Parasites and Infection: Strategies to Control, Diagnose, and Treat Parasitic Diseases

Parasites and Infection: Strategies to Control, Diagnose, and Treat Parasitic Diseases

Guest Editor

Érica S. Martins-Duarte

MDPI

Basel • Beijing • Wuhan • Barcelona • Belgrade • Novi Sad • Cluj • Manchester

Guest Editor
Érica S. Martins-Duarte
Departamento de
Parasitologia
Universidade Federal de
Minas Gerais
Belo Horizonte
Brazil

Editorial Office
MDPI AG
Grosspeteranlage 5
4052 Basel, Switzerland

This is a reprint of the Special Issue, published open access by the journal *Microorganisms* (ISSN 2076-2607), freely accessible at: www.mdpi.com/journal/microorganisms/special_issues/EKV2CA5307.

For citation purposes, cite each article independently as indicated on the article page online and as indicated below:

Lastname, A.A.; Lastname, B.B. Article Title. *Journal Name* **Year**, *Volume Number*, Page Range.

ISBN 978-3-7258-4422-7 (Hbk)
ISBN 978-3-7258-4421-0 (PDF)
https://doi.org/10.3390/books978-3-7258-4421-0

Contents

About the Editor

Érica S. Martins-Duarte

Érica S. Martins-Duarte obtained a degree in Pharmacy, master's in Biological Sciences (Biophysics), and PhD in Biological Sciences from the Federal University of Rio de Janeiro. Doctor Martins-Duarte was a postdoctoral fellow at the University of Georgia in the USA and at the Carlos Chagas Filho Institute of Biophysics at the Federal University of Rio de Janeiro, and has been an Adjunct Professor with the Department of Parasitology at the Federal University of Minas Gerais since 2018. Doctor Martins-Duarte has experience in the areas of parasitology, parasitic chemotherapy, cellular and structural biology (with an emphasis on optical and electron microscopy), and molecular biology.

Preface

Parasitism is an ecological relationship in which an organism, a parasite, lives inside or on another organism, the host, with the former depending on the latter to acquire shelter and essential nutrients, and survive. Thus, it is a relationship that benefits the parasite at the expense of the host, but does not necessarily result in the host's death. In humans, parasitic diseases hamper development and still cause high mortality, especially in children in developing countries. Parasitic infections in poultry, cattle, or swine, for example, are responsible for economic losses in livestock. Indeed, the proximity to infected animals puts humans at risk of zoonosis. Herein are thirteen publications covering basic biology, genetics, new diagnostic tools and treatments, control strategies, disease epidemiology, and the pathogenesis of protozoan and helminth parasites of medical importance.

Érica S. Martins-Duarte
Guest Editor

microorganisms

MDPI

Editorial

Parasites and Infection: Strategies to Control, Diagnose, and Treat Parasitic Diseases

Erica S. Martins-Duarte

Departamento de Parasitologia, Instituto de Ciências Biológicas, Universidade Federal de Minas Gerais, Belo Horizonte 31270-901, MG, Brazil; ericamduarte@ufmg.br

This Special Issue features thirteen publications on the basic biology, genetics, novel diagnostic tools and treatments, control strategies, disease epidemiology, and pathogenesis of medically significant protozoan and helminth parasites. Largely neglected, parasitic diseases contribute to high global morbimortality, particularly in communities lacking adequate sanitation, clean water, and healthcare. Consequently, reduced human work capacity and quality of life are prevalent in endemic areas. Such infections are frequently linked with malnourishment, iron deficiency, anemia, and compromised physical and intellectual development, along with other organic chronic conditions that impede regional growth [1].

However, fighting parasitic diseases presents considerable challenges, as local risk factors are constantly changing and differ widely according to each region's unique environment, ecology, climate, culture, and socioeconomic conditions [2]. In addition, a critical hurdle for eradication in endemic areas is the lack of fast, sensitive, and specific diagnostic tools, effective treatments, and vaccines for most parasitic diseases. Therefore, continuous efforts are essential to thoroughly investigate parasite biology, local genetic variations, modes of transmission, main reservoirs, host immune responses, disease pathogenesis, new treatments, and drug resistance patterns to establish effective and sustainable prevention and control programs.

Helminth infections are common parasitoses in low- and middle-income countries. Among these, *Taenia solium* (pork tapeworm) infection is one of the most prevalent and endemic in sub-Saharan Africa, Asia, Latin America, and the Caribbean regions [3]. Transmission can occur by ingesting larvae in undercooked pork or eggs through the fecal–oral route, leading to various diseases. The ingestion of eggs, whether through self-contamination or the consumption of contaminated food or water, causes neurocysticercosis, a disease that can cause serious encephalic complications [4]. To better understand how oxidative stress contributes to inflammation and the pathophysiology of the disease, Generoso et al. [4] used a rat model that reproduces human extraparenchymal neurocysticercosis, a form of the disease less responsive to treatment and more prone to complications [4]. The obtained results showed a positive correlation between the presence of oxidative stress markers and inflammation intensity. These results encourage future efforts to monitor oxidative stress status during disease as a guide for better clinical intervention decisions for infected patients [5]. Another common helminth infecting humans is *Strongyloides stercoralis*, which causes high morbidity and possibly infects 600 million people globally [6]. It is an intestinal worm, and transmission occurs through skin penetration by larvae that contaminate the environment via shedding in the feces of infected individuals or animals. Based on previous reports of the potential of *Bacillus thuringiensis* crystal protein toxins against *S. stercoralis* and other gastrointestinal helminths, Pommare et al. (2024; [7]) conducted a pilot analysis using

MALDi-TOF and a laboratory genomic pipeline study in stools of *S. stercoralis* infected and non-infected dogs to identify the genes involved in the synthesis of *B. thuringiensis* crystal toxins with possible larvicidal effect [7].

Intestinal parasitic diseases are globally distributed. However, they are most prevalent in developing countries, such as sub-Saharan Africa (SSA), Asia, Latin America, and the Caribbean [8]. These infections are commonly associated with malnourishment, iron deficiency, and anemia in children. For that reason, Cuna et al. [9] investigated whether intestinal parasites were the cause of the high anemia incidence in schoolchildren from a rural area in Bolivia. Although the authors did not find a correlation between anemia and intestinal parasite infections, a high rate of intestinal protozoan and helminths was found in those children. The most prevalent were the protozoans *Entamoeba coli* (48.9%) and *Blastocystis hominis* (40.2%). Still, other parasites of fecal–oral route transmission, such as *Ascaris lumbricoides*, *Giardia intestinalis*, and *Entamoeba histolytica*, were also found, suggesting precarious local sanitary structures [9].

The protozoan *Blastocystis* sp. is a common human intestinal parasite infecting humans and animals worldwide through the fecal–oral route. Usually, infection by this parasite is asymptomatic. However, some patients develop gastrointestinal symptoms, raising questions about whether specific genetic subtypes of the parasite are more likely to cause symptoms [10]. Matovelle et al. [11] explored the subtype diversity in fecal samples obtained from hospitalized patients exhibiting gastrointestinal symptoms from Northern Spain. They found two predominant (ST2 and ST3) and two minor (ST1 and ST4) subtypes. All four of these had been previously identified in patients with gastrointestinal symptoms. Mixed infections with different subtypes of *Blastocystis* were also found. The study of subtype distribution in humans and animals across different countries, along with their phylogenetic analysis and genetic diversity assessment, can help identify the primary sources of parasite transmission and elucidate differences in pathogenicity observed in some patients [11].

Cryptosporidium sp. is one of the world's most frequent causes of diarrhea in humans and animals. Cryptosporidiosis has both anthroponotic and zoonotic spread, and transmission occurs through the fecal–oral route by consuming water or food contaminated with the feces of infected humans or animals [12]. In several countries, it is highly prevalent in children younger than 2–5 years old, and it is recognized as life-threatening for immunosuppressed individuals such as AIDS patients [13]. Diarrhea by *Cryptosporidium* species is also the main cause of mortality in neonatal calves, leading to substantial economic losses in dairy farms [14]. Two works published in this Special Issue examined the incidence of this parasite in animals and humans. In a systematic review and meta-analysis, Tawana et al. [15] showed a high infection rate of *Cryptosporidium* spp. from 1980 to 2020 in South Africa, with *Cryptosporidium parvum* being the prevalent species in both animals and humans. Among humans, HIV+ individuals showed greater exposure to *Cryptosporidium* than HIV- individuals, and higher infection rates were observed in regions where sanitary conditions are poorer. However, they also observed a decline in *Cryptosporidium* spp. infection prevalence from 2011 to 2020 compared to 2001–2010, indicating a possible improvement in local sanitary conditions, medication access, and good animal practice [15]. In another study, Kaduková et al. [16] demonstrated that in calves up to five weeks old in dairy farms in Eastern Slovakia, *C. parvum* is also the main infecting species of *Cryptosporidium*. Genotypic analysis showed that the subtype IIaA17G1R1 was the most prevalent in these calves. Considering that this subtype has already been reported in humans and that calves are the main reservoir of *C. parvum*, these results highlight the importance of preventive actions to reduce infection in those animals, environmental contamination, and the zoonotic spread of the parasite to humans [16].

Costa et al. [17] focused on the genetic diversity of *Toxoplasma gondii* by analyzing the alleles of five virulence factors. Their study examined 103 atypical strains of *T. gondii* isolated from humans, domestic, and wild animals from four different Brazilian States. The results showed a positive relation between the allele 4 of ROP18 and increased virulence and mortality in mice. These findings offer new perspectives for understanding how atypical virulence markers of *T. gondii* modulate the immune system in mice or humans. Further research in this direction could elucidate the higher virulence of atypical strains in mice and their connection to more severe toxoplasmosis outcomes observed Brazil [17].

T. gondii is an important, globally distributed protozoonosis commonly related to sequelae in congenitally infected newborns, ocular lesions, and encephalitis in immunosuppressed individuals [18]. Serological inquiry for this parasite infection is mandatory during early pregnancy to prevent congenital transmission [19]. In a systematic review and meta-analysis, Ribeiro et al. [20] determined the effectiveness of the main treatments and current treatment protocols for gestational and congenital toxoplasmosis, drawing on 56 studies from 16 countries. Their results demonstrated that treating acutely infected pregnant individuals with either spiramycin or triple therapy (sulfadiazine + pyrimethamine + folinic acid) reduces the risk of neonatal vertical infection and sequelae. These findings show that the serological monitoring of susceptible individuals during pregnancy, combined with appropriate treatment, has a good prognosis for reducing congenital toxoplasmosis [20].

The search for novel therapies for toxoplasmosis was the focus of the work of Costa et al. [21]. They employed a drug repositioning strategy to screen 160 drugs or drug-like compounds from the Medicines for Malaria Venture COVID Box to identify against *T. gondii*. A total of 23 molecules inhibited the tachyzoite forms of *T. gondii* in vitro by more than 70% at 1 μM after seven days of treatment. For two of them (apilimod and midostaurin), this was the first reported activity against this parasite. Morphological and ultrastructural analysis by transmission electron microscopy and fluorescence microscopy of tachyzoites treated with nine compounds showed that they all induced alterations in the parasite organelles and cell division [21].

Santos et al. [22] sought to understand the role of SAG2A in the *T. gondii* immune response by the peritoneal exudate cells (PECs) of susceptible and resistant mice. SAG2A is an immunodominant antigen expressed on the surface of the tachyzoite stage and, for that reason, is a promising candidate for vaccine and diagnostic development [23,24]. The results showed that recombinant SAG2A differentially modulates the immune response in the PECs of susceptible and resistant mice. PECs from susceptible mice were more sensitive to modulation by rSAG2A and showed lower parasitism than PECs from resistant mice. These results indicate that developing vaccines and new diagnostic tools for toxoplasmosis requires a better comprehension of the immune response variations from different hosts to *T. gondii* antigens [22].

The immune factors associated with the congenital transmission of the protozoan *Trypanosoma cruzi* were investigated by Herrera Choque et al. [25]. Chagas disease is an endemic zoonosis in South America, associated with poverty in rural areas and precarious housing conditions. It is estimated that *T. cruzi* causes the death of 10,000 individuals worldwide every year [26]. This parasite is primarily transmitted through contact with the feces or urine of blood-sucking triatomine bugs. However, congenital transmission occurs in around 10% of chronically infected pregnant individuals and poses a threat to disease eradication, as it can sustain transmission over time [27]. The immune and genetic factors involved in congenital transmission, however, remain unclear. The authors analyzed IFN-gamma concentration in maternal, placenta, and cord blood samples from both uninfected and infected mothers and their newborns. Higher levels of this cytokine were found in infected, non-transmitting mothers compared to transmitting ones, suggesting a possible role of IFN-gamma in controlling parasite infection and proliferation in the placenta and, consequently, in preventing fetal transmission [25].

Another highly relevant zoonotic disease is visceral leishmaniosis, caused by the *Leishmania infantum* species, especially in Brazil, which accounts for 97% of the total cases in America [28]. In endemic urban areas, dogs serve as the main reservoir for the parasite. Therefore, faster and more accurate diagnosis of infection in these animals is essential to enable timely treatment or, when necessary, euthanasia to prevent further transmission [29]. However, the protocols and diagnostic tools currently available in Brazil have limitations regarding sensitivity and specificity, hindering the detection of early seropositive or asymptomatic infected dogs [30]. To address this gap, Moreira et al. [31] used immunoinformatic tools to discover new peptides with higher sensitivity and specificity for detecting canine visceral leishmaniosis in asymptomatic dogs by ELISA assays. Two peptides were selected and tested, both individually and in combination. The combination of peptides showed 100% specificity in detecting infection in asymptomatic dogs. Additionally, a high specificity (90%) was observed in distinguishing infections caused by other pathogens or in vaccinated dogs. These data offer new possibilities for improving the management of canine visceral leishmaniosis in endemic areas [31].

Lastly, the review of Muh et al. [32] summarized recent findings on the biology, natural hosts and vectors, epidemiology, clinical presentation, and management of the simian malaria parasite *Plasmodium cynomolgi*, an emerging zoonotic pathogen in Southeastern Asia. *P. cynomolgi* was first reported in 2011, in a patient from the east coast of Peninsular Malaysia. Since then, the number of reported cases has increased throughout the region. However, to date, there has been no evidence that *P. cynomolgi* is naturally transmitted from one human to another; instead, proximity to natural reservoir hosts (macaque monkeys) remains the main risk factor for contracting *P. cynomolgi* malaria. The authors highlighted that extensive deforestation has significantly impacted agriculture and human settlements, likely bringing mosquito vectors into closer contact with both humans and reservoir hosts. Further research into the primary vectors, reservoirs, and disease manifestations in humans is essential to better understand this type of zoonotic malaria and to prevent potential, yet unknown, risks to human health [32].

Of the 11 parasites discussed above, 8 are zoonotic—an unsurprising fact given that approximately 60% of infectious diseases are of zoonotic origin, causing millions of deaths annually [33]. Changes in the natural ecosystems, biodiversity, and climate contribute to shifts in the distribution of pathogens, vectors, and hosts. As previously mentioned, population growth over the last and this century has led to the establishment of human settlements in deforested areas, bringing humans closer to wildlife, where the majority of emergent zoonosis originates [34–36]. The risk of zoonotic disease introduction or (re)-emergence is even more critical for the impoverished population, who are already burdened by persistent neglected zoonotic diseases and the most vulnerable to the risk of new zoonotic pathogens [37]. The adoption of a One Health strategy—which seeks to balance human health through investments in food/nutritional security, access to clean water, healthcare services, enhanced diagnostic tools and treatment, and improved housing and education, alongside sustainable animal welfare and agriculture, ecosystem protection, and biodiversity restoration—is crucial for eradicating the existing and preventing the (re)-emergence of parasitic diseases.

Funding: ESMD is a CNPq research productivity fellow (308082/2023-0).

Conflicts of Interest: The author declares no conflicts of interest.

References

1. Celestino, A.O.; Vieira, S.C.F.; Lima, P.A.S.; Rodrigues, L.M.C.L.; Lopes, I.R.S.; França, C.M.; Barreto IDde, C.; Gurgel, R.Q. Prevalence of intestinal parasitic infections in Brazil: A systematic review. *Rev. Soc. Bras. Med. Trop.* **2021**, *54*, e00332021. [CrossRef] [PubMed]
2. Cable, J.; Barber, I.; Boag, B.; Ellison, A.R.; Morgan, E.R.; Murray, K.; Pascoe, E.L.; Sait, S.M.; Wilson, A.J.; Booth, M. Global change, parasite transmission and disease control: Lessons from ecology. *Philos. Trans. R. Soc. B Biol. Sci.* **2017**, *372*, 20160088. [CrossRef] [PubMed]
3. World Health Organization. *Status of Endemicity of Taenia Solium*; WHO: Geneva, Switzerland, 2025. Available online: https://www.Who.Int/Data/Gho/Data/Indicators/Indicator-Details/GHO/Status-of-Endemicity-of-Taenia-Solium (accessed on 22 May 2025).
4. Abanto, J.; Blanco, D.; Saavedra, H.; Gonzales, I.; Siu, D.; Pretell, E.J.; Bustos, J.A.; Garcia, H.H. Mortality in Parenchymal and Subarachnoid Neurocysticercosis. *Am. J. Trop. Med. Hyg.* **2021**, *105*, 176–180. [CrossRef]
5. Generoso, D.; Martins, T.d.C.; Camacho, C.R.C.; Segredo, M.P.d.F.; Batah, S.S.; Fabro, A.T.; Sciutto, E.; Fleury, A.; Hamamoto Filho, P.T.; Zanini, M.A. Oxidative Stress in the Murine Model of Extraparenchymal Neurocysticercosis. *Microorganisms* **2024**, *12*, 1860. [CrossRef]
6. Buonfrate, D.; Bisanzio, D.; Giorli, G.; Odermatt, P.; Fürst, T.; Greenaway, C.; French, M.; Reithinger, R.; Gobbi, F.; Montresor, A.; et al. The Global Prevalence of *Strongyloides stercoralis* Infection. *Pathogens* **2020**, *9*, 468. [CrossRef]
7. Pomari, E.; Orza, P.; Bernardi, M.; Fracchetti, F.; Campedelli, I.; De Marta, P.; Recchia, A.; Paradies, P.; Buonfrate, D. A Pilot Study for the Characterization of *Bacillus* spp. and Analysis of Possible *B. thuringiensis/Strongyloides stercoralis* Correlation. *Microorganisms* **2024**, *12*, 1603. [CrossRef]
8. Ahmed, M. Intestinal Parasitic Infections in 2023. *Gastroenterol. Res.* **2023**, *16*, 127–140. [CrossRef]
9. Cuna, W.R.; Contreras, I.; Rodriguez, A.; Passera, R.; Rodriguez, C. Determinants of Anemia in Schoolchildren in the Highland Bolivia. *Microorganisms* **2024**, *12*, 2491. [CrossRef]
10. Jiménez, P.A.; Jaimes, J.E.; Ramírez, J.D. A summary of *Blastocystis* subtypes in North and South America. *Parasites Vectors* **2019**, *12*, 376. [CrossRef]
11. Matovelle, C.; Quílez, J.; Tejedor, M.T.; Beltrán, A.; Chueca, P.; Monteagudo, L.V. Subtype Distribution of *Blastocystis* spp. in Patients with Gastrointestinal Symptoms in Northern Spain. *Microorganisms* **2024**, *12*, 1084. [CrossRef]
12. Gerace, E.; Lo Presti, V.D.M.; Biondo, C. *Cryptosporidium* Infection: Epidemiology, Pathogenesis, and Differential Diagnosis. *Eur. J. Microbiol. Immunol.* **2019**, *9*, 119–123. [CrossRef] [PubMed]
13. Bouzid, M.; Kintz, E.; Hunter, P.R. Risk factors for *Cryptosporidium* infection in low and middle income countries: A systematic review and meta-analysis. *PLoS Neglected Trop. Dis.* **2018**, *12*, e0006553. [CrossRef] [PubMed]
14. Roblin, M.; Canniere, E.; Barbier, A.; Daandels, Y.; Dellevoet-Groenewegen, M.; Pinto, P.; Tsaousis, A.; Leruste, H.; Brainard, J.; Hunter, P.R.; et al. Study of the economic impact of cryptosporidiosis in calves after implementing good practices to manage the disease on dairy farms in Belgium, France, and the Netherlands. *Curr. Res. Parasitol. Vector-Borne Dis.* **2023**, *4*, 100149. [CrossRef] [PubMed]
15. Tawana, M.; Onyiche, T.E.; Ramatla, T.; Nkhebenyane, S.J.; Grab, D.J.; Thekisoe, O. *Cryptosporidium* Species Infections Detected from Fecal Samples of Animal and Human Hosts in South Africa: Systematic Review and Meta-Analysis. *Microorganisms* **2024**, *12*, 2426. [CrossRef]
16. Kaduková, M.; Schreiberová, A.; Mudroň, P.; Tóthová, C.; Gomulec, P.; Štrkolcová, G. *Cryptosporidium* Infections in Neonatal Calves on a Dairy Farm. *Microorganisms* **2024**, *12*, 1416. [CrossRef]
17. Costa, J.G.L.; Martins-Duarte, É.S.; Pinto, L.V.; Baraviera, R.A.d.C.; Rego, W.M.F.D.; Vitor, R.W.d.A. Investigation of Virulence-Related Markers in Atypical Strains of *Toxoplasma gondii* from Brazil. *Microorganisms* **2025**, *13*, 301. [CrossRef]
18. Kamus, L.; Belec, S.; Lambrecht, L.; Abasse, S.; Olivier, S.; Combe, P.; Bonnave, P.-E.; Vauloup-Fellous, C. Maternal and congenital toxoplasmosis in Mayotte: Prevalence, incidence and management. *PLoS Neglected Trop. Dis.* **2023**, *17*, e0011198. [CrossRef]
19. Lopes, F.M.R.; Gonçalves, D.D.; Mitsuka-Breganó, R.; Freire, R.L.; Navarro, I.T. *Toxoplasma gondii* infection in pregnancy. *Braz. J. Infect. Dis.* **2007**, *11*, 496–506. [CrossRef]
20. Ribeiro, S.K.; Mariano, I.M.; Cunha, A.C.R.; Pajuaba, A.C.A.M.; Mineo, T.W.P.; Mineo, J.R. Treatment Protocols for Gestational and Congenital Toxoplasmosis: A Systematic Review and Meta-Analysis. *Microorganisms* **2025**, *13*, 723. [CrossRef]
21. Costa, A.L.O.; dos Santos, M.; Dantas-Vieira, G.C.; Lopes, R.E.N.; Vommaro, R.C.; Martins-Duarte, É.S. Antiproliferative and Morphological Analysis Triggered by Drugs Contained in the Medicines for Malaria Venture COVID-Box Against *Toxoplasma gondii* Tachyzoites. *Microorganisms* **2024**, *12*, 2602. [CrossRef]
22. dos Santos, T.A.R.; de Oliveira, M.C.; Silva, E.M.d.A.; dos Santos, U.R.; Ferreira, M.M.; Soares, A.L.C.; Silva, N.M.; Mendes, T.A.d.O.; Leal-Sena, J.A.; da Cunha-Júnior, J.P.; et al. Recombinant SAG2A Protein from *Toxoplasma gondii* Modulates Immune Profile and Induces Metabolic Changes Associated with Reduced Tachyzoite Infection in Peritoneal Exudate Cells from Susceptible C57BL/6 Mice. *Microorganisms* **2024**, *12*, 2366. [CrossRef] [PubMed]

23. Macêdo, A.G.; Cunha, J.P.; Cardoso, T.H.; Silva, M.V.; Santiago, F.M.; Silva, J.S.; Pirovani, C.P.; Silva, D.A.; Mineo, J.R.; Mineo, T.W. SAG2A protein from *Toxoplasma gondii* interacts with both innate and adaptive immune compartments of infected hosts. *Parasites Vectors* **2013**, *6*, 163. [CrossRef] [PubMed]
24. Béla, S.R.; Oliveira Silva, D.A.; Cunha-Júnior, J.P.; Pirovani, C.P.; Chaves-Borges, F.A.; Reis de Carvalho, F.; Carrijo de Oliveira, T.; Mineo, J.R. Use of SAG2A recombinant *Toxoplasma gondii* surface antigen as a diagnostic marker for human acute toxoplasmosis: Analysis of titers and avidity of IgG and IgG1 antibodies. *Diagn. Microbiol. Infect. Dis.* **2008**, *62*, 245–254. [CrossRef]
25. Herrera Choque, A.G.; Cuna, W.R.; Gabrielli, S.; Mattiucci, S.; Passera, R.; Rodriguez, C. Parasitic Effects on the Congenital Transmission of *Trypanosoma cruzi* in Mother–Newborn Pairs. *Microorganisms* **2024**, *12*, 1243. [CrossRef]
26. WHO—World Health Organization. *Chagas Disease (American Trypanosomiasis)*; WHO: Geneva, Switzerland, 2025. Available online: https://www.Who.Int/En/News-Room/Fact-Sheets/Detail/Chagas-Disease-(American-Trypanosomiasis) (accessed on 21 May 2025).
27. Howard, E.; Xiong, X.; Carlier, Y.; Sosa-Estani, S.; Buekens, P. Frequency of the congenital transmission of *T rypanosoma cruzi*: A systematic review and meta-analysis. *BJOG Int. J. Obstet. Gynaecol.* **2014**, *121*, 22–33. [CrossRef]
28. PAHO. *Leishmanioses: Informe Epidemiológico Nas Américas*; PAHO: Washington, DC, USA, 2020. Available online: https://iris.paho.org/handle/10665.2/34113 (accessed on 21 May 2025).
29. Quaresma, P.F.; Martins-Duarte, E.S.; Soares Medeiros, L.C. Editorial: One Health Approach in Zoonosis: Strategies to control, diagnose and treat neglected diseases. *Front. Cell. Infect. Microbiol.* **2023**, *13*, 1227865. [CrossRef]
30. Pessoa-e-Silva, R.; Vaitkevicius-Antão, V.; de Andrade, T.A.S.; de Oliveira Silva, A.C.; de Oliveira, G.A.; Trajano-Silva, L.A.M.; Nakasone, E.K.N.; de Paiva-Cavalcanti, M. The diagnosis of canine visceral leishmaniasis in Brazil: Confronting old problems. *Exp. Parasitol.* **2019**, *199*, 9–16. [CrossRef]
31. Moreira, G.; Maia, R.; Soares, N.; Ostolin, T.; Coura-Vital, W.; Aguiar-Soares, R.; Ruiz, J.; Resende, D.; de Brito, R.; Reis, A.; et al. Synthetic Peptides Selected by Immunoinformatics as Potential Tools for the Specific Diagnosis of Canine Visceral Leishmaniasis. *Microorganisms* **2024**, *12*, 906. [CrossRef]
32. Muh, F.; Erwina, A.; Fitriana, F.; Syahada, J.H.; Cahya, A.D.; Choe, S.; Jun, H.; Garjito, T.A.; Siregar, J.E.; Han, J.-H. *Plasmodium cynomolgi*: What Should We Know? *Microorganisms* **2024**, *12*, 1607. [CrossRef]
33. Jones, K.E.; Patel, N.G.; Levy, M.A.; Storeygard, A.; Balk, D.; Gittleman, J.L.; Daszak, P. Global trends in emerging infectious diseases. *Nature* **2008**, *451*, 990–993. [CrossRef]
34. Bird, B.H.; Mazet, J.A.K. Detection of Emerging Zoonotic Pathogens: An Integrated One Health Approach. *Annu. Rev. Anim. Biosci.* **2018**, *6*, 121–139. [CrossRef] [PubMed]
35. Gebreyes, W.A.; Dupouy-Camet, J.; Newport, M.J.; Oliveira, C.J.B.; Schlesinger, L.S.; Saif, Y.M.; Kariuki, S.; Saif, L.J.; Saville, W.; Wittum, T.; et al. The Global One Health Paradigm: Challenges and Opportunities for Tackling Infectious Diseases at the Human, Animal, and Environment Interface in Low-Resource Settings. *PLoS Neglected Trop. Dis.* **2014**, *8*, e3257. [CrossRef] [PubMed]
36. Ahmed, M.M.; Okesanya, O.J.; Othman, Z.K.; Ibrahim, A.M.; Adigun, O.A.; Ukoaka, B.M.; Abdi, M.I.; Lucero-Prisno, D.E. Holistic Approaches to Zoonoses: Integrating Public Health, Policy, and One Health in a Dynamic Global Context. *Zoonotic Dis.* **2025**, *5*, 5. [CrossRef]
37. Technical report of the TDR Disease Reference Group on Zoonoses and Marginalized Infectious Diseases of Poverty. In *Research Priorities for Zoonoses and Marginalized Infections*; WHO: Geneva, Switzerland, 2012.

microorganisms

MDPI

Systematic Review

Treatment Protocols for Gestational and Congenital Toxoplasmosis: A Systematic Review and Meta-Analysis

Sissi Kelly Ribeiro [1], Igor Moraes Mariano [2], Ana Clara Ribeiro Cunha [2], Ana Cláudia Arantes Marquez Pajuaba [1], Tiago Wilson Patriarca Mineo [1] and José Roberto Mineo [1,*]

1 Laboratory of Immunoparasitology, Institute of Biomedical Sciences, Federal University of Uberlândia, Uberlândia 38400-678, MG, Brazil; sissi.ribeiro@ufu.br (S.K.R.); ana.pajuaba@ufu.br (A.C.A.M.P.); tiago.mineo@ufu.br (T.W.P.M.)

2 Laboratory of Cardiorespiratory and Metabolic Physiology, Federal University of Uberlândia, Uberlândia 38405-302, MG, Brazil; igor.mariano@ufu.br (I.M.M.); anacrcunha@ufu.br (A.C.R.C.)

* Correspondence: jrmineo@ufu.br; Tel.: +55-34-3225-8666

Abstract: Toxoplasmosis is a globally prevalent zoonotic parasitic disease. Neonates with congenital infection can develop severe long-term sequelae, which can be mitigated or prevented through early diagnosis and therapeutic approaches. In this context, the main objective of this study was to describe the main treatments and evaluate the effectiveness of the current treatment protocols for gestational and congenital toxoplasmosis to prevent vertical transmission and to reduce clinical manifestations in neonates. This systematic review with a meta-analysis searched digital databases (PUBMED, SCOPUS, WEB OF SCIENCE, EMBASE, and COCHRANE) for observational cohort studies published between 1 January 2013 and 29 January 2025, evaluating treatment effectiveness in gestational and congenital toxoplasmosis. Risk ratios (RRs) were calculated using random effects models to assess infection risk and clinical manifestations in neonates. The study quality was assessed following the Joanna Briggs Institute protocol and fifty-six studies from 16 countries were included, comprising 11,090 pregnant women and 4138 children. Studies were predominantly from Brazil (38%), France, and Italy. Only 9% of the studies indicated knowledge of the serological status of the pregnant woman before the gestational stage. Of 10,148 women with confirmed toxoplasmosis, 8600 received treatment, with 18% of their children infected, compared to a 58% infection rate in untreated mothers' children. Meta-analysis showed that treatment reduced infection risk (RR = 0.34 [0.21; 0.57]) and clinical manifestations (RR = 0.30 [0.17; 0.56]). While spiramycin or triple therapy showed similar effects, triple therapy demonstrated more consistent results (RR: 0.22 [0.15; 0.32]) compared to spiramycin alone (RR: 0.54 [0.06; 4.67]). In conclusion, treatment protocols for congenital or gestational toxoplasmosis have proven to be effective in reducing the risk of infection and clinical manifestations in neonates. Regarding the type of treatment, although they have similar responses, the use of triple therapy shows more consistent responses than isolated spiramycin. It can be also concluded that prevention and mitigation of congenital toxoplasmosis require standardized treatment protocols, improved diagnostic methods, and educational programs for women of childbearing age, as treatment initiation timing and protocol choice are crucial factors in determining outcomes.

Keywords: congenital toxoplasmosis; maternal and newborn treatment; efficacy; therapeutic failure; diagnostics; sequelae

1. Introduction

Toxoplasma gondii is one of the most successful parasitic pathogens worldwide, capable of infecting a remarkable range of avian and mammalian hosts, including humans [1]. This protozoan's heteroxenous life cycle, with felids as definitive hosts, contributes to its widespread distribution and significant public health impact [2,3]. Humans can be infected by several routes, by eating undercooked meat of animals containing tissue cysts, by consuming food or water contaminated with cat feces, by contaminated environmental samples, by blood transfusion or organ transplantation, as well as transplacentally from mother to fetus [1]. While typically asymptomatic in immunocompetent hosts, *T. gondii* infection poses significant health challenges in two critical scenarios: immunocompromised individuals, where it can cause severe neurological complications, and pregnant women, where transplacental transmission puts the fetus at risk of congenital infection [3]. In addition to these implications, socioeconomic factors such as lower socioeconomic status and education of the population play a significant role in the prevalence of toxoplasmosis. Although socioeconomic factors influence toxoplasmosis prevalence, congenital infection can affect all demographic groups, making it a universal concern in maternal healthcare [4].

Considering this complex scenario, the therapeutic approach to toxoplasmosis is a significant challenge [5]. Conventional treatment is limited to the acute phase of the disease, with no effects on latent parasites; consequently, a cure is still not available. Furthermore, considerable toxic effects and long-term therapy contribute to high rates of treatment abandonment [6]. However, evidence from French studies suggests that preventive and therapeutic approaches during pregnancy can reduce the risk of symptoms and sequelae in children [7].

Regarding the type of treatment, anti-*Toxoplasma* chemotherapy consists of several medications that can be used individually or in combination. These include sulfadiazine (SDZ), pyrimethamine (PYR), sulfamethoxazole (SMT), trimethoprim (TMP), and spiramycin [8]. Currently, the gold standard treatment involves a combination of SDZ and PYR, with the option to use the combination of SMT and TMP, clindamycin, and spiramycin, all of which present a synergistic effect affecting the replication of the tachyzoite form of *T. gondii* [9,10]. New research on the impacts of toxoplasmosis highlights the need to increase institutional awareness of infection pathways and implement comprehensive, interdisciplinary actions to control transmission and optimize treatment [11].

Given the complex clinical context and varied therapeutic approaches, this systematic review with a meta-analysis aims to evaluate treatment protocols for gestational and congenital toxoplasmosis, focusing on their effectiveness in preventing vertical transmission and reducing clinical manifestations in newborns.

2. Methods

2.1. Search Strategy and Selection Criteria

This systematic review and meta-analysis was conducted following the Preferred Reporting Items for Systematic Reviews and Meta-Analysis (PRISMA-P) guidelines [12], and the method protocol used was previously published (Available online: https://www.protocols.io/view/protocol-of-a-systematic-review-with-metanalysis-f-e6nvwddp2lmk/v1 accessed on 10 February 2025) and registered on the PROSPERO platform (Available online: https://protocols.io/view/protocol-of-a-systematic-review-with-metanalysis-f-cymnxu5e.pdf accessed on 10 February 2025).

Searches were conducted in PUBMED, SCOPUS, WEB OF SCIENCE, EMBASE, and the COCHRANE Library, manually via Google Scholar, plus with reference lists of key articles, for studies published between 1 January 2013 and 29 January 2025. The search strategy combined three categories of terms (Treatment, Congenital toxoplasmosis, and

Population) using Boolean operators 'OR' within categories and 'AND' between categories, as follows: (Treatment OR Drug OR Drugs OR Spiramycin OR Sulfadiazine OR "Folinic acid" OR Pyrimethamine) and ("Congenital toxoplasmosis") and (Maternal OR Pregnant OR Children OR Child OR Newborn OR Baby OR Babies OR Pediatric OR Infant OR Neonate).

Studies were eligible with the following characteristics: (1) population: human, pregnant, and newborn; (2) intervention: treatment of toxoplasmosis; (3) control/comparator: children/mothers not treated for toxoplasmosis; (4) outcomes of interest: treatment protocols used and their effectiveness in reducing risk of vertical transmission and clinical manifestations in neonates; (5) languages: there was no language restriction; while no design restrictions were applied, only cohort studies (prospective or retrospective) met the inclusion criteria of reporting both *T. gondii* infection status and treatment data; (6) publication dates: within the specific time frame that has been set.

Excluded works included literature reviews, meta-analyses, letters to the editor, animal studies, studies unrelated to toxoplasmosis and treatment, studies that treat other comorbidities in addition to toxoplasmosis, full texts that were not accessible, and texts published before 2013. To avoid bias in the results, studies whose mothers had co-infection were excluded.

Regarding outcome assessment, vertical transmission was confirmed through serological tests (IgM and IgG), PCR in amniotic fluid, or placental examination. Clinical manifestations were documented via ophthalmological examination, neurological assessment, and imaging studies. Treatment effectiveness was evaluated by comparing infection rates and clinical manifestations between treated and untreated groups. These standardized outcome measures were used to ensure consistency in data extraction and analysis across included studies.

First, the titles and abstracts of the research studies were evaluated independently by two researchers. Duplicates and studies that did not meet the inclusion or exclusion criteria were excluded from the analyses. Title and abstract screening were conducted using Rayyan (https://www.rayyan.ai/), accessed on 10 February 2025, a software for organizing and managing systematic reviews independently, where each reviewer performed article selection blindly [13]. During the selection process, all data were cross-checked, and discrepancies were resolved through consensus and, if necessary, by the senior researcher. Finally, the full texts of eligible articles were read by the same researchers to decide on their definitive inclusion. For articles considered relevant to the review but not available in full for reading, authors were contacted to request the availability of the full article.

Data were collected using a standardized Excel spreadsheet (Microsoft Excel®, version 2016). The data extraction included the following: (1) bibliometrics (country where the study was conducted, title, journal, language, DOI, and publication year); (2) time of diagnosis; (3) initiation phase of therapy, therapeutic options; (4) dosage form of drugs; (5) dosage; (6) main clinical manifestations and sequels; (7) duration of treatment; (8) patient follow-up; (9) number of treated and untreated patients; (10) idiosyncrasies of the population or treatment.

After the selection phase, the data were extracted, and the quality assessment was performed using the Joanna Briggs Institute (JBI) critical appraisal checklist for cohort studies, which evaluates key methodological aspects including population selection, exposure measurement, identification of confounding factors, outcome assessment, and follow-up adequacy. Studies were rated on each criterion as 'yes', 'no', 'unclear', or 'not applicable' [14].

2.2. Data Synthesis and Analysis

The data were evaluated using the programming language "R" [15] through the supplements "meta" [16] and "metafor" [17]. The pooled effect estimates were computed from risk ratio (RR) differences between treated and untreated groups using random effects models due to expected clinical and methodological heterogeneity between studies. Statistical heterogeneity among studies was evaluated by Cochran's Q test and I^2 inconsistency test [18]. Data synthesis involved pooling RR for infection and clinical manifestations separately, with subgroup analyses performed by treatment type. Subgroups analysis was conducted through comparing the incidence of cases of congenital toxoplasmosis after different treatments in pregnant women. Forest plots were generated to present the pooled effect and the 95% confidence interval.

3. Results

3.1. Qualitative Results

From the initial selection of 1156 studies, 635 were duplicates. Out of the remaining 521, 465 were excluded for pre-defined reasons, such as studies unrelated to the treatment of congenital toxoplasmosis, case reports, studies conducted on animals, reviews, and presentations at scientific conferences, resulting in the selection of 56 studies that addressed the treatment of congenital or gestational toxoplasmosis (Figure 1). No article was excluded due to its inaccessibility.

All 56 eligible studies contributed to the construction of data to elucidate the evidence, considering the following variables: time of diagnosis; initial phase of therapy, therapeutic options; pharmaceutical form of medications; dosage; main clinical manifestations and sequelae; treatment duration; patient follow-up; number of treated and untreated patients; population or treatment idiosyncrasies. The characteristics of all studies included in the qualitative analysis are shown in Supplementary Table S1. A total of 11,090 pregnant women and 4138 children were enrolled in these studies. Eligible studies were published between 2013 and 2025 in 16 countries. Out of the total number of studies analyzed, 20 were conducted in Brazil, followed by France and Italy, with 6 studies each. Moreover, 85% of the studies were conducted in Referral Centers.

Some studies have shown that the screening period for toxoplasmosis during pregnancy may vary from country to country, but indirect diagnostic methods for detecting IgM and IgG immunoglobulins were unanimous, albeit with variations in serological methodology [19–22]. In this regard, one of the mentioned diagnostic methods was the Sabin Feldman test, considered the gold standard for toxoplasmosis diagnosis. Some studies have demonstrated the use of other immunoglobulins such as IgA and IgE [23–32] to complement the diagnosis of *Toxoplasma* infection. Another methodology addressed in the studies was the avidity test, considered a useful tool for detecting the timing of infection [33]. Ultrasound imaging examinations in pregnant women were also present in all studies, while the PCR technique was present in 50% of the studies.

Regarding the knowledge of the pregnant woman's serological status before the gestational stage, a relevant factor in defining the timing of infection if the infection was acquired during or early in pregnancy, which would facilitate the detection of the moment of seroconversion. However, this analysis was addressed in only 9% of the studies.

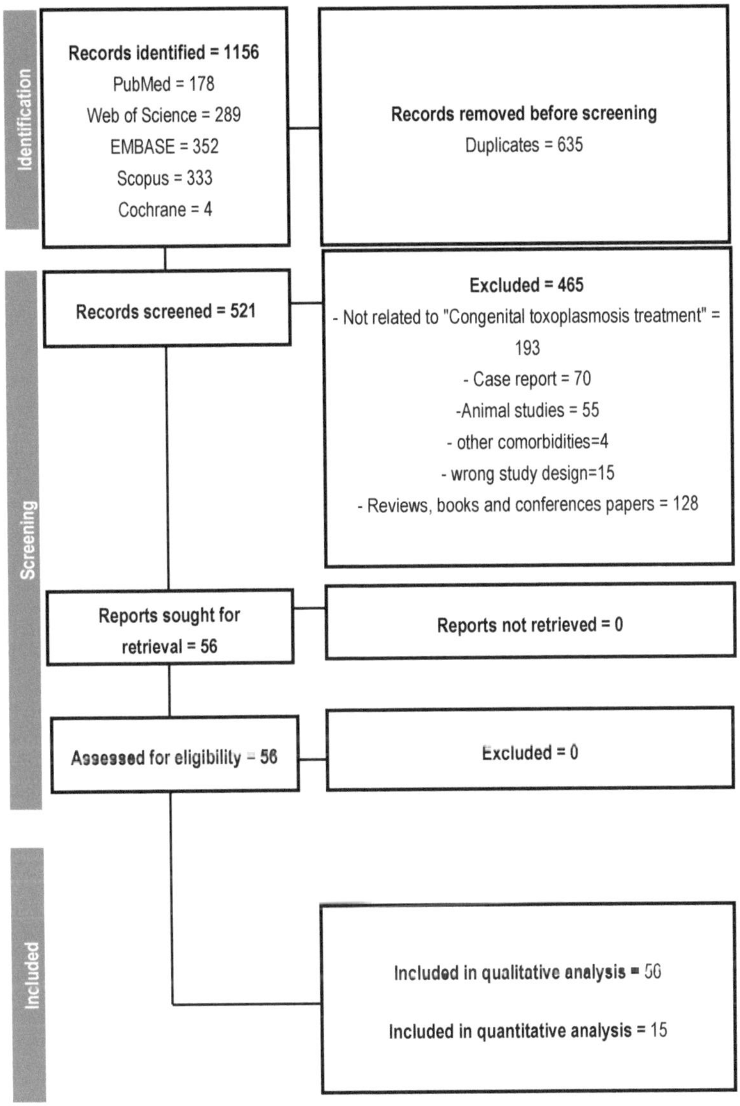

Figure 1. Follow-up flowchart.

A total of 2652 IgG avidity tests were conducted, with 1595 of them detecting low avidity. Additionally, a total of 4284 PCR tests were performed, with 3724 on amniotic fluid and 560 on placenta or umbilical cord, resulting in 442 positive results for a total of 995 children diagnosed with congenital toxoplasmosis with follow-up after one year of age. Additionally, routine follow-up tests for pregnant women considered seronegative for toxoplasmosis during prenatal care are monthly in France and Italy [3,28,31,34–36]. Ultrasounds were conducted on pregnant women as a complementary examination to indirect diagnostic tests, and in children, fundoscopic eye examinations, transfontanelle ultrasounds performed at birth, and physical and neurological examinations were carried out in all studies conducted at reference centers.

According to Damar et al. [20], in Turkey, specifically in the city of Sanlúrfa, there is no established protocol considering that it is not a highly prevalent disease in the country. Similarly, in Japan, the frequency of prenatal screening exams, according to Hijikata et al. [37], is quarterly since it does not have a high incidence. In Vienna, Austria, the frequency of exams is bimonthly [22]. According to Carral [19], in Buenos Aires, Argentina, exams are repeated every trimester and during childbirth, and in Brazil, exams are repeated quarterly [38–40].

Out of a total of 10,148 pregnant women diagnosed with toxoplasmosis, 8600 received some form of treatment and had 1548 (18%) infected children, while 1586 untreated mothers had 922 children infected with *T. gondii* (58%).

In 71% of the studies, the triple therapy regimen (sulfadiazine, pyrimethamine, and folinic acid) associated or not with spiramycin is used as a treatment for gestational and congenital toxoplasmosis, with the following dosage regimens: pyrimethamine 25–50 mg/day, sulfadiazine 3 g/day, folinic acid 25–50 mg twice a week, and spiramycin 3 g/day. For neonates, the established regimen is pyrimethamine 3 mg/kg every 3 days, sulfadiazine 25 mg/kg every 8 hours, and folinic acid 50 mg every 7 days orally [41].

Only five studies cited the pharmaceutical form adopted for the treatment of neonates/children. The most common clinical manifestations in children of treated mothers were ophthalmological, while in children of untreated mothers, the predominant clinical manifestations were neurological. In total, 113 deaths were reported, of which 48% were spontaneous abortions, and 25% were terminations of pregnancy after amniocentesis results [26,42,43], with the remainder being represented by 15% stillbirths, 11% postnatal deaths, and 1% death during adolescence.

3.2. Quantitative Results

To perform the meta-analysis, only studies with groups larger than 10 individuals per group and that did not include only infected individuals were considered. This approach reduces the risk of bias associated with small sample sizes and studies in which both groups would have all individuals infected, which would not allow for infection risk analyses. As shown in Table 1, a total of 15 studies were included in the meta-analysis, some of which contained infection risk data, while others presented clinical manifestation data, leaving 10 for each analysis. However, the timing of maternal infection, a potential important confounder, could not be adequately analyzed as this information was not consistently reported across studies. Although the exact timing of infection was uncertain, studies reported mean treatment initiation at 24 ± 6 gestational weeks.

Regarding the risk of infection, out of the 2923 mothers treated during pregnancy and the 516 who remained untreated, it was observed that a total of 276 (9%) and 231 (45%) had infected newborns, respectively (Figure 2). Therefore, treatment reduces the possibility of vertical transmission (RR = 0.34 [0.21; 0.57]; Figure 2). Additionally, no outlier values were identified regarding the risk of infection.

Regarding the clinical manifestations of newborns from treated or untreated mothers, it was observed that out of 2886 treated mothers, 183 (6%) had children who presented some type of clinical manifestation (Figure 3). In contrast, out of 678 untreated mothers, a total of 334 (49%) of their children presented one of the typical clinical manifestations (Figure 3). However, the study by Conceição et al. [24] was identified as an outlier and influential point without overlapping effects with other studies. By omitting this study, the result changes from RR = 0.37 [0.18; 0.77] (Figure 3) to RR = 0.30 [0.17; 0.56].

Table 1. General characteristics of the studies included in the meta-analysis.

Study	Country	Population (*)	Mother Treatment (Diagnosis Trimester/Week Initiated Protocol)	Newborn Treatment (*)	Conclusions
Avelino et al., 2014 [44]	Brazil	120 treated PW, 115 untreated PW, 162 treated NB, 0 untreated NB	S (2T,25W)	SPF	Treatment of pregnant women with spiramycin reduces the possibility of transmission of infection to the fetus. However, a lack of proper treatment is associated with the onset of the neural–optical form of congenital infection. Primary preventive measures should be increased for all pregnant women during the prenatal period, and secondary prophylaxis through surveillance of seroconversion in seronegative pregnant woman should be introduced to reduce the severity of congenital infection in the environment.
Conceição et al., 2021 [24]	Brazil	82 treated PW, 102 untreated PW, 29 treated NB, 0 untreated NB	Undefined (3T,29W)	SPF-C	High prevalence rates of clinical manifestations were observed in infants with congenital toxoplasmosis after a waterborne toxoplasmosis outbreak, the largest yet described. Cerebral calcifications were higher in infants with ocular abnormalities, and maternal infection during the third pregnancy trimester was associated with a higher rate of congenital toxoplasmosis independent of maternal treatment.
Damar et al., 2023 [20]	Turkey	103 treated PW, 16 untreated PW, 3 treated NB, 2 untreated NB	S (2T,15W)	C-SX	In conclusion, although *Toxoplasma* seroprevalence was found to be high in our region, there was a paucity in diagnosis, follow-up, and treatment. Our findings support that prenatal spiramycin prophylaxis is effective in preventing the transmission of parasites from mother to child.
De Santis et al., 2024 [45]	Italy	537 treated PW, 35 untreated PW, 34 treated NB, 0 untreated NB	S/SPF-Cl/S	Undefined	The study discusses the efficacy of available treatments to reduce the risk of vertical transmission of toxoplasmosis during pregnancy, highlighting the controversy over their effectiveness. Although a large randomized clinical trial would be ideal to validate or modify current clinical practices, randomization against placebo is considered unethical. The authors' experience indicates that maternal treatment with spiramycin and cotrimoxazole, even with negative amniocentesis, can significantly reduce the rate of transmission of congenital toxoplasmosis without causing harm to the mother or fetus.

Table 1. *Cont.*

Study	Country	Population (*)	Mother Treatment (Diagnosis Trimester/Week Initiated Protocol)	Newborn Treatment (*)	Conclusions
Gomes-Ferrari-Strang et al., 2023 [38]	Brazil	48 treated PW, 13 untreated PW, 61 treated NB, 0 untreated NB	S-S/SPF (2T,26W)	SPF	Based on the follow-up of women with acute *T. gondii* infection and their children, through a multidisciplinary team, the availability of anti-*T. gondii* serology and pre- and post-natal treatments reduced the risk of toxoplasmosis transmission.
Guarch-Ibáñez et al., 2024 [46]	Spain	36 treated PW, 18 untreated PW, 0 treated NB, 0 untreated NB	S-SPF-S/SPF	S/SPF	Since cases detected by prenatal screening and treated with SPI and/or PSA presented fewer complications at birth and during follow-up, it is recommended to implement universal screening in Spain and in countries with similar epidemiological data. Long-term follow-up of the REIV-TOXO cohort will provide more information on late complications and the effects of pre- and postnatal treatments.
Guegan et al., 2021 [27]	Frace/Serbia/USA	61 treated PW, 54 untreated PW, 0 treated NB, 0 untreated NB	S-SPF-S/SPF (3T,28W)	Undefined	The sensitivity of PCR for detecting *Toxoplasma* in blood was also reduced by maternal treatment from 39.1% to 23.2%. These results highlight that anti-*Toxoplasma* therapy during pregnancy may set back biological evidence of neonatal infection at birth and underline the need for a careful serological follow-up of infants with normal workup.
Lago et al., 2014 [47]	Brazil	12 treated PW, 16 untreated PW, 59 treated NB, 6 untreated NB	Undefined	Undefined	Even with high sensitivity methods, children with congenital toxoplasmosis can have a negative anti-*Toxoplasma* IgM result at birth. It is important not to interrupt the monitoring of infants with suspected congenital toxoplasmosis simply because they present a negative anti-*Toxoplasma* IgM result.
Losa et al., 2024 [48]	Portugal	63 treated PW, 7 untreated PW, 0 treated NB, 0 untreated NB	S-SPF-S/SPF	S/SPF	The lower incidence observed in the study, compared to Europe, may be related to the reduction in the prevalence of toxoplasmosis, the effectiveness of primary infection prevention measures, and a well-structured prenatal screening program, which allows early initiation of treatment to prevent vertical transmission.

Table 1. *Cont.*

Study	Country	Population (*)	Mother Treatment (Diagnosis Trimester/Week Initiated Protocol)	Newborn Treatment (*)	Conclusions
Mejia-Oquendo et al., 2021 [49]	Colombia	52 treated PW, 47 untreated PW, 0 treated NB, 0 untreated NB	S-SX/P (2T,22W)	Undefined	The study showed that an early detection program for gestational toxoplasmosis implemented at a public health center in Armenia, Quindío, correctly followed evidence-based guidelines. Diagnostic tests were requested in a timely manner, with adequate follow-up of seronegative pregnant women and timely initiation of treatment. Before the implementation of the guidelines, some mothers were not treated, and their children had more ocular and neurological sequelae, something that decreased after the adoption of the recommendations. However, the frequency of infection did not decrease compared to previous studies, and there were failures in the reporting of some IgA results.
Olariu et al., 2019 [21]	USA and Romania	23 treated PW, 164 untreated PW, 0 treated NB, 0 untreated NB	Undefined	Undefined	These findings provide further evidence that anti-parasitic treatment if administered during pregnancy can contribute to better clinical outcomes, even in countries where systematic screening and treatment have not been routinely implemented.
Prusa et al., 2015 [22]	Austria	660 treated PW, 27 untreated PW, 35 treated NB, 4 untreated NB	S-SPF-S/SPF-SZ (3T,28W)	S/SPF	Amniocentesis is indicated in women with acute maternal infection and facilitated targeted therapies in pregnant women and their offspring. In women with late *T. gondii* infection, negative amniotic fluid PCR made treatment of infants unnecessary. Serological and clinical follow-up of infants is important to confirm the infection status of the infant. Recommendations, based on our 17-year experience, to improve the current diagnostic strategies and to reduce unnecessary amniocentesis, are given.
Prusa et al., 2015-2 [50]	Austria	1110 treated PW, 63 untreated PW, 141 treated NB, 0 untreated NB	S/SPF (3T,30W)	S/SPF	Results from the Austrian Toxoplasmosis Register show the efficiency of the prenatal screening program. Our results are of clinical relevance for infants, healthcare systems, and policy makers to consider preventive *T. gondii* screening as a potential tool to reduce the incidence of congenital toxoplasmosis.

Table 1. *Cont.*

Study	Country	Population (*)	Mother Treatment (Diagnosis Trimester/Week Initiated Protocol)	Newborn Treatment (*)	Conclusions
Rodrigues et al., 2014 [51]	Brazil	44 treated PW, 24 untreated PW, 46 treated NB, 0 untreated NB	S (1T,13W)	SPF	The higher proportion of infants without clinical symptoms in group 1 (70.4%) suggests that maternal treatment with spiramycin delays fetal infection, reducing the clinical sequelae of the disease in newborns. Given the low sensitivity of the tests used, when there is suspicion of congenital transmission, several serological and parasitological tests are required in order to confirm or exclude congenital toxoplasmosis in newborns.
Soares et al., 2023 [30]	Brazil	46 treated PW, 33 untreated PW, 79 treated NB, 0 untreated NB	S-SPF-S/SPF (3T,28W)	SPF	A positive advance was observed regarding the care provided for the mother–child binomial affected by *T. gondii*, with a reduction in negative outcomes for the child. However, there are still challenges concerning the diagnosis and proper management of the disease.

(*) PW: pregnant women; NB: newborn; S: spiramycin; SPF: sulfadiazine + pyrimethamine + folinic acid; S/SPF: spiramycin alternate with SPF; SX: sulfadoxine; SM: sulfamethoxazol + trimetropim; AZ: azithromycin; C: corticosteroid; CL: clotrimazol.

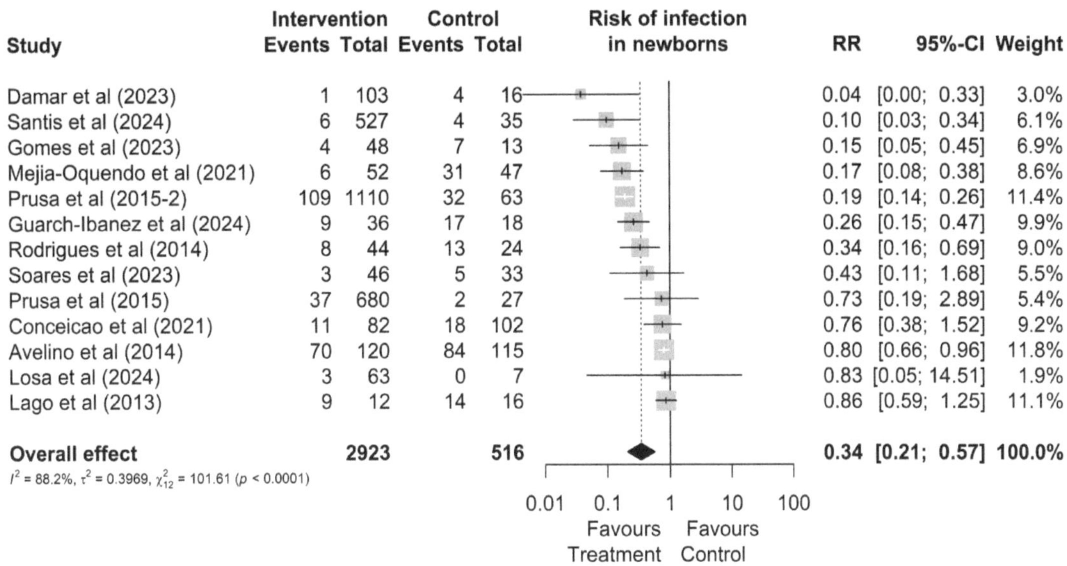

Figure 2. Forest graph of the risk of infection in the fetus after gestational treatment for toxoplasmosis [20,22,24,30,38,44–51].

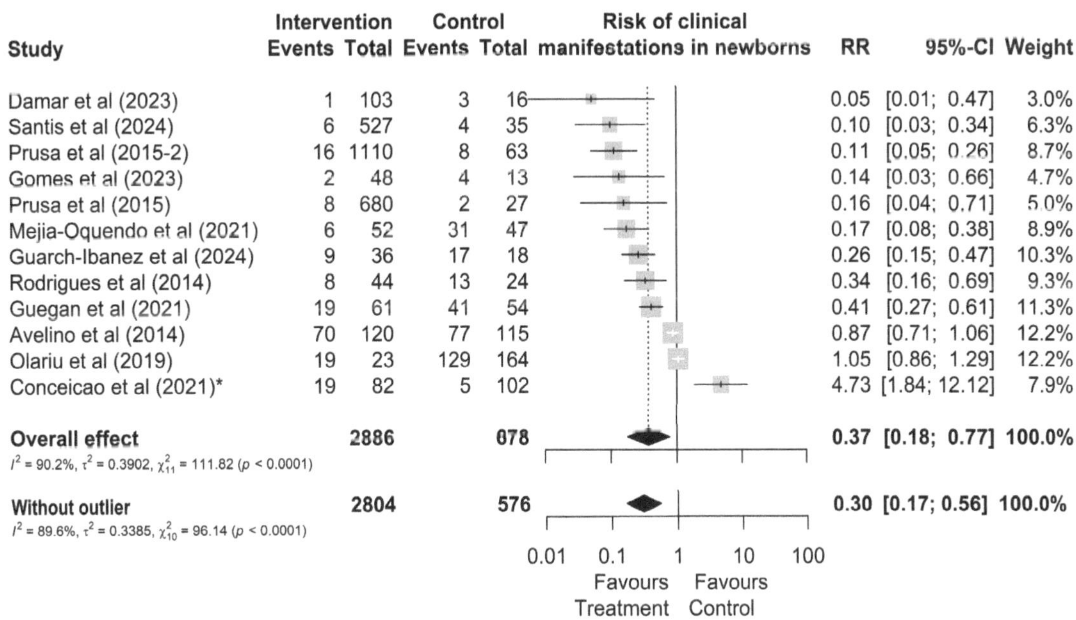

Figure 3. Forest chart of the risk of clinical manifestations in children after gestational treatment for toxoplasmosis. (*) The outlier study [20–22,24,27,38,44–46,49–51].

3.3. Comparative Analysis Between Treatments

As shown in Figure 4, although no significant difference between treatments was observed in the overall analysis ($p = 0.0833$), the subgroup of studies involving mothers treated with the triple regimen demonstrated lower heterogeneity ($I^2 = 23.2\%$, χ^2 $p = 0.18$) and statistically significant results (RR: 0.22 [0.15; 0.32]) compared to the group treated with spiramycin alone ($I^2 = 84.1\%$, χ^2 $p < 0.01$), which showed non-significant results (RR: 0.54; [0.06; 4.67]). These findings suggest greater consistency in treatment effects for mothers who received the triple regimen.

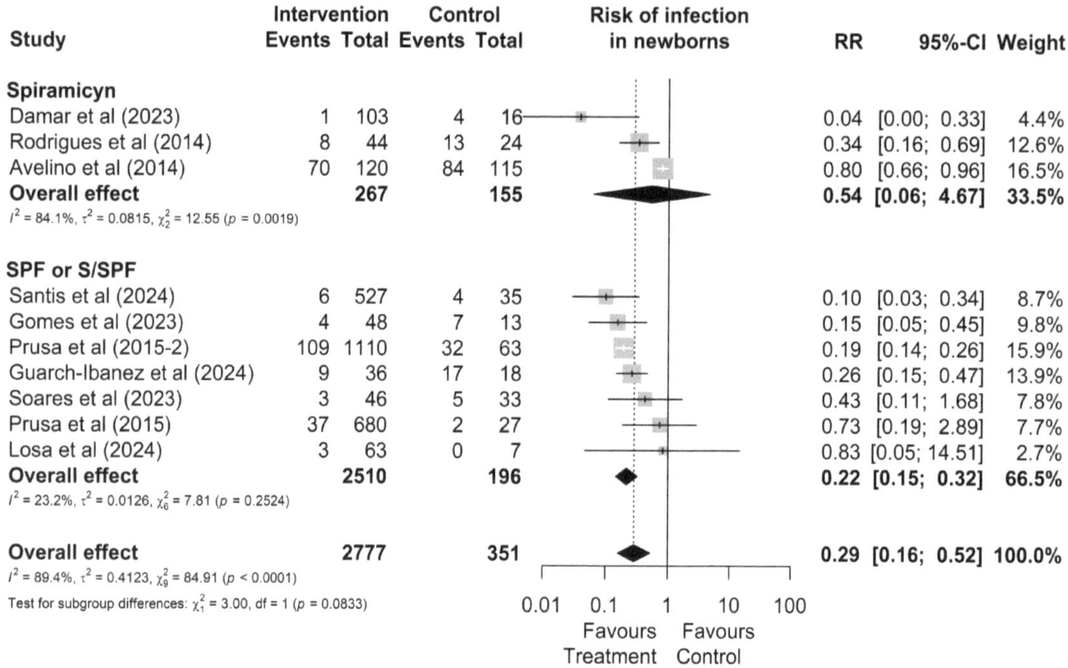

Figure 4. Forest plot of the proportion of infected neonates divided into treatment drug subgroups. CI: confidence interval; studies with more than one drug subgroup; SPF: sulfadiazine + pyrimethamine + folinic acid; W/SPF: spiramycin alternated with SPF [20,22,30,38,44–46,48,50,51].

3.4. Risk of Bias

The graphical summary of the bias risk assessment using the Joanna Briggs Institute critical appraisal tool for all studies included in qualitative and quantitative analyses is shown in Figure 5. Overall, the analyzed studies demonstrated robust methodological quality, with 84% average adherence to the assessment instrument criteria. The risk of bias evaluation revealed that most studies (81%) presented a low risk, while only 11% showed moderate risk, and 8% were classified as high risk. This favorable distribution of bias risk levels substantiates the methodological rigor of the included studies, enhancing the reliability of the findings. No authors reported conflicts of interest. Regarding publication bias, the Egger's test did not show an increased risk; however, funnel plots with the trim-and-fill method suggest the possible omission of three studies regarding the risk of infection and four studies regarding the risk of clinical manifestations (Figure 6).

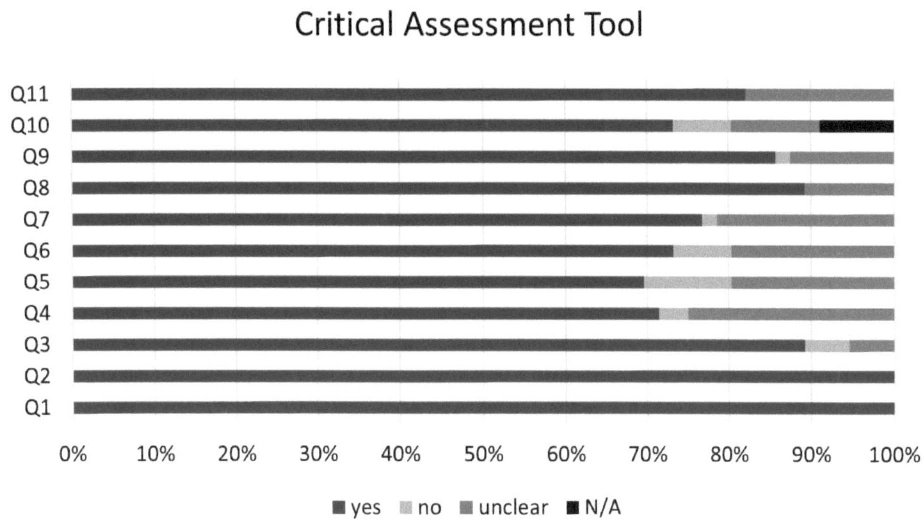

Figure 5. Joanna Briggs Institute Critical Assessment Tool. Q1. Were the two groups similar and recruited from the same population? Q2. Were the exposures measured similarly to assign people to both exposed and unexposed groups? Q3. Was the exposure measured in a valid and reliable way? Q4. Were confounding factors identified? Q5. Were strategies to deal with confounding factors stated? Q6. Were the groups/participants free of the outcome at the start of the study (or at the moment of exposure)? Q7. Were the outcomes measured in a valid and reliable way? Q8. Was the follow up time reported and sufficient to be long enough for outcomes to occur? Q9. Was follow up complete, and if not, were the reasons to loss to follow up described and explored? Q10.Were strategies to address incomplete follow up utilized? Q11. Was appropriate statistical analysis used? N/A—not applicable.

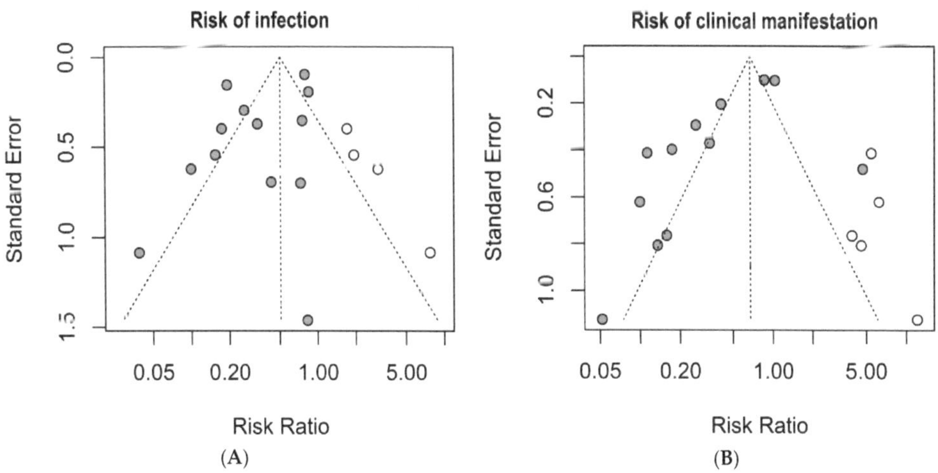

Figure 6. Publication bias risk by the trim-and-fill method and funnel plots. (**A**) risk of infection; (**B**) risk of clinical manifestations. Grey circles: Published studies included in the meta-analysis. White circles: Hypothetical missing studies identified through trim-and-fill analysis. These represent potentially unpublished studies that, if included, would balance the funnel plot and produce a more symmetrical distribution around the estimated true effect size.

4. Discussion

The most effective approach to preventing congenital toxoplasmosis requires comprehensive management from diagnosis through treatment of the mother–child pair. Our meta-analysis demonstrated that maternal treatment significantly reduces both vertical transmission risk (RR = 0.34 [0.21; 0.57) and clinical manifestations in newborns (RR = 0.30 [0.17; 0.56]). While both spiramycin alone and triple therapy (sulfadiazine, pyrimethamine, and folinic acid) showed similar efficacy, triple therapy demonstrated more consistent results with lower heterogeneity between studies. In this systematic review, only 9% of studies reported prior knowledge of women's serological status during the preconception period, highlighting a critical gap in preventive care. These findings emphasize the urgent need to implement protocols promoting early serological screening in women of childbearing age, particularly in countries with high toxoplasmosis incidence, as determining the timing of infection is crucial for optimal therapeutic management [36].

The implementation of monthly serological monitoring represents a critical strategy in preventing congenital toxoplasmosis. This is particularly evident when examining data from certain regions of high-incidence countries like Italy [28] and Brazil [38], where reported rates are 0.12% and 0.15%, respectively. While economic constraints in low- and middle-income countries often challenge the feasibility of monthly testing during prenatal care, the scientific evidence strongly supports that regular monitoring is essential for early detection and appropriate intervention [32]. This is primarily because the effectiveness of antiparasitic treatment significantly decreases once the parasite establishes intracellular infection beyond the initial parasitemia phase.

The critical importance of comprehensive diagnostic approaches is further emphasized by discordant immunoglobulin levels as reported by Fricker-Hidalgo et al. [34], demonstrating that interpretation of current tests extends beyond standard parameters. Misinterpretation or relativization of results outside recommended levels may result in irreversible fetal harm. Beyond standard IgG and IgM measurements, complementary testing for anti-*Toxoplasma* IgA and IgE isotypes [27,29], PCR techniques [36], and proper timing of diagnostic procedures are crucial for accurate diagnosis. The PCR test timing is particularly critical, as it should be performed between the 16th and 30th weeks of gestation and not exceed 4–6 weeks from the estimated infection date to avoid false-negative results. However, the accuracy of the PCR method remains unclear, as the occurrence of true false positive and false negative rates still need to be determined, even considering that diagnosis of congenital toxoplasmosis using a combination of IgG avidity in maternal blood and multiplex nested PCR in amniotic fluid and neonatal blood is helpful to detect a high-risk pregnancy, as well as to diagnose *T. gondii* infection [32].

The number of infected cells and intracellular parasite concentration significantly influence infection pathophysiology, emphasizing the importance of early and accurate diagnosis for optimal prognosis [50]. Variations in PCR diagnostic outcomes may result from multiple factors, including reduced test sensitivity, suboptimal timing of specimen collection relative to infection onset [29,34,50], or treatment initiation effects [27]. These technical challenges underscore why a comprehensive diagnostic approach is essential, incorporating multiple methodologies to ensure accurate detection and monitoring.

While this multi-method diagnostic approach represents the ideal standard of care, its implementation faces significant challenges, particularly in resource-limited settings. Healthcare systems in low- and middle-income countries must balance the substantial costs of comprehensive testing against the potential consequences of delayed or missed diagnoses. This economic consideration becomes particularly relevant given that regular monitoring remains fundamental for successful treatment outcomes, as the parasite's intracellular behavior after initial parasitemia significantly reduces antiparasitic drug efficacy.

The complexity is further compounded by the specificity of acquired immunity to primary *T. gondii* genotype contact. Exposure to different genotypes requires development of new immunological memory, highlighting the importance of continued serological monitoring throughout pregnancy to identify recent infections. This biological characteristic, combined with the time-sensitive nature of effective intervention, reinforces why sustained monitoring, despite its associated costs, remains a cornerstone of effective prevention strategies.

When *T. gondii* infection is confirmed during pregnancy, treatment protocols should follow a clear progression based on gestational timing and diagnostic findings [52]. Initial treatment with spiramycin is indicated until the 16th week of gestation. After this period, the therapeutic approach should be guided by amniocentesis results and ultrasonographic findings. Detection of parasitic DNA through PCR or identification of morphological changes via ultrasound suggests fetal infection, necessitating a transition to triple therapy with sulfadiazine, pyrimethamine, and folinic acid. This protocol modification considers the parasite's complex immunological interactions, as acquired immunity is genotype-specific [53]. The development of new immunological responses is required when encountering different *T. gondii* genotypes, making sustained monitoring essential throughout pregnancy for detecting potential new or reactivated infections. This biological characteristic underscores the importance of a flexible therapeutic approach that can be adjusted based on ongoing diagnostic findings and the specific nature of the infection.

Multiple factors influence the detection of fetal infection, including infection timing, PCR sample collection timing, and treatment initiation, all of which can affect test sensitivity [42]. While imaging studies provide valuable diagnostic information, they cannot definitively rule out fetal infection, particularly retinal involvement. Our meta-analysis demonstrates that maternal treatment reduces both transmission risk (42%) and clinical manifestations (40%), with subgroup analyses indicating more consistent outcomes for combination therapy (SPF or S/SPF) compared to spiramycin alone, aligning with findings from Montoya et al. [4]. These results support transitioning from spiramycin monotherapy to combination treatment after the initial gestational period, as spiramycin alone does not effectively treat fetal infection.

The timing of the immunological response presents additional diagnostic challenges in neonatal cases. The interval between placental and fetal infection allows anti-*Toxoplasma* maternal IgG antibody transfer, which can suppress fetal antibody production. This dynamic may result in negative anti-*Toxoplasma* IgM antibodies at birth despite infection, even with highly sensitive testing methods. Conversely, when maternal infection occurs near delivery, neonates may develop positive serology within weeks after birth, though subsequent confirmatory testing may become negative. These variable serological patterns emphasize the importance of comprehensive follow-up, as demonstrated by Lago et al. [54], who found that initiating treatment within two months of life serves as a protective factor against late-onset retinochoroiditis, which may manifest later in life [55].

Neonatal treatment during the first year of life aims to control infection until the child develops a sufficient immune response to inhibit parasite proliferation. McLeod et al. [56] emphasizes that treatment requires individually compounded formulations based on the child's weight, aligning with World Health Organization guidelines for patient safety and pharmacotherapeutic principles [56,57]. The implementation of standardized syrup formulations is particularly crucial as these medications are typically only available in adult tablet form, helping optimize dosing while minimizing handling errors and inappropriate administration. However, as Trotta et al. [33] note, even with appropriate postnatal treatment, serious complications may occur, especially in cases of first-trimester infection.

The lack of global consensus on toxoplasmosis screening during pregnancy presents an ongoing challenge [58]. While some countries mandate comprehensive maternal monitoring, others do not recommend routine screening [32,59]. Our systematic review provides evidence supporting standardized protocols that include therapeutic regimen modification after the 16th–18th weeks of pregnancy, coupled with monthly serological monitoring. This approach is particularly important given that congenital toxoplasmosis is primarily associated with delayed maternal diagnosis and subsequent treatment initiation [39]. The evidence from our analysis supports improving surveillance of women of childbearing age, pregnant women, and infected children, as such measures have demonstrated reduced incidence of congenital toxoplasmosis [38]. Additionally, our findings reinforce the importance of immediate postnatal treatment initiation and standardized medication preparation protocols [44].

Our meta-analysis presents important limitations that should be considered. The high heterogeneity observed could not be fully explained by our subgroup analyses. Important clinical factors, such as the gestational age at infection, were not consistently reported across studies, limiting our ability to control for this potential confounder. Additionally, the included studies varied in design and diagnostic criteria, contributing to the observed heterogeneity. Consequently, the aggregated results should be interpreted with caution, especially since only observational studies were included. While our findings strongly support monthly monitoring and comprehensive diagnostic approaches, we acknowledge the economic challenges faced by healthcare systems, particularly in low- and middle-income countries. Treatment approaches did not account for potential differences in gestational infection by different genotypes, which could influence treatment resistance and the immune response. Indeed, there are several studies showing that certain genotypes display differential sensitivity to common treatments that are preconized nowadays [60,61]. For instance, type I strains, known for their high virulence in mice, may exhibit resistance to some drugs due to differences in their metabolic pathways or drug target sites; type II and III strains, which are more prevalent in human infections, may show varying degrees of susceptibility based on their enzyme activity related to folate metabolism and drug detoxification; and atypical strains found in specific geographic regions may have unique genetic mutations that affect drug efficacy [60,61]. Understanding these genotype-specific responses is crucial for optimizing treatment strategies and developing more effective therapies against *T. gondii* infections. Future research should focus on developing and validating cost-effective monitoring strategies that maintain clinical efficacy while considering resource constraints in different healthcare settings. Thus, future studies on congenital toxoplasmosis need to concentrate on new diagnostic tools, novel drugs with efficacy against potentially resistant genotypes and fewer side effects, as well as innovative strategies for health education aimed at women of childbearing age.

Overall, the qualitative analysis of this systematic review focused on the main aspects defined as useful to clinical practice and that may serve as guidelines for future research. First, prior knowledge of the serological status of women of childbearing age, and the detection of seroconversion in non-reactive pregnant women or those reinfected by *T. gondii*, should be encouraged monthly. This contributes to more assertive and prompter decision-making regarding treatment and better prognosis for infections caused by this disease. Additionally, we suggest the implementation of adequate formulas specifically prepared for neonates and children to minimize improper manipulation of these medications through home preparations. This allows for the optimization of medication doses, reduces the risk of using these medications in inadequate doses, and promotes the rational use of these medications, contributing to the goal of minimizing clinical manifestations in these children.

In conclusion:

- The occurrence of congenital toxoplasmosis remains a significant health problem in numerous countries. There is no global consensus on the traceability of toxoplasmosis during pregnancy. While some countries advocate monitoring all pregnant women, others do not recommend it, and the choice of treatment conduct is not yet well-established. The frequency of congenital toxoplasmosis is primarily associated with late, inaccurate, or nonexistent diagnosis during pregnancy, leading to delays or the absence of adequate treatment;
- This systematic review and meta-analysis of articles published from 2013 to 2025, selected based on inclusion and exclusion criteria, reveals an urgent need to establish standardization for therapeutic protocols. This is particularly crucial after the 16th or 18th week of pregnancy. Additionally, monthly monitoring of pregnant women with serological tests is recommended as a predictor to reduce the vertical transmission of *Toxoplasma gondii*. Ensuring proper healthcare access and promoting adequate treatment will improve the overall health of pregnant women and their children;
- Future studies on congenital toxoplasmosis that address procedures in the mother–child dyad should concentrate on new diagnostic tools, novel drugs with efficacy against potentially resistant genotypes and fewer side effects, as well as innovative strategies for health education aimed at women of childbearing age.

Supplementary Materials: The following supporting information can be downloaded at: https://www.mdpi.com/article/10.3390/microorganisms13040723/s1, Table S1: Cohort selected studies among 1089 initially screened meeting inclusion but not exclusion criteria, with a focus on the treatment of congenital toxoplasmosis. Refs. [45–49,51,62–81] can be found in Supplementary Materials.

Author Contributions: S.K.R. participated in the study design, data collection and analysis, wrote the original manuscript, and approval its final version; I.M.M. participated in the data selection, statistical analysis, and review of this manuscript and approved its final version; A.C.R.C. participated in the data analysis and review of this manuscript and approved its final version; A.C.A.M.P. participated in the data collection and review and approval of the final version of this manuscript; T.W.P.M. participated in the idealization, planning, review, and approval of the final version; J.R.M. participated in the study design, data collection and analysis, and review and approval of the final version of this manuscript. All authors have read and agreed to the published version of the manuscript.

Funding: This study was financially supported by Brazilian research agencies (Fundação de Amparo à Pesquisa do Estado de Minas Gerais FAPEMIG, Grant # RED-00198-23 and #APQ-01313-14; Conselho Nacional de Desenvolvimento Científico e Tecnológico (CNPq, Brazil) Grant # 312201/2021-4 and # INCT-Toxo 406572/2022-4; Coordenação de Aperfeiçoamento de Pessoal de Nível Superior-CAPES, Grant # AUXPE-02450/09-7).

Institutional Review Board Statement: Not applicable.

Informed Consent Statement: Not applicable.

Data Availability Statement: The original contributions presented in this study are included in the article/Supplementary Materials. Further inquiries can be directed to the corresponding author.

Conflicts of Interest: The authors declare no conflicts of interest.

References

1. Aguirre, A.A.; Longcore, T.; Barbieri, M.; Dabritz, H.; Hill, D.; Klein, P.N.; Lepczyk, C.; Lilly, E.L.; McLeod, R.; Milcarsky, J.; et al. The One Health Approach to Toxoplasmosis: Epidemiology, Control, and Prevention Strategies. *Ecohealth* **2019**, *16*, 378–390. [CrossRef] [PubMed]
2. Vieira, G.C.D.; Duarte, E.S.M. Atypical strains of *Toxoplasma gondii* and its impact on development of toxoplasmosis. *Res. Soc. Dev.* **2023**, *12*, 2. [CrossRef]
3. Kamus, L.; Belec, S.; Lambrecht, L.; Abasse, S.L.; Olivier, S.; Combe, P.L.; Bonnave, P.-E.; Vauloup-Fellous, C.L.; Jaffe, C. Maternal and congenital toxoplasmosis in Mayotte: Prevalence, incidence and management. *PLoS Negl. Trop. Dis.* **2023**, *17*, e0011198. [CrossRef]
4. Montoya, J.G.; Laessig, K.; Fazeli, M.S.; Siliman, G.; Yoon, S.S.; Drake-Shanahan, E.; Zhu, C.Y.; Akbary, A.; McLeod, R. A fresh look at the role of spiramycin in preventing a neglected disease: Meta-analyses of observational studies. *Eur. J. Med. Res.* **2021**, *26*, 143. [CrossRef]
5. Konstantinovic, N.; Guegan, H.; Stäjner, T.; Belaz, S.; Robert-Gangneux, F. Treatment of toxoplasmosis: Current options and future perspectives. *Food Waterborne Parasitol.* **2019**, *15*, e00036. [CrossRef]
6. Deng, Y.; Wu, T.; Zhai, S.Q.; Li, C.H. Recent progress on anti-Toxoplasma drugs discovery: Design, synthesis and screening. *Eur. J. Med. Chem.* **2019**, *183*, 111711. [CrossRef]
7. Wallon, M.; Peyron, F. Effect of Antenatal Treatment on the Severity of Congenital Toxoplasmosis. *Clin. Infect. Dis.* **2015**, *62*, 811–812. [CrossRef]
8. Dunay, I.R.; Gajurel, K.; Dhakal, R.; Liesenfeld, O.; Montoya, J.G. Treatment of Toxoplasmosis: Historical Perspective, Animal Models, and Current Clinical Practice. *Clin. Microbiol. Rev.* **2018**, *31*, e00057-17. [CrossRef]
9. Alday, P.H.; Doggett, J.S. Drugs in development for toxoplasmosis: Advances, challenges, and current status. *Drug Des. Devel. Ther.* **2017**, *11*, 273–293. [CrossRef]
10. Chorlton, S.D. Adjunctive bradyzoite-directed therapy for reducing complications of congenital toxoplasmosis. *Med. Hypotheses* **2019**, *133*, 109376. [CrossRef]
11. Dos Santos, D.A.; Souza, H.F.S.; Silber, A.M.; Souza, T.D.A.C.B.D.; Ávila, A.R. Protein kinases on carbon metabolism: Potential targets for alternative chemotherapies against toxoplasmosis. *Front. Cell. Infect. Microbiol.* **2023**, *13*, 1175409. [CrossRef]
12. Page, M.; McKenzie, J.; Bossuyt, P.; Boutron, I.; Hoffmann, T.; Mulrow, C.; Shamseer, L.; Tetzlaff, J.; Akl, E.; Brennan, S.; et al. The PRISMA 2020 statement: An updated guideline for reporting systematic reviews. *Syst. Rev.* **2021**, *10*, 89. [CrossRef] [PubMed]
13. Ouzzani, M.; Hammady, H.; Fedorowicz, Z.; Elmagarmid, A. Rayyan—A web and mobile app for systematic reviews. *Syst. Rev.* **2016**, *5*, 210. [CrossRef] [PubMed]
14. The Joanna Briggs. The Joanna Briggs Institute Reviewers'. In *Manual 2015: Methodology for JBI Scoping Reviews*; Joanne Briggs Institute: Adelaide, Australia, 2015; pp. 1–24.
15. R Core Team. R: A Language and Environment for Statistical Computing [Internet]. Vienna R Foundation for Statistical Computing. 2019. Available online: https://www.r-project.org (accessed on 10 February 2025).
16. Balduzzi, S.; Rücker, G.; Schwarzer, G. How to perform a meta-analysis with R: A practical tutorial. *Evid. Based Ment. Health* **2019**, *22*, 153–160. [CrossRef]
17. Viechtbauer, W. Conducting meta-analyses in R with the metafor package. *J. Stat. Softw.* **2010**, *36*, 1–48. [CrossRef]
18. Higgins, J.P.T.; Thompson, S.G.; Spiegelhalter, D.J. A re-evaluation of random-effects meta-analysis. *J. R. Stat. Soc. Ser. A Stat. Soc.* **2009**, *172*, 137–159. [CrossRef]
19. Carral, L.; Kaufer, F.; Olejnik, P.; Freuler, C.; Durlach, R. Prevention of congenital toxoplasmosis in a Buenos Aires hospital. *Medicina* **2013**, *73*, 238–242.
20. Damar Çakırca, T.; Can, İ.N.; Deniz, M.; Torun, A.; Akçabay, Ç.; Güzelçiçek, A. Toxoplasmosis: A Timeless Challenge for Pregnancy. *Trop. Med. Infect. Dis.* **2023**, *8*, 63. [CrossRef]
21. Olariu, T.R.; Press, C.; Talucod, J.; Olson, K.; Montoya, J.G. Toxoplasmose congénitale aux États-Unis: Observations cliniques et sérologiques chez les nourrissons nés de mères traitées pendant la grossesse. *Parasite* **2019**, *26*, 13. [CrossRef]
22. Prusa, A.-R.; Kasper, D.C.; Pollak, A.; Olischar, M.; Gleiss, A.; Hayde, M. Amniocentesis for the detection of congenital toxoplasmosis: Results from the nationwide Austrian prenatal screening program. *Clin. Microbiol. Infect. Off. Publ. Eur. Soc. Clin. Microbiol. Infect. Dis.* **2015**, *21*, 191.e1–191.e8. [CrossRef]
23. Buonsenso, D.; Pata, D.; Colonna, A.T.; Iademarco, M.; de Santis, M.; Masini, L.; Conti, G.; Molle, F.; Baldascino, A.; Acampora, A.; et al. Spyramicine and Trimethoprim-Sulfamethoxazole Combination to Prevent Mother-To-Fetus Transmission of *Toxoplasma gondii* Infection in Pregnant Women: A 28-Years Single-center Experience. *Pediatr. Infect. Dis. J.* **2022**, *41*, E223–E227. [CrossRef] [PubMed]
24. Conceição, A.R.; Belucik, D.N.; Missio, L.; Gustavo Brenner, L.; Henrique Monteiro, M.; Ribeiro, K.S.; Costa, D.F.; Valadão, M.C.D.S.; Commodaro, A.G.; de Oliveira Dias, J.R.; et al. Ocular Findings in Infants with Congenital Toxoplasmosis after a Toxoplasmosis Outbreak. *Ophthalmology* **2021**, *128*, 1346–1355. [CrossRef] [PubMed]

25. De Araújo, T.E.; Gomes, A.O.; Coelho-dos-Reis, J.G.; Carneiro, A.C.A.V.; Machado, A.S.; Andrade, G.M.Q.; Vasconcelos-Santos, D.V.; Januário, J.N.; Peruhype-Magalhães, V.; Teixeira-Carvalho, A.; et al. Long-term impact of congenital toxoplasmosis on phenotypic and functional features of circulating leukocytes from infants one year after treatment onset. *Clin. Immunol.* **2021**, *232*, 108859. [CrossRef] [PubMed]
26. Findal, G.; Helbig, A.; Haugen, G.; Jenum, P.A.; Stray-Pedersen, B. Management of suspected primary *Toxoplasma gondii* infection in pregnant women in Norway: Twenty years of experience of amniocentesis in a low-prevalence population. *BMC Pregnancy Childbirth* **2017**, *17*, 1–9. [CrossRef]
27. Guegan, H.; Stajner, T.; Bobic, B.; Press, C.; Olariu, R.T.; Olson, K.; Srbljanovic, J.; Montoya, J.G.; Djurković-Djaković, O.; Robert-Gangneux, F. Maternal Anti-Toxoplasma Treatment during Pregnancy Is Associated with Reduced Sensitivity of Diagnostic Tests for Congenital Infection in the Neonate. *J. Clin. Microbiol.* **2021**, *59*, e01368-20. [CrossRef]
28. Piffer, S.; Lauriola, A.L.; Pradal, U.; Collini, L.; Dell'Anna, L.; Pavanello, L. *Toxoplasma gondii* infection during pregnancy: A ten-year observation in the province of Trento, Italy. *Le Infez. Med.* **2020**, *28*, 603–610.
29. Olariu, T.R.; Remington, J.S.; McLeod, R.; Alam, A.; Montoya, J.G. Severe congenital toxoplasmosis in the United States: Clinical and serologic findings in untreated infants. *Pediatr. Infect. Dis. J.* **2011**, *30*, 1056–1061. [CrossRef]
30. Soares, J.A.S.; Holzmann, A.P.F.; Alves, B.B.S.; Lima, C.F.Q.; Caldeira, A.P. Profile of pregnant women and children accompanied due to *T. gondii* exposure at a referred healthcare center: What has changed in 10 years? *Rev. Bras. Saude Matern. Infant.* **2023**, *23*, e20220225. [CrossRef]
31. Valentini, P.; Buonsenso, D.; Barone, G.; Serranti, D.; Calzedda, R.; Ceccarelli, M.; Speziale, D.; Ricci, R.; Masini, L. Spiramycin/cotrimoxazole versus pyrimethamine/sulfonamide and spiramycin alone for the treatment of toxoplasmosis in pregnancy. *J. Perinatol.* **2015**, *35*, 90–94. [CrossRef]
32. Yamada, H.; Tanimura, K.; Deguchi, M.; Tairaku, S.; Morizane, M.; Uchida, A.; Ebina, Y.; Nishikawa, A. A cohort study of maternal screening for congenital *Toxoplasma gondii* infection: 12 years' experience. *J. Infect. Chemother.* **2019**, *25*, 427–430. [CrossRef]
33. Trotta, M.; Trotta, A.; Spataro, E.; Giache, S.; Borchi, B.; Zammarchi, L.; Campolmi, I.; Galli, L.; Pasquini, L. Primary toxoplasmosis acquired during early pregnancy: Is it currently overestimated? *Eur. J. Obstet. Gynecol. Reprod. Biol.* **2021**, *267*, 285–289. [CrossRef] [PubMed]
34. Fricker-Hidalgo, H.; Cimon, B.; Chemla, C.; Darde, M.L.; Delhaes, L.; L'Ollivier, C.; Godineau, N.; Houze, S.; Paris, L.; Quinio, D.; et al. Toxoplasma seroconversion with negative or transient immunoglobulin M in pregnant women: Myth or reality? A French multicenter retrospective study. *J. Clin. Microbiol.* **2013**, *51*, 2103–2111. [CrossRef] [PubMed]
35. Vimercati, A.; Chincoli, A.; de Gennaro, A.C.; Calvario, A.; Amendolara, M.; Del Gaudio, G.; Laforgia, N.; Carbonara, S. Congenital toxoplasmosis and proposal of a new classification for the likelihood of primary maternal infection: Analysis of 375 cases in Southeast Italy. *J. Matern. Neonatal Med.* **2020**, *33*, 3746–3751. [CrossRef] [PubMed]
36. Wallon, M.; Peyron, F.; Cornu, C.; Vinault, S.; Abrahamowicz, M.; Kopp, C.; Binquet, C. Congenital toxoplasma infection: Monthly prenatal screening decreases transmission rate and improves clinical outcome at age 3 years. *Clin. Infect. Dis.* **2013**, *56*, 1223–1231. [CrossRef]
37. Hijikata, M.; Morioka, I.; Okahashi, A.; Nagano, N.; Kawakami, K.; Komatsu, A.; Kawana, K.; Ohyama, S.; Fujioka, K.; Tanimura, K.; et al. A prospective cohort study of newborns born to mothers with serum *Toxoplasma gondii* immunoglobulin M positivity during pregnancy. *J. Infect. Chemother.* **2022**, *28*, 486–491. [CrossRef]
38. Gomes-Ferrari-Strang, A.G.; Ferrar, R.G.; Falavigna-Guilherme, A.L. Gestational toxoplasmosis treatment changes the child's prognosis: A cohort study in southern Brazil. *PLoS Negl. Trop. Dis.* **2023**, *17*, e0011544. [CrossRef]
39. Righi, N.C.; Hermes, L.; Piccini, A.D.; Branco, J.C.; Skupien, J.A.; Weinmann, A.R.M.; Valadao, M.C.D.; Schuch, N.J. Epidemiological profile of gestational and congenital toxoplasmosis cases arising out of the population outbreak. *Sci. Med.* **2021**, *31*, e40108. [CrossRef]
40. Tibúrcio, J.D.; Vasconcelos-Santos, D.V.; Vasconcelos, G.C.; Carellos, E.V.M.; Romanelli, R.M.d.C.; Januario, J.N.; Andrade, G.M.Q. Psychometric properties of CVFQ7-BR-toxo to evaluate vision-related quality of life in children with congenital toxoplasmosis in Brazil. *Arq. Bras. Oftalmol.* **2022**, *85*, 46–58. [CrossRef]
41. Bartholo, B.B.G.R.; Monteiro, D.L.M.; Rodrigues, N.C.P.; Trajano, A.J.B.; de Jesus, N.R.; Cardoso, F.F.O.; de Souza, F.M.; Werner, H.; Araujo Júnior, E. Treatment of Acute Toxoplasmosis in Pregnancy: Influence in the Mother-to-Child Transmission. *J. Obstet. Gynaecol. Can.* **2020**, *42*, 1505–1510. [CrossRef]
42. Avci, M.E.; Arslan, F.; Çiftçi, S.; Ekiz, A.; Tüten, A.; Yildirim, G.; Madazli, R. Role of spiramycin in prevention of fetal toxoplasmosis. *J. Matern. Neonatal Med.* **2016**, *29*, 2073–2076. [CrossRef]
43. Mandelbrot, L.; Kieffer, F.; Sitta, R.; Laurichesse-Delmas, H.; Winer, N.; Mesnard, L.; Berrebi, A.; Le Bouar, G.; Bory, J.-P.P.; Cordier, A.-G.G.; et al. Prenatal therapy with pyrimethamine plus sulfadiazine vs spiramycin to reduce placental transmission of toxoplasmosis: A multicenter, randomized trial. *Am. J. Obstet. Gynecol.* **2018**, *219*, e1–e386. [CrossRef]
44. Avelino, M.M.; Amaral, W.N.; Rodrigues, I.M.; Rassi, A.R.; Gomes, M.B.; Costa, T.L.; Castro, A.M. Congenital toxoplasmosis and prenatal care state programs. *BMC Infect. Dis.* **2014**, *14*, 33. [CrossRef]

45. De Santis, M.; Tartaglia, S.; Apicella, M.; Visconti, D.; Noia, G.; Valentini, P.; Lanzone, A.; Santangelo, R.; Masini, L. The prevention of congenital toxoplasmosis using a combination of Spiramycin and Cotrimoxazole: The long-time experience of a tertiary referral centre. *Trop. Med. Int. Health* **2024**, *29*, 697–705. [CrossRef] [PubMed]
46. Guarch-Ibáñez, B.; Carreras-Abad, C.; Frick, M.A.; Blázquez-Gamero, D.; Baquero-Artigao, F.; Fuentes, I. REIV-TOXO Project: Results from a Spanish cohort of congenital toxoplasmosis (2015–2022). The beneficial effects of prenatal treatment on clinical outcomes of infected newborns. *PLoS Neglected Trop. Dis.* **2024**, *18*, e0012619. [CrossRef]
47. Lago, E.G.; Oliveira, A.P.; Bender, A.L. Presence and duration of anti-*Toxoplasma gondii* immunoglobulin M in infants with congenital toxoplasmosis. *J. Pediatr.* **2014**, *90*, 363–369. [CrossRef]
48. Losa, A.; Carvalho, I.; Sousa, B.; Ashworth, J.; Guedes, A.; Carreira, L.; Pinho, L.; Godinho, C. Congenital Toxoplasmosis Diagnosis: Challenges and Management Outcomes. *Cureus* **2024**, *16*, e52971. [CrossRef]
49. Mejia-Oquendo, M.; Marulanda-Ibarra, E.; Gomez-Marin, J.E. Evaluation of the impact of the first evidence-based guidelines for congenital toxoplasmosis in Armenia (Quindío) Colombia: An observational retrospective analysis. *Lancet Reg. Health—Am.* **2021**, *1*, 100010. [CrossRef]
50. Prusa, A.-R.; Kasper, D.C.; Pollak, A.; Gleiss, A.; Waldhoer, T.; Hayde, M. The Austrian toxoplasmosis register, 1992–2008. *Clin. Infect. Dis.* **2015**, *60*, e4–e10. [CrossRef]
51. Rodrigues, I.M.; Costa, T.L.; Avelar, J.B.; Amaral, W.N.; Castro, A.M.; Avelino, M.M. Assessment of laboratory methods used in the diagnosis of congenital toxoplasmosis after maternal treatment with spiramycin in pregnancy. *BMC Infect. Dis.* **2014**, *14*, 349. [CrossRef]
52. Felín, M.S.; Wang, K.; Moreira, A.; Grose, A.; Leahy, K.; Zhou, Y.; Clouser, F.A.; Siddiqui, M.; Leong, N.; Goodall, P.; et al. Building Programs to Eradicate Toxoplasmosis Part I: Introduction and Overview. *Curr. Pediatr. Rep.* **2022**, *10*, 57–92. [CrossRef]
53. Ferguson, D.J. Use of molecular and ultrastructural markers to evaluate stage conversion of *Toxoplasma gondii* in both the intermediate and definitive host. *Int. J. Parasitol.* **2004**, *34*, 347–360. [CrossRef] [PubMed]
54. Lago, E.G.; Endres, M.M.; Scheeren, M.F.D.C.; Fiori, H.H. Ocular Outcome of Brazilian Patients with Congenital Toxoplasmosis. *Pediatr. Infect. Dis. J.* **2021**, *40*, e21–e27. [CrossRef] [PubMed]
55. Wallon, M.; Garweg, J.G.; Abrahamowicz, M.; Cornu, C.; Vinault, S.; Quantin, C.; Bonithon-Kopp, C.; Picot, S.; Peyron, F.; Binquet, C. Ophthalmic outcomes of congenital toxoplasmosis followed until adolescence. *Pediatrics* **2014**, *133*, e601–e608. [CrossRef]
56. McLeod, R.; Lykins, J.; Gwendolyn Noble, A.; Rabiah, P.; Swisher, C.N.; Heydemann, P.T.; McLone, D.; Frim, D.; Withers, S.; Clouser, F.; et al. Management of Congenital Toxoplasmosis. *Curr. Pediatr. Rep.* **2014**, *2*, 166–194. [CrossRef]
57. World Health Organization. *Promoting Safety of Medicines for Children*; WHO Press: Geneva, Switzerland, 2007; 59p, ISBN 978-92-4-156343-7.
58. Koçak, Ö.; Kan, Ö.; Kocak, O.; Kan, O. Results of the Toxoplasmosis Screening in 9311 Pregnant Women in a Tertiary Center in Turkey. *Flora* **2020**, *25*, 332–338. [CrossRef]
59. Kalem, A.K.; Hasanoğlu, I.; Ayhan, M.; Kayaaslan, B.; Eser, F.; Oğuz, Y.; Avşar, F.Y.; Güner, R. Toxoplasmosis in pregnancy: Test, treatment and outcome. *Eur. Res. J.* **2022**, *8*, 296–303. [CrossRef]
60. Montazeri, M.; Mehrzadi, S.; Sharif, M.; Sarvi, S.; Tanzifi, A.; Aghayan, S.A.; Daryani, A. Drug Resistance in *Toxoplasma gondii*. *Front. Microbiol.* **2018**, *9*, 2587.
61. Brito, R.M.d.M.; Bessa, G.d.L.; Bastilho, A.L.; Dantas-Torres, F.; de Andrade-Neto, V.F.; Bueno, L.L.; Fujiwara, R.T.; Magalhães, L.M.D. Genetic diversity of *Toxoplasma gondii* in South America: Occurrence, immunity, and fate of infection. *Parasites Vectors* **2023**, *16*, 461.
62. Andrade, F.M.; Santana, E.F.M.; Júnior, E.A.; Andrade, S.G.A.; Bortoletti Filho, J.; Amed, A.M.; Moron, A.F. Polymerase chain reaction analysis of amniotic fluid for diagnosis of fetal toxoplasmosis. *Clin. Exp. Obstet. Gynecol.* **2019**, *46*, 593–595. [CrossRef]
63. Boudaouara, Y.; Aoun, K.; Maatoug, R.; Souissi, O.; Bouratbine, A.; Ben Abdallah, R. Congenital toxoplasmosis in Tunisia: Prenatal and neonatal diagnosis and postnatal follow-up of 35 cases. *Am. J. Trop. Med. Hyg.* **2018**, *98*, 1722–1726. [CrossRef]
64. Bouhlel, S.; Ben Abdallah, R.; Aoun, K.; Maatoug, R.; Souissi, O.; Bouratbine, A. Management of Toxoplasmic Seroconversion in the Third Trimester of Pregnancy in Tunisia. *Bull. De La Soc. De Pathol. Exot. (1990)* **2018**, *111*, 269–274. [CrossRef] [PubMed]
65. Capobiango, J.D.; Breganó, R.M.; Navarro, I.T.; Neto, C.P.R.; Casella, A.M.B.; Mori, F.M.R.L.; Pagliari, S.; Inoue, I.T.; Reiche, E.M.V. Congenital toxoplasmosis in a reference center of Paraná, Southern Brazil. *Braz. J. Infect. Dis.* **2014**, *18*, 364–3671. [CrossRef] [PubMed]
66. Carellos, E.V.M.; de Andrade, J.Q.; Romanelli, R.M.C.; Tibúrcio, J.D.; Januário, J.N.; Vasconcelos-Santos, D.V.; Figueiredo, R.M.; de Andrade, G.M.Q. High Frequency of Bone Marrow Depression During Congenital Toxoplasmosis Therapy in a Cohort of Children Identified By Neonatal Screening in Minas Gerais, Brazil. *Pediatr. Infect. Dis. J.* **2017**, *36*, 1169–1176. [CrossRef] [PubMed]
67. Caro-Garzón, J.D.; Gómez-Henck, C.; Jaramillo-Giraldo, T.; Cifuentes-Botero, J.M.; Gómez-Marín, J.E. Evaluation of the avidity test for the follow up on children treated for congenital toxoplasmosis during the first year of life. *Iatreia* **2021**, *34*, 25–32. [CrossRef]

68. De Paula, H.L.; de Lucca, L.; Vendrame, S.A.; Wess, L.C.; Stein, C.d.S.; Moresco, R.N.; Beck, S.T.; Gonçalves, T.d.L. Delta-aminolevulinate dehydratase enzyme activity and the oxidative profile of pregnant women being treated for acute toxoplasmosis. *Microb. Pathog.* **2022**, *164*, 105455. [CrossRef]
69. Diesel, A.A.; Zachia, S.d.A.; Müller, A.L.L.; Perez, A.V.; Uberti, F.A.d.F.; Magalhães, J.A.d.A. Follow-up of Toxoplasmosis during Pregnancy: Ten-Year Experience in a University Hospital in Southern Brazil. *Rev. Bras. Ginecol. Obstet.* **2019**, *41*, 539–547. [CrossRef]
70. Donadono, V.; Saccone, G.; Maruotti, G.M.; Berghella, V.; Migliorini, S.; Esposito, G.; Sirico, A.; Tagliaferri, S.; Ward, A.; Mazzarelli, L.L. Incidence of toxoplasmosis in pregnancy in Campania: A population-based study on screening, treatment, and outcome. *Eur. J. Obstet. Gynecol. Reprod. Biol.* **2019**, *240*, 316–321. [CrossRef]
71. Donadono, V.; Saccone, G.; Sarno, L.; Esposito, G.; Mazzarelli, L.L.; Sirico, A.; Guida, M.; Martinelli, P.; Zullo, F.; Maruotti, G.M. Association between lymphadenopathy after toxoplasmosis seroconversion in pregnancy and risk of congenital infection. *Eur. J. Clin. Microbiol. Infect. Dis.* **2022**, *41*, 45–51. [CrossRef]
72. Evangelista, F.F.; Mantelo, F.M.; de Lima, K.K.; Marchioro, A.A.; Beletini, L.F.; de Souza, A.H.; Santana, P.L.; Riedo, C.d.O.; Higa, L.T.; Guilherme, A.L.F. Prospective evalution of pregnant women with suspected acute toxoplasmosis treated in a reference prenatal care clinic at a university teaching hospital in Southern Brazil. *Rev. Inst. Med. Trop. Sao Paulo* **2020**, *62*, 46. [CrossRef]
73. Ludwig, A.; Fernandes, F.D.; Guerra, R.R.; Braünig, P.; Ramos, L.S.; Pacheco, L.S.; Sangioni, L.A.; Vogel, F.S.F. Molecular detection of *Toxoplasma gondii* in placentas of women who received therapy during gestation in a toxoplasmosis outbreak. *Infect. Genet. Evol.* **2022**, *97*, 105145. [CrossRef]
74. Mueller, R.A.S.; Frota, A.C.C.; Barreto, D.D.M.; Vivacqua, D.P.F.; Loria, G.B.; Lebreiro, G.P.; Martins, M.G.; Potsch, M.V.; Maia, P.D.; Coutinho, R.L.M.; et al. Congenital Toxoplasmosis: Missed Opportunities for Diagnosis and Prevention. *J. Trop. Pediatr.* **2021**, *67*, fmaa069. [CrossRef] [PubMed]
75. Pengsaa, K.; Hattasingh, W. Congenital toxoplasmosis: An uncommon disease in Thailand. *Paediatr. Int. Child Health* **2015**, *35*, 56–60. [CrossRef] [PubMed]
76. Prasil, P.; Sleha, R.; Kacerovsky, M.; Bostik, P. Comparison of adverse reactions of spiramycin versus pyrimethamine/sulfadiazine treatment of toxoplasmosis in pregnancy: Is spiramycin really the drug of choice for unproven infection of the fetus? *J. Matern.-Fetal Neonatal Med.* **2023**, *36*, 2215377. [CrossRef] [PubMed]
77. Saghrouni, F.; Khammari, I.; Ben Abdeljelil, J.; Yaacoub, A.; Meksi, S.G.; Ach, H.; Garma, L.; Fathallah, A.; Ben Saïd, M. La toxoplasmose congénitale: à propos de 21 cas. *J. Pédiatr. Puéricult.* **2013**, *26*, 83–89. [CrossRef]
78. Teil, J.; Dupont, D.; Charpiat, B.; Corvaisier, S.; Vial, T.; Leboucher, G.; Wallon, M.; Peyron, F. Treatment of Congenital Toxoplasmosis: Safety of the Sulfadoxine-Pyrimethamine Combination in Children Based on a Method of Causality Assessment. *Pediatr. Infect. Dis. J.* **2016**, *35*, 634–638. [CrossRef]
79. Villar, B.B.D.L.F.; Neves, E.d.S.; Louro, V.C.; Lessa, J.F.; Rocha, D.N.; Gomes, L.H.F.; Junior, S.C.G.; Pereira, J.P.; Moreira, M.E.L.; Guida, L.d.C. Toxoplasmosis in pregnancy: A clinical, diagnostic, and epidemiological study in a referral hospital in Rio de Janeiro, Brazil. *Braz. J. Infect. Dis.* **2020**, *24*, 517–523. [CrossRef]
80. Yamamoto, L.; Targa, L.S.; Sumita, L.M.; Shimokawa, P.T.; Rodrigues, J.C.; Kanunfre, K.A.; Okay, T.S. Association of Parasite Load Levels in Amniotic Fluid With Clinical Outcome in Congenital Toxoplasmosis. *Obstet. Gynecol.* **2017**, *130*, 335–345. [CrossRef]
81. Zuluaga, L.M.; Hernández, J.C.; Castaño, C.F.; Donado, J.H. Effect of antenatal spiramycin treatment on the frequency of retinochoroiditis due to congenital toxoplasmosis in a Colombian cohort. *Biomedica* **2017**, *37* (Suppl. S1), 86–91. [CrossRef]

microorganisms

MDPI

Communication

Investigation of Virulence-Related Markers in Atypical Strains of *Toxoplasma gondii* from Brazil

Júlia Gatti Ladeia Costa [1,†], Érica Santos Martins-Duarte [2,†], Lorena Velozo Pinto [2], Ramon Araujo de Castro Baraviera [2], Wagner Martins Fontes do Rego [2] and Ricardo Wagner de Almeida Vitor [2,*]

1 Departamento de Ciências Biológicas, Universidade do Estado de Minas Gerais, Ibirité 32412-190, Brazil; julia.costa@uemg.br

2 Departamento de Parasitologia, Universidade Federal de Minas Gerais, Belo Horizonte 31270-901, Brazil; emartinsduarte@gmail.com (É.S.M.-D.); nena.twist@hotmail.com (L.V.P.); ramonbaraviera@gmail.com (R.A.d.C.B.); wagner.fontesrego@gmail.com (W.M.F.d.R.)

* Correspondence: ricardovitor@icb.ufmg.br

† These authors contributed equally to this work.

Abstract: *Toxoplasma gondii* is an obligate intracellular protozoan parasite distributed worldwide that infects a wide range of warm-blooded animals, including humans. Recent studies sought to clarify the relationship between the alleles GRA15, ROP5, ROP16, ROP17, and ROP18 and the virulence of *T. gondii* isolates in mice. This work aims to analyze the variability of genes that express *T. gondii* virulence proteins of 103 strains. Most strains were virulent for mice (76/103–73.79%); within these, 30 were 100% lethal, and 46 caused a cumulative mortality range from 20% to 93%. For the GRA15 marker, most strains presenting allele 2 were non-lethal. For the ROP17 marker, allele 4 was associated with mortality, compared to allele 1. For the ROP18 marker, alleles 1 and 4 were associated with mortality, compared to alleles 2 and 3. A combined analysis of alleles showed low cumulative mortality when the strains presented alleles 3 and 1 for ROP18 and ROP16, respectively. On the other hand, allele 4 of ROP17 was a determinant for virulence when associated with ROP18 allele 3 and ROP16 allele 1. Our analysis shows that ROP18 is the primary determinant of the virulence of atypical strains in mice. Additionally, ROP17 genotyping should not be overlooked, as it has proven critical to enhance this prediction.

Keywords: toxoplasmosis; rhoptry proteins; genotyping; cumulative mortality; PCR-RFLP

1. Introduction

Toxoplasmosis is a worldwide zoonosis that presents various clinical conditions, from asymptomatic infection to severe systemic manifestations. In immunocompetent individuals, the disease, in general, is asymptomatic. However, latent toxoplasmosis may be associated with neurological disorders such as schizophrenia [1]. In immunocompromised individuals, sequelae caused by *Toxoplasma gondii* have been observed in individuals with acquired immunodeficiency syndrome (AIDS), in treatment with immunosuppressive drugs, or in those who have received an organ or bone marrow transplant. In congenital toxoplasmosis, the parasite can cause abortion or the birth of children with ocular and neurological disorders [2,3].

Studies carried out with strains of *T. gondii* isolated from patients or animals from South America have shown a high rate of genetic polymorphism [4]. Recent studies sought to clarify the relationship between the alleles identified by polymerase chain reaction–restriction fragment length polymorphism (PCR-RFLP) of the protein virulence markers, such as dense granule protein 15 (GRA15), rhoptry protein 5 (ROP5), ROP16, ROP17, and ROP18, with the virulence of *T. gondii* isolates in mice [5]. Among these proteins, ROP16 and ROP17 were associated with the modulation of the host's immune response by the parasite, and ROP18 and ROP5 were highlighted as the main determinants of *T. gondii* virulence in mice [5]. Rêgo et al. [6] demonstrated that in Brazilian atypical strains of *T. gondii*, the ROP18 protein, analyzed alone or in combination with ROP5, effectively determined the parasite's virulence for mice.

Different mouse strains, life cycle stages of the parasite, and routes of inoculation are used in *T. gondii* virulence assays. These limit comparisons considering genotype and virulence in mice and the integration of data [7]. Here, we analyze a large number of atypical isolates obtained in Brazil, including those isolated from humans, with a standardized methodology to evaluate the variability of genes that express *T. gondii* virulence proteins.

2. Materials and Methods

2.1. *Toxoplasma gondii Strains*

All 103 strains used in this work were previously isolated by our group (Supplementary Table S1) and have been maintained cryopreserved in N2 banks at the Laboratory of Toxoplasmosis at the Universidade Federal de Minas Gerais—UFMG, Brazil. According to Brazilian law, access to Brazilian genetic heritage was approved by SisGen protocols A3F9195, AD3C00F, AA14CC9, AB9C3CF, A375FCF, AC52DBB, and AD3745E.

2.2. *PCR-RFLP Genotyping of T. gondii*

PCR-RFLP protocols for GRA15, ROP16, and ROP18 were performed according to Dubey et al., 2014 [8]. PCR-RFLP protocols for ROP5 and ROP17 were performed according to Shwab et al. [5].

Genomic DNA samples from *T. gondii* strains were obtained according to Carneiro et al. [9] and provided by the Laboratory of Toxoplasmosis (UFMG). Briefly, DNA extraction of the tachyzoites was performed using the Promega Wizard genomic DNA purification kit following the manufacturer's instructions. Genomic DNA stocks were stored at −20 °C to 4 °C before analysis. The RH88 (type I), ME49 (type II), PTG (II), VEG (type III), and MAS (atypical) strains were used as controls and references. The negative control was distilled water in the presence of primers. For each locus, PCR was performed on parasite genomic DNA using external primers, followed by a second round of PCR with nested primers using external primer products as templates. Primer sequences and restriction enzymes for PCR-RFLP genotyping of rhoptry gene loci ROP18, ROP5, ROP16, and ROP17 were the same as previously described [5,8]. The amplified products were digested using restriction endonucleases (New England BioLabs, Ipswich, MA, USA) specific for each marker according to the manufacturer's instructions. The DNA of the digested products was purified by extraction with an equal volume of phenol/chloroform (1:1), subjected to polyacrylamide gel (5%) electrophoresis, stained with silver nitrate, and photographed.

2.3. *Parasite Virulence Determination*

Strain virulence information (Supplementary Table S1) was obtained from previously published data from bioassays in mice [6,9–15]. Five female BALB/c mice, six to eight weeks old, were inoculated intraperitoneally (i.p.) with 10, 100, or 1000 tachyzoites of each strain in 0.2 mL of PBS (pH 7.2). A total of 15 mice per strain were used. Five animals inoculated i.p. with PBS were maintained as negative controls. All efforts were made to minimize animal suffering during the study. Mouse mortality was observed for a period of 30 days. ELISA (anti-*T. gondii* IgG) was performed on all surviving mice to confirm infection. The survivor mice that did not seroconvert were excluded from the experiment.

RH (virulent) and ME49 (nonvirulent) strains were used as references for comparison. According to Saraf et al. [7], cumulative mortality was calculated based on the number of mice that died divided by the number of mice infected, analyzing three sequential inoculation dosages, with the lowest dose resulting in only partial infection. The Animal Ethics Committee (CEUA) of the Universidade Federal de Minas Gerais (UFMG), Minas Gerais, Brazil, approved all experiments and procedures, including euthanasia and blood collection (CEUA Protocols: 013/2007, 128/2010, 266/2012, 067/2016, and 048/2018).

2.4. Statistical Analyses

The analysis of the correlation between allele types (GRA15, ROP5, ROP16, ROP17, and ROP18) and the median mouse cumulative mortality was performed using the non-parametric Mann–Whitney U test (individual comparisons) and the Kruskal–Wallis test (group analysis) followed by Dunn's test for multiple comparisons. Statistical correlation between lethality and genotype was analyzed using Pearson chi^2. Analyses were performed using GraphPad Prism version 5.0 (GraphPad Software, La Jolla, CA, USA), and $p < 0.05$ was considered significant.

3. Results

In total, we studied 103 atypical strains isolated from four different Brazilian states: Espírito Santo state (ES) and Minas Gerais state (MG), both located in Southeast Brazil, with a predominantly tropical climate, and Piauí state (PI) and Rio Grande do Norte state (RN), both located in the Northeast region of Brazil with a predominantly semi-arid climate (Figure 1A). Considering the hosts, 29 strains were obtained from humans (28.16%), 29 (28.16%) from chickens, 21 (20.39%) from pigs, 10 (9.71%) from goats, 8 (7.77%) from dogs, and 6 (5.83%) from wild birds (Supplementary Table S1). Most strains showed some virulence level for mice (76/103–73.79%); within these, 30 were 100% lethal, and 46 caused a cumulative mortality ranging from 20% to 93%. Within the 103 strains, previous studies showed that 39 different genotypes were identified, with 26 strains of unique genotypes, 17 identified as ToxoDB #163, 14 as #11(BrII), 7 as #108, 6 as #146, 6 as #206, 5 as #8 (BrIII), 5 as #109, 4 as #13, 4 as #24, 3 as #57, 2 as #6 (BrI), 2 as #41, and 2 as #19 (Supplementary Table S1).

The atypical strains presented diverse combinations of virulence factors. Representative gel images of PCR-RFLP genotyping are presented in Supplementary Figure S1. There was no difference between the accumulated mortality from strains obtained in different geographic regions ($p = 0.2083$). The diversity of alleles found for the studied markers GRA15, ROP5, ROP16, ROP17, and ROP18 in strains obtained from different geographic regions is shown in Figure 1B. To validate the correlation of the allele types for GRA15, ROP5, ROP16, ROP17, ROP18, and combinations of ROP18/ROP5 with the virulence of *T. gondii* strains in mice, we plotted allele types against the cumulative mortality (Figure 2). The cumulative mortality was calculated from published data (Supplementary Table S1). Despite a large variation in the percentage of cumulative mortality within allele types, the differences between the medians were significant for some virulence factor types. For the GRA15 marker, most strains presenting allele 2 showed cumulative mortality of 0%, and the median cumulative mortality in isolates that presented alleles 1 or 3 was around 90% (Figure 2A). For the ROP17 marker, we observed that the median cumulative mortality of the isolates presenting allele 4 was significantly higher than that of allele 1 (Figure 2B). For the ROP16 and ROP5 markers, there were no differences in the cumulative mortality within the different alleles (Figure 2C,E).

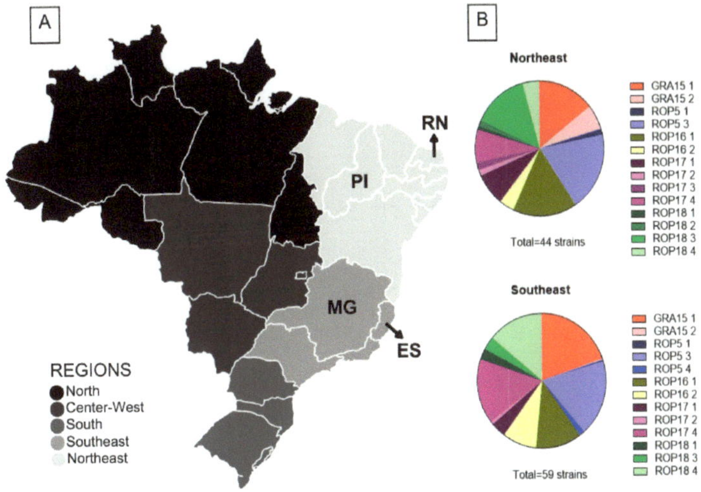

Figure 1. Distribution of *Toxoplasma gondii* strains according to the Brazilian regions from which they were isolated. (**A**) Map of Brazil showing the four states where the 103 *Toxoplasma gondii* strains were isolated: ES (Espírito Santo state), MG (Minas Gerais state), PI (Piauí state), and RN (Rio Grande do Norte state). In total, 5 strains were obtained from chickens from ES; 54 strains were obtained from MG (29 from humans—25 from peripheral blood of newborns with congenital toxoplasmosis and 4 from amniotic fluid; 8 from dogs; 11 from chickens; and 6 from free-living wild birds); 25 strains were obtained from PI (16 from pigs and 9 from goats); and 19 strains were obtained from RN (13 from chickens, 5 from pigs, and 1 from goat). (**B**) Distribution of alleles of the virulence protein genes GRA15, ROP5, ROP16, ROP17, and ROP18 identified by PCR-RFLP in 103 strains isolated in the Northeast and Southeast regions of Brazil.

The isolates that presented allele 1 of the ROP18 marker had 100% cumulative mortality, a significantly different value from those that had the 2 or 3 ($p < 0.001$) allele (Figure 2D). The median cumulative mortality for the isolates with alleles 1 or 4 of the ROP18 marker was higher than that of isolates with alleles 2 or 3. Equally, analysis of the combined alleles of ROP18/ROP5 with mouse virulence indicated that the combination had an overall stronger correlation. Six different combinations were identified (Figure 2F). Some combination segregation with high or low mouse virulence was statistically significant.

The relation of alleles 4 and 3 of ROP18 with high or low cumulative mortality, respectively, was evident when a combined analysis of alleles of ROP18 and ROP16 was performed. While no significant statistical difference was seen between the combinations of ROP18 allele 4 with ROP16 alleles 1 or 2 (Figure 3B), a significant difference was seen when the strains presented alleles 3 and 1 for ROP18 and ROP16, respectively, compared to those with ROP18 allele 4 (Figure 3B). The relation of allele 4 of ROP18 with cumulative mortality in strains presenting ROP16 allele 2 was irrespective of the ROP17 allele, as no significant difference was seen between the presence of ROP17 alleles 4 and 3 (Figure 3E). However, allele 4 of ROP17 was a determinant for the virulence of strains with ROP18 allele 3 and ROP16 allele 1 (Figure 3D). In the strains harboring alleles 3 and 1 of ROP18 and ROP17, respectively, the presence of alleles 1/3 or 2 of GRA15 did not influence virulence (Figure 3C).

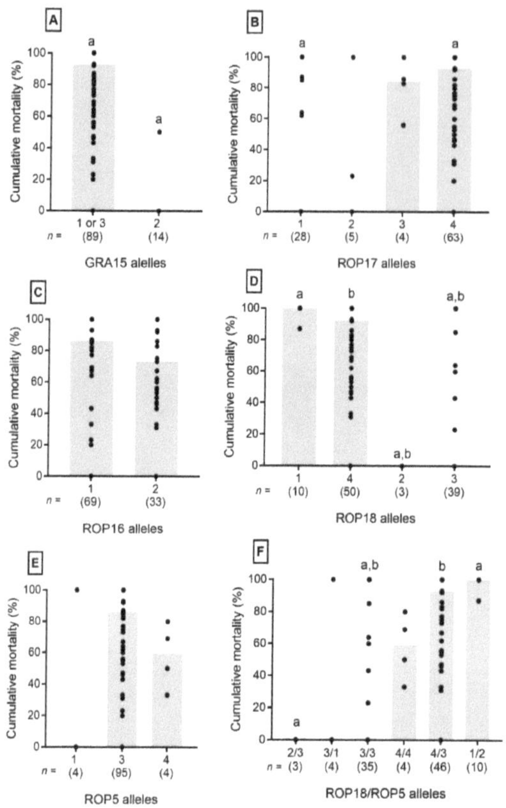

Figure 2. Virulence of *T. gondii* strains in mice, categorized by dense granule GRA15 gene and rhoptry genes ROP5, ROP16, ROP17, and ROP18 allele types. Cumulative mouse mortality for 103 *Toxoplasma gondii* strains with published bioassay results was categorized for alleles of GRA15 (**A**), ROP17 (**B**), ROP16 (**C**), ROP18 (**D**), ROP5 (**E**), and the ROP18/ROP5 (**F**) combinations. The percentage of mouse cumulative mortality was plotted per allele type, with each dot representing the percent mortality reported for an individual parasite strain. The gray bars indicate the median cumulative mortality for each allele type. The number (n) of strains representing each allele type is in parentheses below allele identification. Different letters above allele data (a or b) indicate statistical differences as determined by Mann–Whitney U tests. Differences between data are significant at $p < 0.001$ for GRA15 and ROP18, $p = 0.003$ for ROP17, and $p < 0.05$ for the ROP18/ROP5 allele combinations.

To identify alleles that could be possible molecular markers of lethal strains, we also analyzed the frequency of alleles in two groups of strains according to their lethality (Table 1). Only 2 isolates from 50 that presented ROP18 allele 4 were non-lethal. ROP18 allele 1 and allele 2 were only found in lethal and non-lethal strains, respectively. Allele 4 of ROP5 and allele 3 of ROP17 were identified only in lethal strains, but all of them also harbored ROP18 allele 4. Strains harboring ROP17 allele 4 also showed a higher frequency of lethal phenotype. A higher frequency of non-lethal phenotype was observed in strains presenting GRA15 allele 2, ROP18 allele 3, and ROP5 allele 1 (Table 1). However, all strains with ROP5 allele 1 also harbored ROP18 allele 3, and the non-lethal phenotype is possibly associated with ROP18 alleles. Equally, most strains with ROP16 allele 2 presented a lethal phenotype; however, those also harbored ROP18 allele 4. In conclusion, our analysis shows that ROP18 is the major determinant of the virulence of atypical strains in mice.

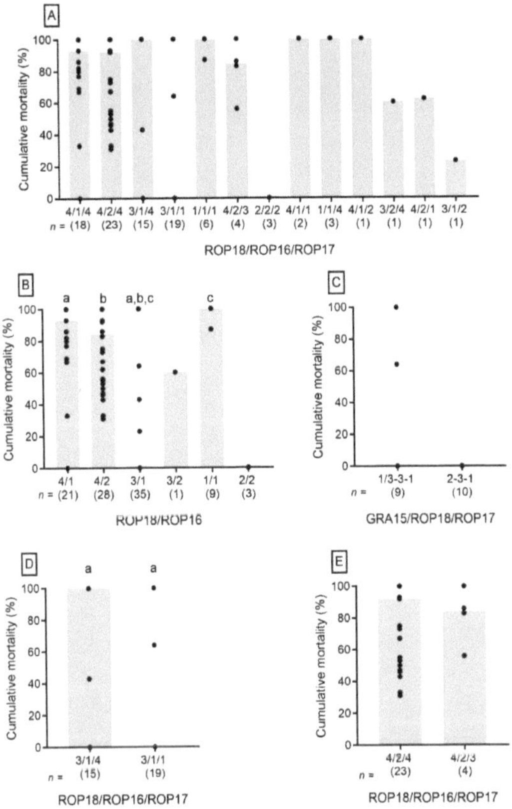

Figure 3. Cumulative mouse mortality for 103 *Toxoplasma gondii* strains categorized for ROP18/ROP17/ROP16 (**A**,**D**,**E**) ROP18/ROP16 (**B**), and GRA15/ROP18/ROP17 (**C**) allele combinations. Percentage of mouse cumulative mortality is plotted against allele type, with each individual dot representing the percentage of mortality reported for an individual parasite strain. The gray bars indicate the median cumulative mortality for each allele type. The number (n) of strains representing each allele type is given in parentheses below allele identification. Different letters above allele data (a, b, or c) indicate statistical differences as determined by Mann–Whitney U tests. Differences between data are significant at $p < 0.0001$ for the ROP18/ROP16 allele combinations (**B**) and at $p = 0.0004$ for the ROP18/ROP16/ROP17 allele combinations (**D**).

Table 1. Distribution of alleles of the virulence protein genes GRA15, ROP5, ROP16, ROP17, and ROP18 identified by PCR-RFLP in Brazilian atypical strains according to lethality in mice.

		Lethal Strains	Non-Lethal Strains		
GRA15		n (*f*)	n (*f*)	Total	*p* value *
	1 or 3	75 (98.7%)	14 (51.9%)	89 (86.4%)	<0.001
	2	1 (1.3%)	13 (48.1%)	14 (13.6%)	
	Total	76	27	103	
ROP5					
	1	1 (1.3%)	3 (11.1%)	4 (3.9%)	
	3	71 (93.4%)	24 (88.9%)	95 (92.2%)	0.04
	4	4 (5.3%)	0 (0.0%)	4 (3.9%)	
	Total	76	27	103	

Table 1. *Cont.*

		Lethal Strains	Non-Lethal Strains		
ROP16					
	1	46 (60.5%)	23 (88.5%)	69 (67.6%)	0.007
	2	30 (39.5%)	3 (11.5%)	33 (32.4%)	
	Total	76	26	102 [a]	
ROP17					
	1	11 (15.1%)	17 (63.0%)	28 (28.0%)	<0.001
	2	2 (2.7%)	3 (11.1%)	5 (5.0%)	
	3	4 (5.5%)	0 (0.0%)	4 (4.0%)	
	4	56 (76.7%)	7 (25.9%)	63 (63.0%)	
	Total	73	27	100 [b]	
ROP18					
	1	10 (13.3%)	0 (0.0%)	10 (9.8%)	<0.001
	2	0 (0.0%)	3 (11.1%)	3 (2.9%)	
	3	17 (22.7%)	22 (81.5%)	39 (38.2%)	
	4	48 (64.0%)	2 (7.4%)	50 (49.0%)	
	Total	75	27	102 [a]	

[a] incomplete ROP16 and ROP18 genotyping for one strain. [b] incomplete ROP17 genotyping for three strains. * Pearson's chi^2.

4. Discussion

Toxoplasma gondii is a successful protozoa parasite due to its ability to manipulate host cell behavior [16]. Rhoptry proteins (ROPs) function at the parasitophorous vacuole to inhibit the recruitment of immunity-related GTPases (IRGs) and guanylate-binding proteins (GBPs), playing a critical role in host–parasite interactions [17,18]. ROP5 and ROP18 work synergistically to block innate immune mechanisms triggered by IFN-γ in murine hosts and serve as predictors of strain virulence in mice [5,19]. Consistent with our findings, previous studies have shown that the combined analysis of ROP18 and ROP5 can predict high virulence in mice from Brazilian strains isolated from pigs, goats, and even cases of human congenital toxoplasmosis [6,20]. However, most strains analyzed in this study exhibited ROP5 allele 3, which made it difficult to assess the contribution of this allele to the observed virulence.

In this work, we found ROP18 allele 4 in 49 isolates; in contrast, only 9 isolates had allele 1. Additionally, the combination of ROP18 allele 2/ROP5 allele 3 had not been previously reported, and strains carrying these alleles showed low cumulative mortality. The ROP18 allele 4/ROP5 allele 4 combination was first identified in *T. gondii* samples from Misiones province, Argentina, leading the authors to propose a possible regionalization of the ROP18/ROP5 profile [21]. However, in our study, we analyzed many strains and detected the same ROP18 allele 4/ROP5 allele 4 combination in four strains from Minas Gerais, Brazil. This finding underscores the importance of analyzing isolates from different regions to validate the predictive value of the ROP18/ROP5 profile for *T. gondii* virulence in mice, especially considering that standardized bioassays in mice are time-consuming, costly, and not feasible in all laboratories [7,22].

ROP5 forms complexes with the ROP18 and ROP17 kinases, working together to regulate acute virulence in mice [23]. ROP16 allele 1/2 can phosphorylate STAT3 and STAT6, leading to the alternative activation of macrophages (M2), which promotes parasite replication within host cells [19]. When analyzed together, our results indicate that ROP18 alleles 1 and 4, irrespective of the ROP5 or ROP16 background, are strongly associated with the virulence of Brazilian strains, showing a high predictive value for cumulative mortality in mice.

In contrast, ROP18 allele 3 is associated with lower virulence and may be influenced by the ROP17 background. Etheridge et al. [23] showed that while ROP17 is not directly responsible for mortality in mice, it contributes to the enhanced virulence associated with the ROP18/ROP5 combination. Hamilton et al. [24] suggested that ROP17 may play a role in determining virulence, which could explain cases where strains predicted to be non-lethal based on the ROP18/ROP5 allele combination are found to be virulent in mouse bioassays. This makes ROP17 a priority gene for genotyping analysis [25]. Thus, determining virulence may be more complex when examining only a few alleles.

GRA15 allele 2 emerged as an important marker in the non-lethal strains identified in this study. In clonal type II strains, it facilitates NF-kB nuclear translocation, induces macrophages toward a classically activated (M1) phenotype, and enhances the host's innate immunity against *T. gondii* infection [26]. Notably, the only lethal strain with GRA15 allele 2 also harbored ROP18 allele 4 and ROP17 allele 4, both key virulence markers, despite the presence of ROP16 allele 2.

Like the current study, most research has focused on the mechanisms by which ROP proteins confer virulence in mice. However, how those virulence factors would impact the outcome of *T. gondii* infection in humans still needs to be discovered [27]. However, type I and atypical strains are frequently associated with symptomatic and severe forms of toxoplasmosis [28–32], and some studies shown the possible association of ROP18 with congenital and ocular diseases in humans [33–35].

To date, strain virulence cannot be reliably assessed using PCR-RFLP genotyping alone, except for the three archetypal strains [36]. Virulence protein polymorphisms were accessed in our study using the PCR-RFLP technique. This methodology limits the detection of single nuclear polymorphisms (SNPs) that are recognized by restriction enzymes and may not be sufficient to detect all polymorphisms in key genes that contribute to phenotypic and pathogenic differences in *T. gondii* strains [16]. So, a future perspective is to sequence the virulence proteins of isolates that presented alleles related to virulence but were non-lethal to mice or vice versa. However, a growing body of research, including our study, indicates that the combination of ROP18 and ROP5 loci offers a strong predictive capacity for virulence in mice, as proposed by Shwab et al. [5].

Our analysis shows that ROP18 is the major determinant of the virulence of Brazilian atypical strains in mice, and highlights the higher prevalence of strains with allele 4. However, how that allele would modulate the immune system in mice or humans has yet to be found. Additionally, ROP17 genotyping should not be overlooked, as it has proven to be a key factor in enhancing virulence prediction. Establishing the fundamental roles of these genetic markers in identifying the virulence of *T. gondii* will allow, in the future, the identification of virulent strains circulating in the environment, especially in livestock. Consequently, more appropriate measures to control the parasite will be adopted.

A limitation of our study is that we evaluated 103 isolates of *T. gondii* of only 39 genotypes. Thus, it is necessary to study more strains, with different genotypes obtained from both animals and humans, from other Brazilian regions and worldwide, to confirm our findings. We expect this work to stimulate other research groups to include similar or other analyses with more strains of identical or different genotypes from different world regions. Thus, over time, this analysis will gain strength and will support whether ROP18 is the major virulence determinant of *T. gondii* in mice and, potentially, in humans.

Supplementary Materials: The following supporting information can be downloaded at https://www.mdpi.com/article/10.3390/microorganisms13020301/s1: Table S1: GRA15, ROP18, ROP16, ROP5, and ROP17 genotypes of *Toxoplasma gondii* strains, according to lethality and cumulative mortality for mice. Figure S1: Representative gel images of PCR-RFLP genotyping (markers GRA15, ROP5, ROP16, ROP17, and ROP18). Sample IDs are at the top of the gel images; genotype results are at the bottom. MK is the DNA size marker.

Author Contributions: Conceptualization, J.G.L.C. and R.W.d.A.V.; methodology, J.G.L.C.; validation, W.M.F.d.R.; formal analysis, J.G.L.C., W.M.F.d.R. and É.S.M.-D.; investigation, J.G.L.C., L.V.P. and R.A.d.C.B.; resources, R.W.d.A.V.; data curation, J.G.L.C. and É.S.M.-D.; writing—original draft preparation, J.G.L.C.; writing—review and editing, R.W.d.A.V. and É.S.M.-D.; visualization, J.G.L.C. and R.W.d.A.V.; supervision, R.W.d.A.V.; project administration, J.G.L.C. and R.W.d.A.V.; funding acquisition, R.W.d.A.V. All authors have read and agreed to the published version of the manuscript.

Funding: This work was supported by the Conselho Nacional de Desenvolvimento e Pesquisa (CNPq) (grants 305574/2021-3 and 308082/2023-0), and Fundação de Amparo à Pesquisa de Minas Gerais (FAPEMIG) (grant APQ-00537-17). J.G.L.C. was supported by a post-doctoral fellowship from CNPq (grant 152652/2016-7). R.W.d.A.V and É.S.M.-D. are Research Fellows at CNPq.

Institutional Review Board Statement: The animal study protocol was approved by the Animal Ethics Committee (CEUA) of the Universidade Federal de Minas Gerais (UFMG), Minas Gerais, Brazil, Protocols: 013/2007, 128/2010, 266/2012, 067/2016, and 048/2018.

Informed Consent Statement: Not applicable.

Data Availability Statement: The raw data supporting the conclusions of this article will be made available by the authors on request.

Acknowledgments: The authors would like to thank Rosalida Estevan Nazar Lopes for the valuable technical assistance and Pró-Reitoria de Pesquisa of the Universidade Federal de Minas Gerais for support.

Conflicts of Interest: The authors declare no conflicts of interest.

References

1. Flegr, J. How and Why *Toxoplasma* Makes Us Crazy. *Trends Parasitol.* **2013**, *29*, 156–163. [CrossRef]
2. Deganich, M.; Boudreaux, C.; Benmerzouga, I. Toxoplasmosis Infection during Pregnancy. *Trop. Med. Infect. Dis.* **2022**, *8*, 3. [CrossRef]
3. Layton, J.; Theiopoulou, D.-C.; Rutenberg, D.; Elshereye, A.; Zhang, Y.; Sinnott, J.; Kim, K.; Montoya, J.G.; Contopoulos-Ioannidis, D.G. Clinical Spectrum, Radiological Findings, and Outcomes of Severe Toxoplasmosis in Immunocompetent Hosts: A Systematic Review. *Pathogens* **2023**, *12*, 543. [CrossRef] [PubMed]
4. Pardini, L.; Bernstein, M.; Carral, L.A.; Kaufer, F.J.; Dellarupe, A.; Gos, M.L.; Campero, L.M.; Moré, G.; Messina, M.T.; Schneider, M.V.; et al. Congenital Human Toxoplasmosis Caused by Non-Clonal *Toxoplasma gondii* Genotypes in Argentina. *Parasitol. Int.* **2019**, *68*, 48–52. [CrossRef]
5. Shwab, E.K.; Jiang, T.; Pena, H.F.J.; Gennari, S.M.; Dubey, J.P.; Su, C. The ROP18 and ROP5 Gene Allele Types Are Highly Predictive of Virulence in Mice across Globally Distributed Strains of *Toxoplasma gondii*. *Int. J. Parasitol.* **2016**, *46*, 141–146. [CrossRef]
6. Rêgo, W.M.F.; Costa, J.G.L.; Baraviera, R.C.A.; Pinto, L.V.; Bessa, G.L.; Lopes, R.E.N.; Vitor, R.W.A. Association of ROP18 and ROP5 Was Efficient as a Marker of Virulence in Atypical Isolates of *Toxoplasma gondii* Obtained from Pigs and Goats in Piauí, Brazil. *Vet. Parasitol.* **2017**, *247*, 19–25. [CrossRef] [PubMed]
7. Saraf, P.; Shwab, E.K.; Dubey, J.P.; Su, C. On the Determination of *Toxoplasma gondii* Virulence in Mice. *Exp. Parasitol.* **2017**, *174*, 25–30. [CrossRef] [PubMed]
8. Dubey, J.P.; Van Why, K.; Verma, S.K.; Choudhary, S.; Kwok, O.C.H.; Khan, A.; Behinke, M.S.; Sibley, L.D.; Ferreira, L.R.; Oliveira, S.; et al. Genotyping *Toxoplasma gondii* from Wildlife in Pennsylvania and Identification of Natural Recombinants Virulent to Mice. *Vet. Parasitol.* **2014**, *200*, 74–84. [CrossRef] [PubMed]
9. Carneiro, A.C.A.V.; Andrade, G.M.; Costa, J.G.L.; Pinheiro, B.V.; Vasconcelos-Santos, D.V.; Ferreira, A.M.; Su, C.; Januário, J.N.; Vitor, R.W.A. Genetic Characterization of *Toxoplasma gondii* Revealed Highly Diverse Genotypes for Isolates from Newborns with Congenital Toxoplasmosis in Southeastern Brazil. *J. Clin. Microbiol.* **2013**, *51*, 901–907. [CrossRef]

10. Brandão, G.P.; Ferreira, A.M.; Melo, M.N.; Vitor, R.W.A. Characterization of *Toxoplasma gondii* from Domestic Animals from Minas Gerais, Brazil. *Parasite* **2006**, *13*, 143–149. [CrossRef]
11. Ferreira, A.d.M.; Vitor, R.W.A.; Gazzinelli, R.T.; Melo, M.N. Genetic Analysis of Natural Recombinant Brazilian *Toxoplasma gondii* Strains by Multilocus PCR-RFLP. *Infect. Genet. Evol.* **2006**, *6*, 22–31. [CrossRef] [PubMed]
12. Clementino Andrade, M.M.; Pinheiro, B.V.; Cunha, M.M.; Carneiro, A.C.a.V.; Andrade Neto, V.F.; Vitor, R.W.A. New Gentotypes of *Toxoplasma gondii* Obtained from Farm Animals in Northeast Brazil. *Res. Vet. Sci.* **2013**, *94*, 587–589. [CrossRef]
13. Ferreira, T.C.R.; Buery, J.C.; Moreira, N.I.B.; Santos, C.B.; Costa, J.G.L.; Pinto, L.V.; Baraviera, R.C.d.A.; Vitor, R.W.A.; Fux, B. *Toxoplasma gondii*: Isolation, Biological and Molecular Characterisation of Samples from Free-Range Gallus gallus Domesticus from Countryside Southeast Brazil. *Rev. Bras. Parasitol. Vet.* **2018**, *27*, 384–389. [CrossRef] [PubMed]
14. Rêgo, W.M.F.; Costa, J.G.L.; Baraviera, R.C.A.; Pinto, L.V.; Bessa, G.L.; Lopes, R.E.N.; Silveira, J.a.G.; Vitor, R.W.A. Genetic Diversity of *Toxoplasma gondii* Isolates Obtained from Free-Living Wild Birds Rescued in Southeastern Brazil. *Int. J. Parasitol. Parasites Wildl.* **2018**, *7*, 432–438. [CrossRef] [PubMed]
15. Pinheiro, B.V.; Noviello, M.d.L.M.; Cunha, M.M.; Tavares, A.T.; Carneiro, A.C.A.V.; Arantes, R.M.E.; Vitor, R.W.A. Pathological Changes in Acute Experimental Toxoplasmosis with *Toxoplasma gondii* Strains Obtained from Human Cases of Congenital Disease. *Exp. Parasitol.* **2015**, *156*, 87–94. [CrossRef]
16. Cheng, W.; Wang, C.; Xu, T.; Liu, F.; Pappoe, F.; Luo, Q.; Xu, Y.; Lu, F.; Shen, J. Genotyping of Polymorphic Effectors of *Toxoplasma gondii* Isolates from China. *Parasit. Vectors* **2017**, *10*, 580. [CrossRef] [PubMed]
17. Jensen, K.D.C.; Camejo, A.; Melo, M.B.; Cordeiro, C.; Julien, L.; Grotenbreg, G.M.; Frickel, E.-M.; Ploegh, H.L.; Young, L.; Saeij, J.P.J. *Toxoplasma gondii* Superinfection and Virulence during Secondary Infection Correlate with the Exact ROP5/ROP18 Allelic Combination. *mBio* **2015**, *6*, e02280. [CrossRef]
18. Kim, E.W.; Nadipuram, S.M.; Tetlow, A.L.; Barshop, W.D.; Liu, P.T.; Wohlschlegel, J.A.; Bradley, P.J. The Rhoptry Pseudokinase ROP54 Modulates *Toxoplasma gondii* Virulence and Host GBP2 Loading. *mSphere* **2016**, *1*, e00045-16. [CrossRef]
19. Saeij, J.P.J.; Boyle, J.P.; Coller, S.; Taylor, S.; Sibley, L.D.; Brooke-Powell, E.T.; Ajioka, J.W.; Boothroyd, J.C. Polymorphic Secreted Kinases Are Key Virulence Factors in Toxoplasmosis. *Science* **2006**, *314*, 1780–1783. [CrossRef]
20. de Melo, R.P.B.; Wanderley, F.S.; Porto, W.J.N.; Pedrosa, C.d.M.; Hamilton, C.M.; de Oliveira, M.H.G.S.; Ribeiro-Andrade, M.; Rêgo, R.C.d.S.; Katzer, F.; Mota, R.A. Description of an Atypical *Toxoplasma gondii* Isolate from a Case of Congenital Toxoplasmosis in Northeastern Brazil. *Parasitol. Res.* **2020**, *119*, 2727–2731. [CrossRef] [PubMed]
21. Bernstein, M.; Rudzinski, M.; Schneider, V.; Messina, M.; Gos, M.L.; Helman, E.; Dellarupe, A.; Unzaga, J.M.; Venturini, M.C.; Moré, G.; et al. Genetic Characterization of *Toxoplasma gondii* from Human and Chicken Isolates from Argentina. *Parasitol. Res.* **2024**, *123*, 129. [CrossRef]
22. Calero-Bernal, R.; Fernández-Escobar, M.; Katzer, F.; Su, C.; Ortega-Mora, L.M. Unifying Virulence Evaluation in *Toxoplasma gondii*: A Timely Task. *Front. Cell. Infect. Microbiol.* **2022**, *12*, 868727. [CrossRef]
23. Etheridge, R.D.; Alaganan, A.; Tang, K.; Lou, H.J.; Turk, B.E.; Sibley, L.D. The *Toxoplasma* Pseudokinase ROP5 Forms Complexes with ROP18 and ROP17 Kinases That Synergize to Control Acute Virulence in Mice. *Cell Host Microbe* **2014**, *15*, 537–550. [CrossRef] [PubMed]
24. Hamilton, C.M.; Black, L.; Oliveira, S.; Burrells, A.; Bartley, P.M.; Melo, R.P.B.; Chianini, F.; Palarea-Albaladejo, J.; Innes, E.A.; Kelly, P.J.; et al. Comparative Virulence of Caribbean, Brazilian and European Isolates of *Toxoplasma gondii*. *Parasit. Vectors* **2019**, *12*, 104. [CrossRef] [PubMed]
25. Bernstein, M.; Pardini, L.; Bello Pede Castro, B.; Unzaga, J.M.; Venturini, M.C.; Moré, G. ROP18 and ROP5 Alleles Combinations Are Related with Virulence of *T. gondii* Isolates from Argentina. *Parasitol. Int.* **2021**, *83*, 102328. [CrossRef]
26. Rosowski, E.E.; Lu, D.; Julien, L.; Rodda, L.; Gaiser, R.A.; Jensen, K.D.C.; Saeij, J.P.J. Strain-Specific Activation of the NF-kappaB Pathway by GRA15, a Novel *Toxoplasma gondii* Dense Granule Protein. *J. Exp. Med.* **2011**, *208*, 195–212. [CrossRef] [PubMed]
27. Niedelman, W.; Gold, D.A.; Rosowski, E.E.; Sprokholt, J.K.; Lim, D.; Farid Arenas, A.; Melo, M.B.; Spooner, E.; Yaffe, M.B.; Saeij, J.P.J. The Rhoptry Proteins ROP18 and ROP5 Mediate *Toxoplasma gondii* Evasion of the Murine, but Not the Human, Interferon-Gamma Response. *PLoS Pathog.* **2012**, *8*, e1002784. [CrossRef] [PubMed]
28. Gilbert, R.E.; Freeman, K.; Lago, E.G.; Bahia-Oliveira, L.M.G.; Tan, H.K.; Wallon, M.; Buffolano, W.; Stanford, M.R.; Petersen, E. European Multicentre Study on Congenital Toxoplasmosis (EMSCOT) Ocular Sequelae of Congenital Toxoplasmosis in Brazil Compared with Europe. *PLoS Negl. Trop. Dis.* **2008**, *2*, e277. [CrossRef]
29. Khan, A.; Jordan, C.; Muccioli, C.; Vallochi, A.L.; Rizzo, L.V.; Belfort, R.; Vitor, R.W.A.; Silveira, C.; Sibley, L.D. Genetic Divergence of *Toxoplasma gondii* Strains Associated with Ocular Toxoplasmosis, Brazil. *Emerg. Infect. Dis.* **2006**, *12*, 942–949. [CrossRef] [PubMed]
30. de Lima Bessa, G.; de Almeida Vitor, R.W.; Dos Santos Martins-Duarte, E. *Toxoplasma gondii* in South America: A Differentiated Pattern of Spread, Population Structure and Clinical Manifestations. *Parasitol. Res.* **2021**, *120*, 3065–3076. [CrossRef]

31. Switaj, K.; Master, A.; Borkowski, P.K.; Skrzypczak, M.; Wojciechowicz, J.; Zaborowski, P. Association of Ocular Toxoplasmosis with Type I *Toxoplasma gondii* Strains: Direct Genotyping from Peripheral Blood Samples. *J. Clin. Microbiol.* **2006**, *44*, 4262–4264. [CrossRef]
32. Vallochi, A.L.; Muccioli, C.; Martins, M.C.; Silveira, C.; Belfort, R.; Rizzo, L.V. The Genotype of *Toxoplasma gondii* Strains Causing Ocular Toxoplasmosis in Humans in Brazil. *Am. J. Ophthalmol.* **2005**, *139*, 350–351. [CrossRef] [PubMed]
33. Hernández-de-Los-Ríos, A.; Murillo-Leon, M.; Mantilla-Muriel, L.E.; Arenas, A.F.; Vargas-Montes, M.; Cardona, N.; de-la-Torre, A.; Sepúlveda-Arias, J.C.; Gómez-Marín, J.E. Influence of Two Major *Toxoplasma gondii* Virulence Factors (ROP16 and ROP18) on the Immune Response of Peripheral Blood Mononuclear Cells to Human Toxoplasmosis Infection. *Front. Cell. Infect. Microbiol.* **2019**, *9*, 413. [CrossRef]
34. Torres-Morales, E.; Taborda, L.; Cardona, N.; De-la-Torre, A.; Sepulveda-Arias, J.C.; Patarroyo, M.A.; Gomez-Marin, J.E. Th1 and Th2 Immune Response to P30 and ROP18 Peptides in Human Toxoplasmosis. *Med. Microbiol. Immunol.* **2014**, *203*, 315–322. [CrossRef] [PubMed]
35. Sánchez, V.; de-la-Torre, A.; Gómez-Marín, J.E. Characterization of ROP18 Alleles in Human Toxoplasmosis. *Parasitol. Int.* **2014**, *63*, 463–469. [CrossRef] [PubMed]
36. Rico-Torres, C.P.; Valenzuela-Moreno, L.F.; Luna-Pastén, H.; Cedillo-Peláez, C.; Correa, D.; Morales-Salinas, E.; Martínez-Maya, J.J.; Alves, B.F.; Pena, H.F.J.; Caballero-Ortega, H. Genotyping of *Toxoplasma gondii* Isolates from México Reveals Non-Archetypal and Potentially Virulent Strains for Mice. *Infect. Genet. Evol.* **2023**, *113*, 105473. [CrossRef] [PubMed]

microorganisms

MDPI

Article

Antiproliferative and Morphological Analysis Triggered by Drugs Contained in the Medicines for Malaria Venture COVID-Box Against *Toxoplasma gondii* Tachyzoites

Andréia Luiza Oliveira Costa [1,†], Mike dos Santos [1,†], Giulia Caroline Dantas-Vieira [1], Rosálida Estevam Nazar Lopes [1], Rossiane Claudia Vommaro [2] and Érica S. Martins-Duarte [1,*]

[1] Laboratório de Quimioterapia de Protozoários Egler Chiari, Departamento de Parasitologia—ICB, Universidade Federal de Minas Gerais, Belo Horizonte 31270-901, MG, Brazil; andreialuizaaa594@gmail.com (A.L.O.C.); mikesnts@outlook.com (M.d.S.); giuliacdv@outlook.com (G.C.D.-V.); rosalidaenlopes@gmail.com (R.E.N.L.)

[2] Laboratório de Ultraestrutura Celular Hertha Meyer, Instituto de Biofísica Carlos Chagas Filho, Centro de Pesquisa em Medicina de Precisão, Universidade Federal do Rio de Janeiro, Rio de Janeiro 21941-902, RJ, Brazil; vommaro@biof.ufrj.br

* Correspondence: ericamduarte@icb.ufmg.br; Tel.: +55-(31)-3409-2847

† These authors contributed equally to this work.

Abstract: *Toxoplasma gondii* is a protozoan, and the etiologic agent of toxoplasmosis, a disease that causes high mortality in immunocompromised individuals and newborns. Despite the medical importance of toxoplasmosis, few drugs, which are associated with side effects and parasite resistance, are available for its treatment. Here, we show a screening of molecules present in COVID-Box to discover new hits with anti-*T. gondii* activity. COVID-Box contains 160 molecules with known or predicted activity against SARS-CoV-2. Our analysis selected 23 COVID-Box molecules that can inhibit the tachyzoite forms of the RH strain of *T. gondii* in vitro by more than 70% at 1 μM after seven days of treatment. The inhibitory curves showed that most of these molecules inhibited the proliferation of tachyzoites with IC_{50} values below 0.80 μM; Cycloheximide and (-)-anisomycin were the most active drugs, with IC_{50} values of 0.02 μM. Cell viability assays showed that the compounds are not toxic at active concentrations, and most are highly selective for parasites. Overall, all 23 compounds were selective, and for two of them (apilimod and midostaurin), this is the first report of activity against *T. gondii*. To better understand the effect of the drugs, we analyzed the effect of nine of them on the ultrastructure of *T. gondii* using transmission electron microscopy. After treatment with the selected drugs, the main changes observed in parasite morphology were the arrestment of cell division and organelle alterations.

Keywords: toxoplasmosis; drug repositioning; cycloheximide; bortezomib; midostaurin; (-)-anisomycin; almitrine

1. Introduction

Toxoplasma gondii is the etiologic agent of toxoplasmosis, a zoonosis with a high proportion of seropositive individuals worldwide [1]. The importance of this disease for humans is related to the high morbidity of immunocompromised individuals, such as AIDS patients and newborns [2]. *T. gondii* infection is asymptomatic in 80% of infected individuals. However, primary infection in AIDS patients can cause the cerebral form of the disease [3].

Congenitally infected newborns can develop neurological problems and eye diseases [4]. In Europe, a risk of eye damage cases of 0.3% to 1% has been observed in adults one or two years after acquiring the infection, and retinochoroiditis results in higher damage in America than in Europe or North America [5,6]. It is believed that 400 to

4000 children are born with toxoplasmosis each year in the United States [7]. In Brazil, the incidence of congenital transmission can reach 1:770 live births [4], and the estimated prevalence of toxoplasmosis is 42 to 92%, depending on the region of the country [8].

Despite the medical importance of toxoplasmosis, few drugs are available for its treatment [9]. The combination of pyrimethamine (PYR), sulfadiazine (SDZ), and folinic acid has been administered for approximately 70 years. This association is still the first choice for all clinical conditions of the disease [10]. However, this therapeutic scheme has some critical flaws, such as frequent and non-tolerated side effects, which lead to the abandonment of treatment. In addition, there are reports of therapeutic failures with *T. gondii* strains possibly resistant to these drugs, as well as low absorption and the need for high therapeutic doses or a long-period treatment, making it challenging to manage the treatment [11–13].

In this context, drug repositioning has been a promising strategy for new treatments for infectious and neglected diseases. This strategy consists of redirecting an active pharmaceutical compound with commercial use or in development and reusing it for a new therapeutic indication. The development and discovery of medicines are long and high-investment processes [14,15]. Based on this, the Medicines for Malaria Venture (MMV) develops and offers several promising drug libraries with potential repositioning for treating neglected diseases and infections that can cause epidemics. The first drug box provided by MMV was in 2014, Malaria-Box. The anti-*T. gondii* activity of the compounds available in Malaria-Box has already been described in the literature [16–18]. Several compounds in other MMV boxes, such as Pathogen and Pandemic, have also shown anti-*T. gondii* activity [19–22].

In 2020, with the advent of the COVID-19 pandemic, MMV made available a new box with 160 drugs and compounds, COVID-Box. The drugs and compounds in this box have structurally different therapeutic classes, selected by experts and initially tested against the new coronavirus (SARS-CoV-2). As COVID-Box compounds are in various stages of pharmacological research and development, it is interesting to identify new potential treatments for toxoplasmosis. Previous work demonstrated the activity of those compounds in *T. gondii*, confirming the potential of COVID-Box [23,24]. Thus, in this work, we aimed to expand the study of the activity of those compounds against the tachyzoites of *T. gondii* and characterize the mode of action of the best ones by morphological analysis using fluorescence and transmission electron microscopy.

2. Materials and Methods

2.1. Drugs and Compounds of COVID-Box

The 160 drugs and compounds were provided free of charge by MMV: https://www.mmv.org/mmv-open/covid-box/covid-box-supporting-information (accessed on 12 August 2024). The compounds were provided solubilized in DMSO at 10 mM in two 96-well plates (A and B) containing 80 compounds each. For storage in this box, the drugs and compounds were dissolved in reserves of 2 mM with 100% DMSO (Merck, Darmstadt, Germany).

2.2. Parasites

Toxoplasma gondii tachyzoites of the RH strain were used. The parasites were maintained in vitro through serial passages in 25 cm^2 culture flasks containing confluent neonatal normal human dermal fibroblastic cells (NHDF; Lonza, kindly donated by Dr. Sheila Nardelli, Fiocruz Paraná) in RPMI 1640 (Gibco) medium supplemented with 2% fetal bovine serum (Gibco), penicillin/streptomycin, amphotericin B (Life Technologies, Eugene, OR, USA), and 2 mM glutamine (complete medium). Cells and parasites were maintained at 37 °C in a humid atmosphere of 5% CO_2.

2.3. Antiproliferative Assays Against Tachyzoite Stage

For assessing the proliferation of tachyzoites, we used the classical plaque assay. For preliminary evaluation of the 160 drugs and compounds in the COVID-Box, 6-well plates with monolayers of NHDF cells in complete RPMI-1640 medium were infected with 1000 newly egressed tachyzoites of *T. gondii* and treated with 1 µM of each drug. For antiproliferative curves, 12-well plates were used, and cells were infected with 600 tachyzoites. After adding the parasites, decreasing concentrations of drugs and compounds (1–0.0078 µM in two-fold serial dilutions) were added to each well. In the control wells, 0.1% DMSO was added. During the treatment period, the plates were maintained stable and without disturbing to not spread the egressed parasite from the infection focus (plaques). After 7 days of treatment, the cells were fixed with 70% ethanol and stained with crystal violet. Stained plates were imaged with ChemiDoc MP Imaging System (Bio-Rad, Hercules, CA, USA), and then plaque areas were analyzed with the ImageJ® software (version 1.52e). The destruction areas of treated cells were quantified and compared with the untreated (control) to determine the percentage of proliferation and inhibition of *T. gondii*. For the calculation of the inhibitory concentration of 50% (IC_{50}), the growth inhibition percentage was plotted as a function of drug concentration by fitting the values to the standard curve analysis. The regression analyses were performed using GraphPad Prism 8 Software (GraphPad Inc., San Diego, CA, USA).

2.4. Cytotoxicity Assay in NHDF Cell

The cytotoxic effect of COVID-Box drugs and compounds against NHDF cells was evaluated by the MTS assay [25,26]. For this, 96-well tissue plates containing NHDF cells were treated with different concentrations of drugs and compounds for seven days. At the end of the treatment, the cells were washed with phosphate-buffered saline (PBS, pH 7.4), each well was filled with 100 µL of 10 mM glucose in PBS, and 20 µL of MTS reagent (Promega, Madison, WI, USA) was added. The absorbance was read at 490 nm after three hours of incubation at 37 °C. Cytotoxicity was calculated as the percentage of viable cells versus untreated cells (control). The cytotoxic concentration of 50% (CC_{50}) for the host cells was calculated as for IC_{50}. The selective index (SI) was calculated as the ratio of CC_{50}/IC_{50}.

2.5. Druggability Analysis of Drugs and Compounds

For the in silico analysis, the chemical structures of the simplified molecular line input system (SMILES) code from all drugs were obtained from the plate map available at https://www.mmv.org/mmv-open/covid-box/covid-box-supporting-information and loaded into the online programs pkCSM (https://biosig.lab.uq.edu.au/pkcsm/prediction) and SwissADME (http://www.swissadme.ch/) for physicochemical, toxicity and pharmacokinetics properties analysis of the drugs and compounds of COVID-Box. Pyrimethamine (PYR), sulfadiazine (SDZ), clindamycin (CLI), azithromycin (AZT), and atovaquone (ATO) were analyzed as reference drugs.

2.6. Twenty-Four Hours Antiproliferative Assay

To validate the antiproliferative effect of the compounds with the highest activity present in COVID-Box, 24-well plates containing NDHF cell monolayer coverslips were infected at a 5:1 ratio (tachyzoite/cell) with fresh infusions of the RH strain for two hours. The cells were then washed twice with PBS pH 7.4 to remove non-invaded parasites and incubated for another two hours with fresh medium. At the end of this time, the cells were treated with 1 µM almitrine, 62.5–125 nM (-)-anisomycin, 0.5–1 µM apilimod, 31.2–62.5 nM bortezomib, 31.2–125 nM cycloheximide, 1 µM ivermectin, 1.5 and 2.0 µM merimepodib, 0.25–1 µM midostaurin, 125–250 nM mycophenolic acid, and 31.2–250 nM salinomycin for 24 h. At the end of this time, the washing, fixation, and cell counting steps were followed as described above [21]. The proliferation index is the product of the total number of tachyzoites and the total number of cells divided by the percentage of infected cells [27].

2.7. Transmission Electron Microscopy (TEM)

NHDF cultures infected with *T. gondii* were treated with compounds and fixed with 2.5% glutaraldehyde (EM Grade—Ted Pella Inc., Redding, CA, USA) in 0.1 M sodium cacodylate buffer pH 7.4 (Ted Pella Inc.). Cells were post-fixed with 1% osmium tetroxide, 1.25% potassium ferrocyanide, and 5 mM $CaCl_2$, in 0.1 M sodium cacodylate buffer (pH 7.4). The sample fixation and processing for microscopy were performed as previously described [28]. Samples were dehydrated in alcohol solutions of increasing concentrations (35–100%) and embedded in epoxy resin (Polybed 812, Polysciences). Ultrathin sections were collected in copper grids 300 mesh (Ted Pella Inc.), stained with uranyl acetate and lead citrate, and then observed in a Fei TecNai G2 120 kV Spirit Electron Microscope (FEI Company, Eindhoven, The Netherlands).

2.8. Immunofluorescence Microscopy

For the immunofluorescence microscopy, NHDF cells were infected with newly egressed tachyzoites of *T. gondii* in the ratio 5:1 and treated for 24 h with 1 μM ivermectin, 0.25–0.5 nM midostaurin, 125 nM salinomycin, 1.5 μM merimepodib, 62.5–125 nM (-)-anisomycin, 31.2–62.5 nM bortezomib, and 1 μM almitrine. When completing the treatment, the cells were prepared as previously described [28]. Rabbit anti-IMC1 (1:1000) and anti-ARO (1:1000) (kindly provided by Dr. Dominique Soldati, Universite de Geneve, Switzerland) were used to label the mother cell and daughter cell pellicles throughout the endodyogeny process and rhoptries, respectively. Mouse anti-SAG1 (1:200; kindly donated by Dr. Tiago Mineo, Universidade Federal de Uberlândia) was used to label the parasite plasma membrane. DAPI (2 μg/mL; Sigma-Aldrich, St. Louis, MO, USA) was used to label the DNA. Goat anti-rabbit IgG conjugated to Alexa-488 and goat anti-mouse IgG conjugated to Alexa 546 were used as secondary antibodies (Life Technologies, Eugene, OR, USA). After labeling with antibodies, the coverslips were mounted onto slides using Prolong gold (Life Technologies), and samples were examined on a Zeiss Axio Observer Z1 microscope (Baden-Württemberg, Germany) or Olympus BX60 microscope (Shinjuku City, Tokyo, Japan).

2.9. Statistical Analysis

Data were analyzed using GraphPad Prism 8.0 software (GraphPad Inc., San Diego, CA, USA). IC_{50} and CC_{50} calculations were performed by fitting the values of proliferation/viability in percentage to a non-linear curve followed by dose-response inhibition analysis through log(inhibitor) vs. normalized response. One-way ANOVA and *t*-test were used for the quantitative analysis.

3. Results

3.1. Drugs of COVID-Box Showed High Activity and Selectivity Against T. gondii Tachyzoites

The antiproliferative effect of 1 μM of each of the 160 drugs and compounds present in the COVID-Box (Supplementary Table S1) was screened against tachyzoites of the RH strain of *T. gondii* for seven days of treatment. Thirty compounds inhibited parasite proliferation by at least 70% (Supplementary Figure S1). Of the thirty best drugs, seven were excluded from the studies. Two (amiodarone and proscillaridin) were initially excluded because they presented signs of cytotoxicity for the host cells during the initial screening. Five other drugs (itraconazole, doxycycline, cyclosporine, doxorubicin, and digitoxin) were excluded as their activity against *T. gondii* has been extensively studied before [24,29–34]. Details of the plaque assay of the best selected 23 compounds and drugs are in the Supplementary Figure S2. The chosen drugs (Supplementary Figure S3) were studied for the IC_{50} determination and cytotoxicity analysis (Supplementary Figures S4 and S5 and Table 1).

The drugs Cycloheximide, Bortezomib, Digoxin, and (-)-Anisomycin were the most active, inhibiting the proliferation of *T. gondii* with IC_{50}s values lower than or equal to 30 nM (Table 1). Salinomycin, mycophenolic acid, abemaciclib, midostaurin, emetine, and LY2228820 inhibited *T. gondii* proliferation with IC_{50} lower than 100 nM (Table 1). The drugs ivermectin, almitrine, apilimod, bemcentinib, niclosamide, regorafenib, and merimepodib were also highly active against *T. gondii*, presenting IC_{50} in the range of 0.15–0.48 μM (Table 1).

Table 1. IC_{50} values and cytotoxicity of drugs and compounds of COVID-Box.

Plate Position	Trivial Name	Disease Area	Target	Status [a]	IC_{50} (μM) in Tachyzoites	Cytotoxicity for NHDF		Anti-*T. gondii* Activity [c]
						CC_{50} (μM)	SI [b]	
AA02	Niclosamide	Antiparasitic	Anaerobic phosphorylation inhibitor	Approved	0.36 ± 0.02	3.05 ± 0.08	8	[23,35]
AA04	Bemcentinib	Immune agent	Tyrosine kinase	Phase II	0.15 ± 0.03	0.97 ± 0.15	6	[23]
AB03	Apilimod	Antitumor	PIKfyve inhibitor	Phase II	0.22 ± 0.06	2.12 ± 0.12	10	New
AB04	Regorafenib	Antitumor	Tyrosine kinase	Approved	0.25 ± 0.04	3.03 ± 0.01	12	[23]
AC10	LY2228820	Antitumor agent	p38 mitogen-activated protein kinase inhibitor	Phase II, discontinued	0.04 ± 0.00	nd	nd	[23]
AD02	Digoxin	Antiarrhythmic	Na^+/K^+-ATPase inhibitor	Approved	0.03 ± 0.01	0.34 ± 0.32	11	[24]
AE06	Emetine	Antiparasitic	Unknown	Approved	0.05 ± 0.01	1.02 ± 0.19	20	[36]
AF05	Ivermectin	Antiparasitic	Agonist of glutamate-gated Cl^- channels	Approved	0.21 ± 0.01	nd	nd	[37]
AF09	Sorafenib	Antitumor	Tyrosine kinase inhibitor	Approved	0.56 ± 0.03	nd	nd	[23]
AG03	Manidipine	Antihypertensive	Calcium channel blocker	Phase III	0.74 ± 0.09	nd	nd	[23]
AG04	Almitrine	Respiratory system	Mitochondrial ATP synthase	Approved *	0.33 ± 0.04	nd	n.d	[23,24]
AG08	Midostaurin	Antitumor	Tyrosine kinase	Approved	0.08 ± 0.00	0.24 ± 0.02	3	New
AG11	Abemaciclib	Antitumor	Cyclin-dependent kinase	Approved	0.09 ± 0.01	#	#	[23]
AH03	Tetrandrine	Antitumor	P-glycoprotein	Preclinical	0.35 ± 0.07	4.77 ± 8.25	14	[23]
BA07	Ponatinib	Antitumor	Tyrosine kinase	Approved	0.33 ± 0.03	4.84 ± 0.89	15	[34]
BA09	Berbamine	Antitumor	CAMKII inhibitor	Preclinical	0.31 ± 0.03	4.61 ± 7.52	15	[23]
BB10	Mycophenolic acid	Immunosuppressant	Inosine monophosphate dehydrogenase	Approved	0.07 ± 0.01	23.81 ± 31.97	340	[38,39]
BD02	Salinomycin	Anti-microbial agent	Alkali ion carrier	Approved	0.07 ± 0.01	1.14 ± 0.17	16	[40]
BD08	Merimepodib	Antiviral agent	Inosine monophosphate dehydrogenase	Phase II	0.48 ± 0.08	nd	nd	[23]
BD11	Cycloheximide	Agricultural agent–Fungicide	Protein synthesis inhibitor	Research	0.02 ± 0.00	6.08 ± 4.63	304	[41,42]
BF06	(-)-Anisomycin	Anti-infective agent	Protein synthesis	Approved	0.02 ± 0.00	0.25 ± 0.13	13	[43]
BG06	Bortezomib	Antitumor	Proteasome inhibitor	Approved	0.03 ± 0.00	0.72 ± 0.24	24	[24,34]
BG07	Pimozide	Antipsychotic	Dopamine receptor	Approved	0.64 ± 0.15	nd	nd	[24,33]

nd = not determined. Concentrations up to 3 μM did not affect NHDF cell proliferation. # There was not enough to perform this assay. [a] Approved: FDA-approved drug; Phase I or II or III: Clinical candidate drug in Phase 1, 2, or 3 clinical trials; * Almitrine was withdrawn in some countries. [b] Selectivity index. [c] We searched PubMed for the trivial name of each compound plus "*Toxoplasma gondii*". If available, the reference for a previous study was included. The term "New" indicates when no results were retrieved from the search.

The cytotoxicity assay showed that most compounds were highly selective against *T. gondii*, and the SI ranged from 3 to 304 (Table 1). Drugs with IC_{50}s less than 30 nM (cycloheximide and bortezomib) had the highest SI. Overall, all 23 compounds were selective, and for two of them (apilimod and midostaurin), this is the first report of activity against *T. gondii* (Table 1).

3.2. COVID-Box Drugs Show Potential Oral Druggability

Through the SwissADME [44] platform, it was possible to obtain information about the physical-chemical properties of the drugs that showed the best activity against *T. gondii*. From these analyses, it was possible to predict whether these compounds are by the predictors of Lipinski's rule of five (RO5) and Veber (Table 2). We also compared PYR, SDZ,

CLI, AZT, and ATO, which are currently used for treating toxoplasmosis. The RO5 states that drugs with more than 5H-bond donors, more than 10H-bond acceptors, a molecular weight (MW) greater than 500, and a calculated LogP (a measure of lipophilicity) greater than five are less likely to have good oral absorption and permeation. In addition to the RO5 of Lipinski et al. (1997) [45], the two predictors of Veber et al. (2002) [46] also indicate that compounds with total polar surface area (TPSA) equal to or <140 Å^2 and with ten or less rotating bonds have a greater chance of success in oral bioavailability.

Table 2. Physical-chemical properties of drugs and compounds of COVID-Box according to Lipinski's RO5 and predictors of Veber.

Identification	LogP [a]	H-Bond Donors	H-Bond Receptors	MW [b]	Violations Lipinski	TPSA (Å^2) [c]	n° Rotations	Violations Veber
Pyrimethamine	2.84	2	4	248.7	0	77.83	2	0
Sulfadiazine	0.86	3	5	250.3	0	98.57	3	0
Clindamycin	0.39	4	7	424.9	0	102.25	7	1
Azithromycin	2.50	5	14	749.0	2	198.54	7	1
Atovaquone	5.34	1	3	366.8	1	54.70	2	0
Niclosamide	3.85	2	4	327.1	0	128.62	3	0
Bemcentinib	4.88	2	8	506.7	2	92.35	4	0
Apilimod	3.08	1	8	418.5	0	84.77	8	0
Regorafenibe	4.39	3	7	482.8	0	92.35	5	0
LY2228820	5.51	2	5	420.5	1	85.41	5	0
Digoxin	2.22	6	14	780.9	3	203.06	7	1
Emetine	3.04	1	6	480.7	0	52.20	7	0
Ivermectin	4.37	3	14	875.1	2	170.06	8	1
Sorafenib	4.10	3	7	464.8	0	92.35	9	0
Manidipine	4.04	1	8	610.7	1	116.94	12	1
Almitrine	4.34	2	6	477.6	0	69.21	10	0
Midostaurin	4.05	1	4	570.6	1	77.73	4	0
Abemaciclib	4.04	1	8	506.6	1	75.00	7	0
Tetrandrine	5.49	0	8	622.7	2	61.86	4	0
Ponatinib	4.30	1	8	532.6	1	65.77	6	0
Berbamine	5.13	1	8	608.7	2	72.86	3	0
Mycophenolic acid	2.72	2	6	320.3	0	93.07	6	0
Salinomycin	4.98	4	11	751.0	2	161.21	12	2
Merimepodib	2.36	3	7	452.5	0	123.96	11	1
Cycloheximide	1.23	2	4	281.4	0	83.47	3	0
Anisomycin	1.00	2	5	265.3	0	67.79	5	0
Bortezomib	0.22	4	6	384.2	0	124.44	11	1
Pimozide	5.67	1	5	461.5	1	41.29	7	0

[a] LogP: octanol–water partition coefficient; [b] MW: molecular weight; [c] TPSA: total polar surface area.

The analyses were carried out for the 23 drugs from the COVID-Box selected in the antiproliferative assay and those used as the gold standard (PYR and SDZ) and alternatives (AZT, CLI, and ATO) for toxoplasmosis. As expected, PYR and SDZ results agreed with Lipinski's RO5 and Veber's predictors. Among the drugs used as alternative treatments, only AZT violates two of Lipinski's rules (MW and 5H-bond donors) and one Veber predictor (TPSA < 140 Å^2) (Table 2). Of the 23 selected from the COVID-Box, 10 (niclosamide, apilimod, regorafenib, emetine, sorafenib, mycophenolic acid, merimepodib, cycloheximide, (-)-anisomycin, and bortezomib) showed compliance with the RO5 and Veber's predictors (Table 2). The other drugs and compounds showed at least one or more non-compliances with the RO5 and Veber (Table 2).

Information on the pharmacokinetic properties of selected drugs from the COVID-Box and drugs already used in treating toxoplasmosis was obtained using the platform pkCSM (Supplementary Table S2). Caco-2 permeability values (log Papp at 10^{-6} cm/s) above 0.90 predict high intestinal permeability. The drugs bemcentinib, apilimod, bortezomib, manidipine, almitrine, midostaurin, abemaciclib, and ponatinib had a log Papp at 10^{-6} cm/s above 0.90, and regorafenib, emetine, sorafenib, tetrandrine, and mycophenolic acid showed values of log Papp at 10^{-6} cm/s > 0.70– < 0.90, from which it can be inferred that these also have the potential to present high intestinal permeability [47].

The other drugs showed values below 0.60 log Papp at 10^{-6} cm/s (Table S2). It should also be noted that even the compound cycloheximide, presenting a Caco-2 value below 0.90 (0.553 log Papp at 10^{-6} cm/s, intestinal absorption (human) = 69.8%) (Table S2) was the compound selected in the in vitro tests that most inhibited parasite proliferation with an IC_{50} value = 0.02 μM (Table 1).

The volume of distribution value (VDss) predicts the drug distribution in tissue. It is known that the lower the interaction of drugs with plasma proteins, the faster they will be absorbed, and, therefore, the faster they will be directed to their site of action. Thus, the higher the VDss value above 0.45 log L/kg, the more the drug is distributed in tissues than in plasma, and values below −0.15 log L/kg are considered poorly distributed [47]. Seven drugs or compounds selected from the COVID-Box (bemcentinib, emetine, ivermectin, manidipine, almitrine, abemaciclib, and ponatinib) showed high distribution. Two drugs used for the treatment of toxoplasmosis (SDZ and atovaquone) and 0.329), and the COVID-Box drugs niclosamide, LY2228820, digoxin, sorafenib, salinomycin, merimepodib, cycloheximide, (-)-anisomycin, and pimozide showed VDss between −0.15 and 0.45 log L/kg (Table S2). Fraction unbound analyses showed that (-)-anisomycin and cycloheximide were the drugs with a higher proportion of free state in the plasma (Table S2).

The central nervous system (CNS) is a common site of infection of *T. gondii*; thus, we evaluated the predictors for CNS and BBB permeability of the COVID-Box drugs (Table S2). For CNS permeability, compounds with logPS > −2 are predicted to penetrate but with logPS < −3 are unable to penetrate. From the COVID-Box, six drugs (niclosamide, LY2228820, sorafenib, midostaurin, rapamycin, and pimozide) showed a prediction of CNS penetration, and twelve drugs presented predicted logPS between −3 and −2 and have the potential for penetration too (Table S2).

For BBB permeability, LogBB values above 0.3 predict that a compound could readily cross the BBB, and the ones with <−1 are poorly permeable [47]. None of the drugs and compounds selected from COVID-Box and most of the current ones used for toxoplasmosis treatment showed prediction for a high crossing into the brain. However, thirteen (niclosamide, bemcentinib, emetine, manidipine, almitrine, midostaurin, tetrandrine, ponatinib, berbamine, mycophenolic acid, cycloheximide, (-)-anisomycin, and pimozide) showed values > −1. Of the drugs already used in the treatment of toxoplasmosis, the only one with a value above 0.3 was ATO (0.401 log BB). PYR presented a value close to the expected value (0.278 log BB) (Table S2).

Using the SwissADME platform, we obtained the boiled-Egg graph, which also predicts if the drugs have the potential to cross the BBB and have high gastrointestinal absorption (HIA). The BBB permeability data provided by pkCSM were compared with the data provided in the Swiss ADME boiled-Egg plot [44]. Of the analyzed drugs used for toxoplasmosis treatment, only PYR and ATO showed characteristics with potential BBB permeability at the points drawn above the egg yolk in the graph (yellow color) (Supplementary Figure S6). The pkCSM program predicted that only ATO could cross the BBB (Supplementary Table S2). On the same graphic, those that are plotted in the egg white region (SDZ and CLI) would be more easily absorbed in the gastrointestinal tract by passive transport than the compounds that were plotted in the gray area of the graph (AZT) (Figure S6A). In addition, the graph provides information such as whether the drugs are glycoprotein inhibitors. Only PYR, SDZ, and ATO drugs are not P-gp substrates (marked with red dots in the graph) (Figure S6A). Compounds that are P-gp inhibitors show increased absorption, while P-gp substrates reduce their absorption [48].

Of COVID-Box, almitrine (Figure S6B), emetine, and ponatinib (Figure S6C) presented characteristics with potential permeability through the BBB (points drawn in the upper part of the yolk of the graph) (Figure S6B,C); this information complies with the results obtained for BBB permeability presented in Table S2. According to the results obtained in this analysis, none of the others are predictable for readily crossing the natural protections of the CNS [49]. The drugs niclosamide, bemcentinib, apilimod, LY 2228820, digoxin, manidipine, midostaurin, abemaciclib, tetrandrine, berbamine, mycophenolic acid, merimepodib, cycloheximide, (-)-anisomycin, bortezomib, and pimozide (Figure S6B,C) show potential for easier absorption in the gastrointestinal tract by passive transport. Of the COVID-Box drugs, niclosamide, apilimod, regorafenib, sorafenib, almitrine, tetrandrine, berbamine, mycophenolic acid, salinomycin, cycloheximide, and (-)-anisomycin are predicted as non-P-gp substrates (marked with red dots in the graph) (Figure S6B). Bemcentinib, LY 2228820, digoxin, emetine, ivermectin, manidipine, abemaciclib, ponatinib, merimepodib, bortezomib, and pimozide are P-gp substrates (marked with blue dots in the graph Figure S6C).

In agreement with the ADME in silico analysis, pharmacokinetics in in vivo studies showed that the selected drugs are absorbed and available in the plasma (Table S3). Brain availability was confirmed for abemaciclib, (-)-anisomycin, apilimod, bemcentinib, emetine, ivermectin, ponatinib, regorafenib, salinomycin, sorafenib, and tetrandrine (Table S3).

3.3. *Twenty-Four-Hour Antiproliferative Assay*

For the drugs with the best IC_{50} value (cycloheximide, bortezomib, (-)-anisomycin, mycophenolic acid, and salinomycin), good predictable druggability in the in silico analysis (almitrine and merimepodib), or first reported for anti-*T. gondii* activity (apilimod and midostaurin), we carried out a proliferation assay to verify the inhibitory capacity of the compounds after only 24 h of treatment (Figure 1). As a positive control, we used PYR (Figure 1A). The treatment with 125 nM (-)-anisomycin (Figure 1B), 31.2 nM bortezomib (Figure 1F), 0.25 μM midostaurin (Figure 1D), and 62.5 nM cycloheximide (Figure 1B) inhibited the proliferation around 75–80%. The treatment with 250 nM salinomycin and 62.5 nM bortezomib reduced the proliferation of the parasite by more than 90% (Figure 1E,F). Treatment with 1.5 and 2.0 μM merimepodib (Figure 1C), 1000 nM almitrine, 250 nM mycophenolic acid, and 1000 nM ivermectin (Figure 1E) inhibited the parasite proliferation around 55–60%. Apilimod only showed a modest inhibition after 24 h of treatment (Figure 1E).

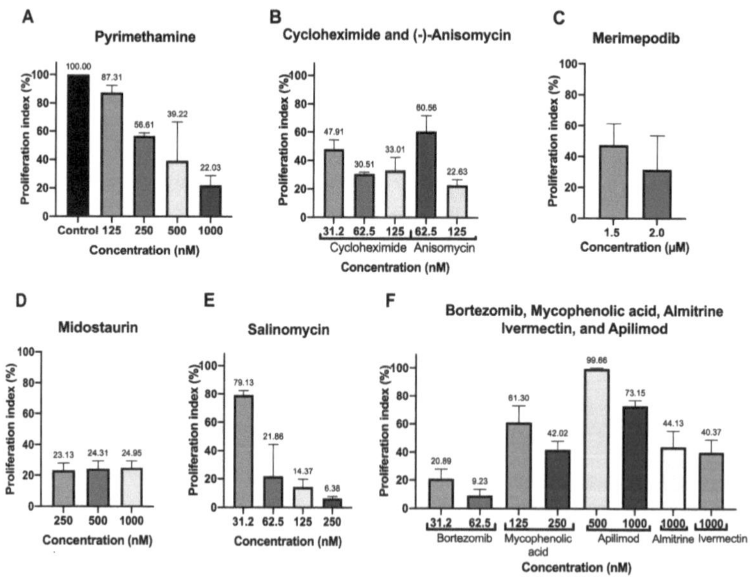

Figure 1. Proliferation index of the best drugs of the COVID-Box after 24 h of treatment with different concentrations of (**A**) Pyrimethamine, (**B**) Cycloheximide, Anisomycin, (**C**) Merimepodib, (**D**) Midostaurin, (**E**) Salimomycin, (**F**) Bortezomib, Mycophenolic acid, Apilimod, Almitrine, and Ivermectin. Values represent mean ± SD of three experiments, except for merimepodib, salinomycin, and pyrimethamine (two experiments).

3.4. Analysis of the Effect of the Drugs and Compounds of COVID-Box on the Parasite Morphology by Transmission Electron Microscopy (TEM) and Fluorescence Microscopy

To verify the direct effect and to determine the mode of action on *T. gondii* of the drugs, the effect on the ultrastructure of *T. gondii* after treatment with those with best IC_{50} value or good performance in the in silico analysis or first reported for anti-*T. gondii* activity was analyzed by TEM and immunofluorescence microscopy (Figures 2–8).

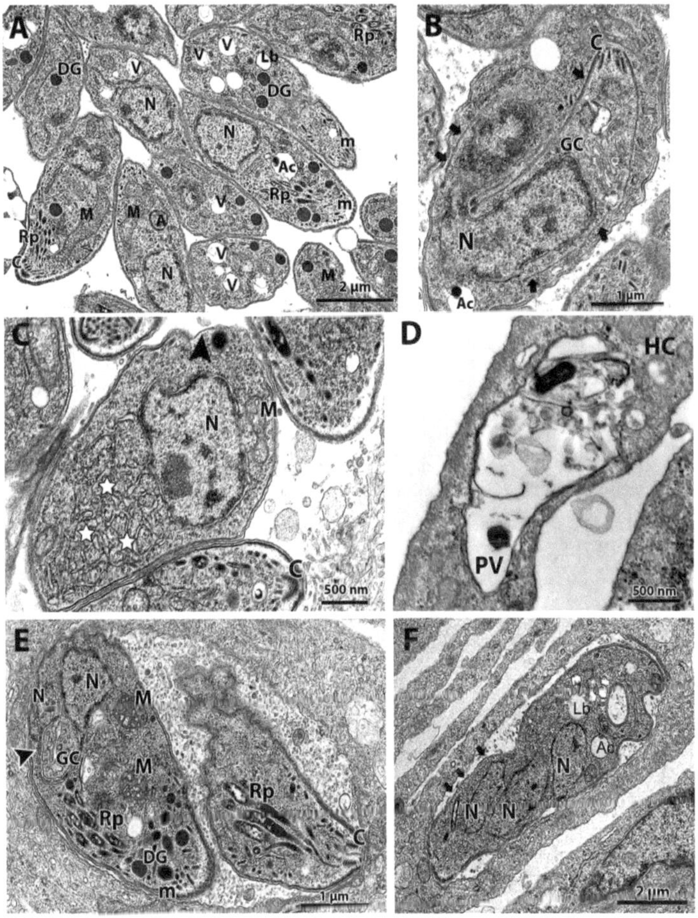

Figure 2. (**A–F**). Transmission electron microscopy and analysis of the ultrastructure of tachyzoites after treatment with cycloheximide and bortezomib. The parasites were treated with the compounds for 48 h. (**A**,**B**) Untreated parasites showed typical morphology (**A**,**B**) and division process by endodyogeny (arrows in (**B**)). (**C**) Parasites treated with the drug 62.5 nM cycloheximide showed an increase in the endoplasmic reticulum area (stars) and alterations in the structure of the plasma membrane, with regions with a lack of inner membrane complex (black arrowhead). (**D**) Parasites treated with 125 nM cycloheximide were destroyed; it is possible to observe parasite content spread through the PV. (**E**,**F**) Parasites treated with 62.5 nM bortezomib showed cell division alterations, as seen by the Golgi complex surrounded by the nucleus envelope (arrowhead) and a parasite presenting three nucleus profiles without constructing new daughter cells. Mitochondrial swelling (M) and regions of parasite devoid IMC were also observed (arrows). A—apicoplast, Ac—acidocalcisome, C—conoid; DG—dense granules, GC—Golgi complex, Lb—lipid body, M—mitochondrion, m—micronemes, N—nucleus, Rp—rhoptries, PV—parasitophorous vacuole, V—vacuolar compartment.

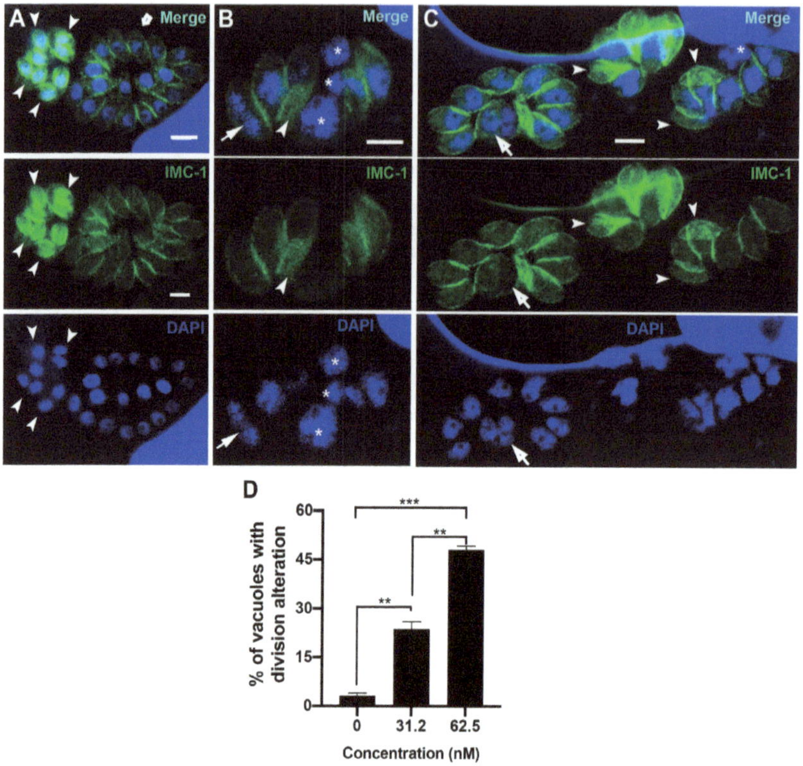

Figure 3. Fluorescence microscopy of untreated parasites (**A**) or after treatment with 31.2 nM (**B**) and 62.5 nM (**C**) of bortezomib. Parasites were labeled with anti-IMC1 for inner membrane complex (IMC, green) and DAPI for DNA (blue). (**A**) Untreated parasites showed typical morphology (arrow) and division process (arrowhead). (**B**,**C**) treated parasites showed an aberrant cell division process with large parasites harboring two or more nuclei (arrow), daughter cells without nuclei (arrowheads), and regions of the cells without IMC coverage (asterisks). (**D**) Quantitative analysis of the number of PVs presenting parasites with aberrant cell division. Results in (**D**) are the mean ± SD of two independent experiments. * $p < 0.05$; ** $p < 0.01$; *** $p < 0.001$. Bars = 2.5 μm.

TEM analysis of the untreated control showed a parasitophorous vacuole (PV) with parasites presenting normal morphology and ultrastructural organization (Figure 2A,B). It is possible to observe structures of the apical complex rhoptries (Rp), conoid (C), and micronemes (m). The nucleus (N), dense granules (DG), apicoplast (A), acidocalcisome (Ac), Golgi complex (GC), lipid body (Lb), vacuolar compartment (V), and mitochondria (M) are also evident (Figure 2A,B). Tachyzoites showed a typical division process by endodyogeny (Figure 2B), with the nucleus presenting a horseshoe shape involved by constructing two new daughter cells delimited by the inner membrane complex (IMC; arrows).

Tachyzoites treated with 62.5 nM cycloheximide showed an increased endoplasmic reticulum area (stars) and alterations on the parasite plasma membrane structure, as evidenced by the presence of regions with only a single pellicle (arrowhead) instead of the three-membrane structure composed by the plasmalemma and IMC (Figure 2C). When tachyzoites were treated with 125 nM cycloheximide, it was observed that vacuoles containing parasites were completely lysed; the asterisk evidences a disrupted parasite and its content spread inside the PV (Figure 2D).

Treatment with 62.5 nM bortezomib affected the parasite cell division, with parasites presenting a nucleus with altered morphology, as evidenced by the enclosure of the Golgi complex by the nucleus (arrow in Figure 2E). The arrestment of the division process was also evidenced as single parasites presenting multiple nucleus profiles without the construction of new daughter cells were observed (Figure 2F). Treatment with bortezomib also caused mitochondrial swelling (M in Figure 2E) and affected the pellicle of the parasite, where it is possible to observe regions devoid of the IMC coverage (arrows in Figure 2F).

To confirm the effect of bortezomib on the parasite cell division, we analyzed tachyzoites treated with 31.2 and 62.5 nM for 24 h after labeling with IMC1 and DAPI by immunofluorescence microscopy (Figure 3A–D). Untreated cells (Figure 3A) showed typical morphology (arrow 3) and division process (arrowheads), where it is possible to observe the construction of two new daughter cells delimited by the IMC, with each containing a divided nucleus (arrowheads in Figure 3A). The effect of bortezomib on cell division was observed even when parasites were treated with 31.2 and 62.5 nM (Figure 3B–D). Treated parasites showed large cells with mitotic nuclei without the construction of new daughter cells (arrows). Areas without IMC cover (asterisks) and daughter cells without nuclei (arrowheads) were also observed (Figure 3B–D). Quantification analysis showed that the effect on parasite division is significantly concentration-dependent, as 23.6% and 48.0% of the PVs had parasites with aberrant division after treatment with 31.2 and 62.5 nM, respectively (Figure 3D).

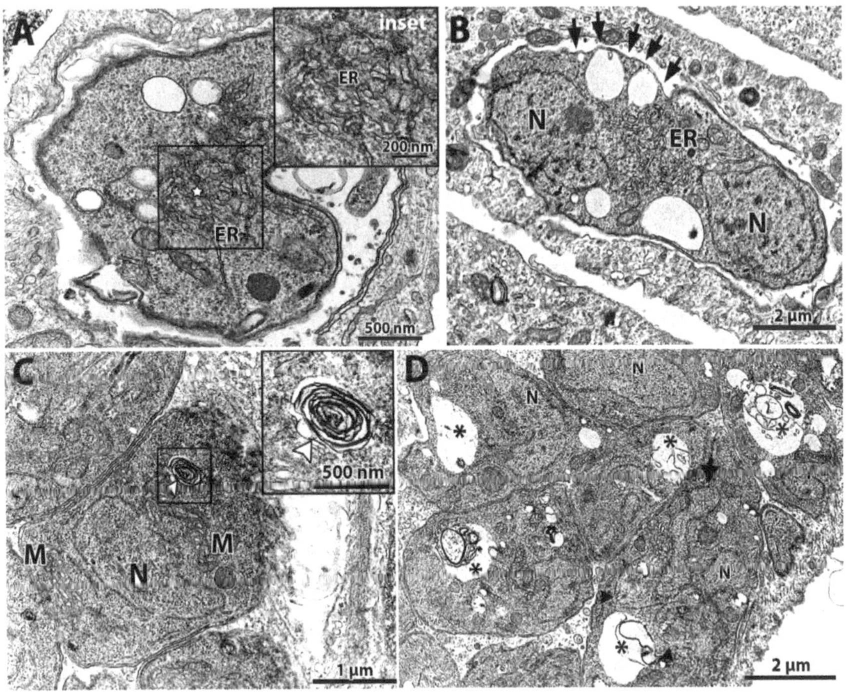

Figure 4. Transmission electron microscopy of *T. gondii* after treatment for 48 h with (-)-anisomycin (**A**,**B**) and ivermectin (**C**,**D**). (**A**) Treatment with 100 nM (-)-anisomycin induced changes in the parasite's endoplasmic reticulum (star in inset) and (**B**) 100 nM (-)-anisomycin also induced impairment of the cell division, making it possible to observe a single parasite with two nuclei and causing discontinuation of the inner membrane complex (black arrows). (**C**) Parasites treated with 1 μM ivermectin induced the formation of myelin-like figures (inset—white arrowhead). (**D**) In this figure, it is also possible to observe an intense vacuolization process in parasites treated with 1 μM ivermectin (asterisks). M—mitochondria; N—nucleus; GC—Golgi complex; ER—endoplasmic reticulum.

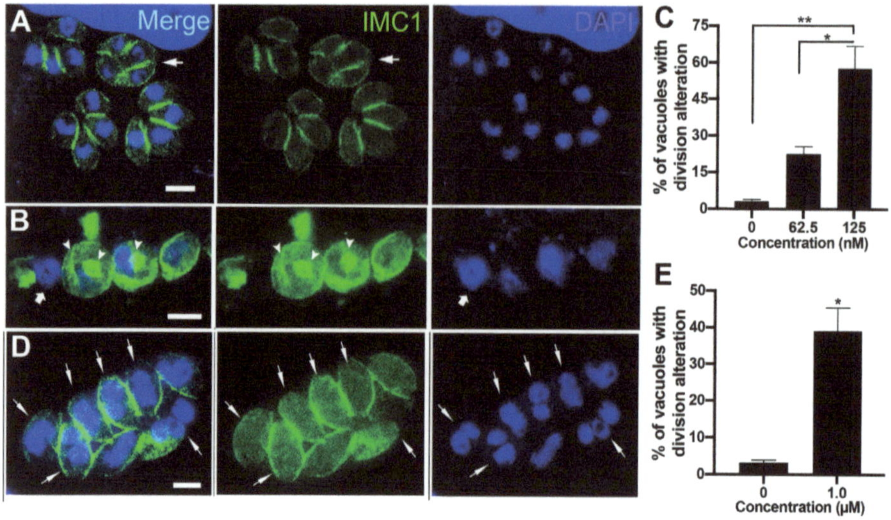

Figure 5. Fluorescence microscopy analysis of tachyzoites treated with 62.5 nM and 125 nM (-)-anisomycin (**A–C**) and 1 µM ivermectin (**D**,**E**). Parasites were labeled with anti-IMC1 for inner membrane complex (IMC, green) and DAPI for DNA (blue). (**A**) Parasites treated with 62.5 nM (-)-anisomycin showed daughter cells' budding arrestment, forming a large mass of tethered daughter cells (arrow). (**B**) Treatment with 125 nM (-)-anisomycin led to a large round mass of cells with a nucleus of increased size and disorganized profiles of IMC (arrowheads). The arrow points to a parasite region without the IMC coverage. (**C**) Quantitative analysis of the number of PVs presenting parasites with aberrant cell division after treatment with (-)-anisomycin. (**D**) Parasites treated with 1 µM ivermectin showed a divided nucleus without the construction of daughter cells (arrows). (**E**) Quantitative analysis of the number of PVs presenting parasites with aberrant cell division after treatment with ivermectin. Results in (**C**,**E**) are the mean ± SD of two independent experiments. * $p < 0.05$; ** $p < 0.01$. Bars = 2 µm.

The tachyzoites treated with 100 nM (-)-anisomycin for 48 h showed alteration of the ER architecture (Figure 4A,B). Figure 4A,B show that the ER has a disorganized architecture extending through a large cytoplasmic area. The treatment with 100 nM (-)-anisomycin also arrested parasite division and the plasma membrane structure (Figure 4B), as seen by a PV containing a single large tachyzoite with two divided nuclei (N) without constructing new daughter cells and the lack of IMC at portions of the plasma membrane (arrows), respectively (Figure 4B).

Treatment with 1 µM ivermectin (Figure 4C,D) induced the formation of myelin-like structures [49,50] (arrowhead in Figure 4C and inset), resembling a process of cell death by autophagy. In Figure 4D, tachyzoites show large vacuoles containing membranous material (asterisks) and a lobular nuclear shape. In one parasite, it is possible to observe the construction of two daughter cells (arrowheads in Figure 4D), but a nucleus lobule (arrow) is not involved by one of the daughter cells.

The effects of (-)-anisomycin and Ivermectin on tachyzoite cell division were analyzed with immunofluorescence microscopy after labeling with IMC1 and DAPI (Figure 5A–D). Treatment with 62.5 nM (-)-anisomycin significantly increased the number of PVs containing parasites with aberrant division (22.5%; Figure 5C). The arrow in Figure 5A points to a mass of tachyzoite containing four non-budded daughter cells. The effect of 125 nM (-)-anisomycin was more drastic (Figure 5B,C), leading to the complete arrestment of cell division with PVs containing a sizeable round mass of cells with a nucleus of increased size and with disorganized profiles of IMC through the cytoplasm (arrowheads). As seen by TEM (Figure 4B), parasites presenting regions lacking IMC cover (arrow in Figure 5B) were also observed. Parasites treated with 1 μM ivermectin also presented significant cell division arrestment (Figure 5D,E), with PVs presenting tachyzoites containing a divided nucleus (arrows in Figure 5D) without the formation of new daughter cells.

Treatment of the tachyzoites with 1 μM almitrine also induced myelin-like structures (Figure 6A and inset) and impairment of the parasite's cell division (Figure 6B). Figure 6B shows the mass of the mother cell with a large vacuole (asterisks) containing two non-budded daughter cells. Large vacuoles with membranous material were also observed (asterisks in Figure 6B). Treatment with 250 μM midostaurin caused drastic effects on the parasite ultrastructure (Figure 6C,D). Parasite masses with daughter cells and IMC profiles spread in the cytoplasm (arrowheads) were observed (Figure 6C). Parasites presenting alterations suggestive of the cell death process, such as the fragmented nucleus (large arrow in Figure 6C), structures resembling the autophagy process (asterisk in Figure 6C), and large masses of parasites with a disrupted cell division process (arrowheads) and in an advanced vacuolization process (asterisks in Figure 6D) were also observed.

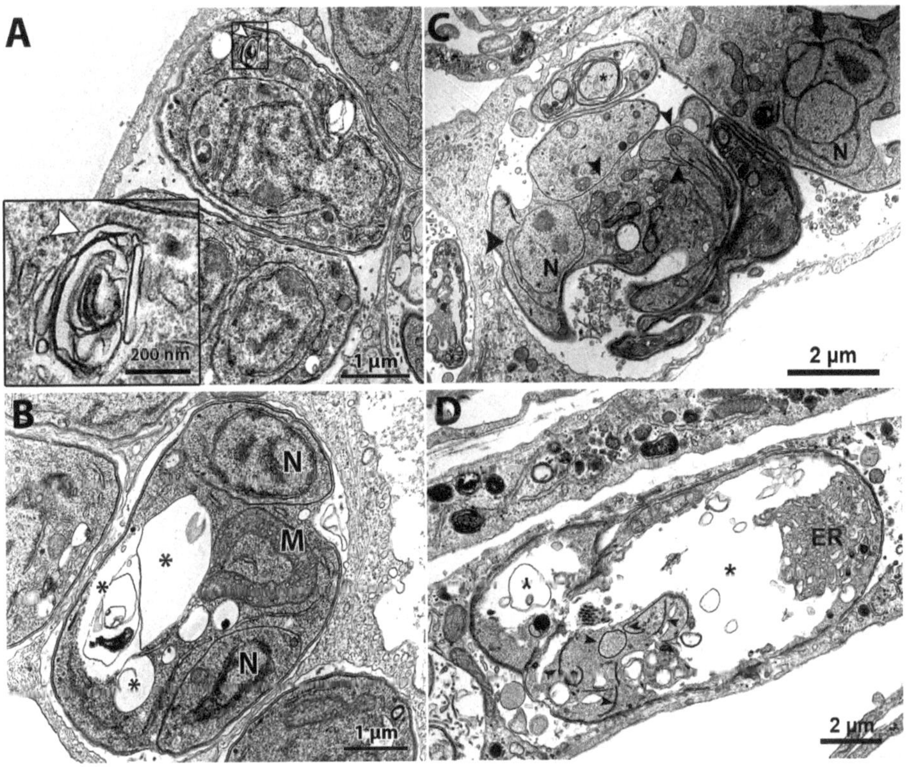

Figure 6. Transmission electron microscopy of *T. gondii* tachyzoites after treatment with almitrine and midostaurin for 48 h. (**A**) Parasites treated with 1 μM almitrine showed myelin-like structures (arrowhead in inset). (**B**) Treatment with one μM almitrine also induced the formation of large vacuoles containing membranous material (asterisks) and disruption of cell division, as seen by a large mother mass harboring two non-budded daughter cells (asterisks). (**C**,**D**) Parasites treated with 250 nM midostaurin for 48 h. (**C**) Vacuole containing a mass of tachyzoite with several arrested daughter cells and IMC profiles through the cytoplasm (arrowheads). A parasite presenting a fragmented nucleus (large arrow), and a process similar to autophagy (asterisks) was observed too.

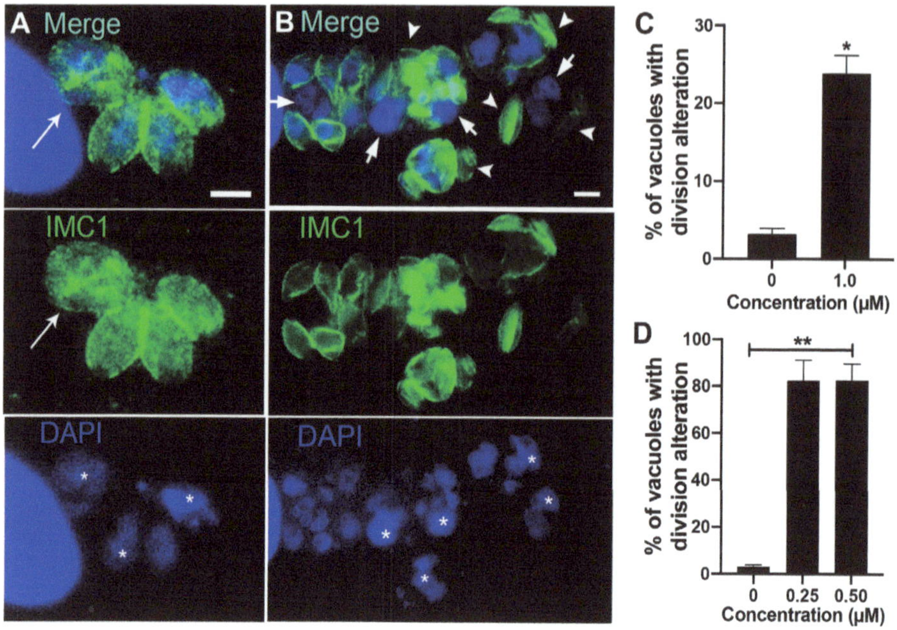

Figure 7. Fluorescence microscopy analysis of tachyzoites treated with almitrine and midostaurin. Parasites were labeled with anti-IMC1 for inner membrane complex (IMC, green) and DAPI for DNA (blue). (**A**) Parasites treated with 1 µM almitrine showed cell division alteration with tachyzoites presenting large nuclei (asterisks) and masses with incomplete division process (arrow). (**B**) Treatment with 0.25 µM midostaurine caused a large round mass of cells with a nucleus of increased size (asterisk), tachyzoites showing regions without the IMC cover (arrows), and daughter cells without a nucleus (arrowheads). (**C**) Quantitative analysis of the number of PVs presenting parasites with aberrant cell division after treatment with almitrine. (**D**) Quantitative analysis of the number of PVs presenting aberrant parasites after treatment with midostaurin. Results in (**C**,**D**) are the mean ± SD of two independent experiments. * $p < 0.05$; ** $p < 0.01$. Bars = 2.5 µm.

The effect on cell division after treatment with almitrine was confirmed after analysis of treated parasites labeled for the IMC and DNA (arrow in Figure 7A). Quantification analysis showed that 23.9% of the vacuoles contained tachyzoites with cell division alteration (Figure 7C). Treatment with Midostaurin was even more drastic (Figure 7B), as more than 80% of vacuoles had aberrant parasites (Figure 7D). Large tachyzoites containing increased-size nuclei (asterisks), regions of the parasite body without IMC cover (arrows), or daughter cells without a nucleus (arrowheads) were observed (Figure 7B).

Treatment with 1.5 µM merimepodib induced Golgi complex fragmentation (GC in Figure 8A), as the stacked membranes were absent and replaced by numerous vesicular structures. Treatment also caused alteration in rhoptry organization and morphology (Rp in Figure 8A and inset). Tachyzoites presenting large vacuoles containing membranous material were also observed after treatment with merimepodib (asterisks in Figure 8B). To better characterize the effect of merimepodib on the rhoptry morphology, parasites treated with 1.5 µM for 24 h were labeled with the antibody against the rhoptry protein ARO, and fluorescence images of six–ten different focal planes (Z) were acquired and analyzed (Figure 8C,D). While untreated parasites showed vacuoles with parasites harboring rhoptries with typical morphology, around 25% of PVs (Figure 8D) showed parasites with rhoptry alteration (arrowheads in Figure 8C) after treatment with merimepodib.

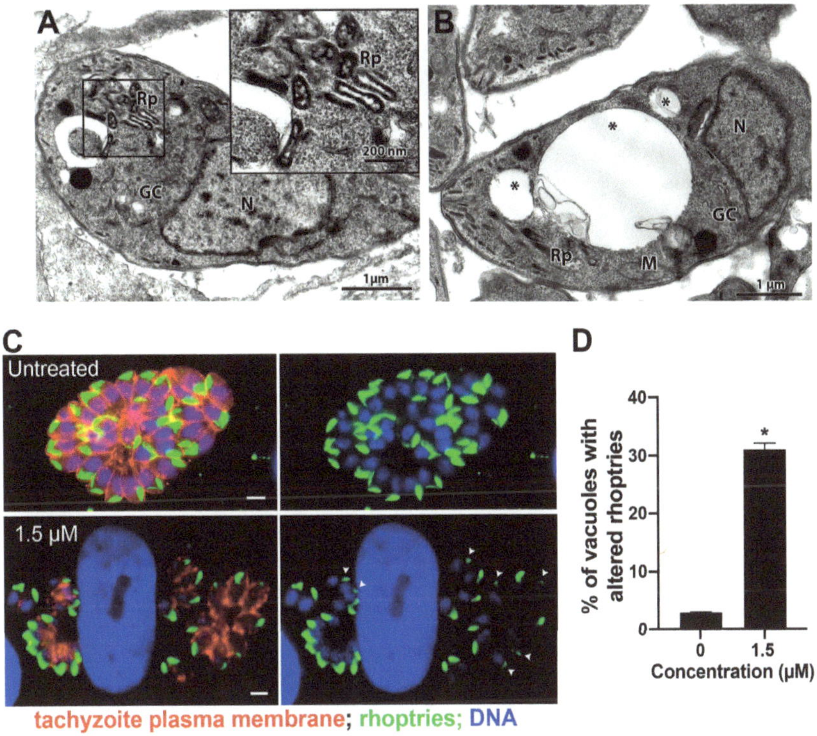

Figure 8. Morphological analysis of tachyzoites of *T. gondii* after treatment with 1.5 µM merimepodib. (**A**) Treatment with 1.5 µM induced Golgi complex fragmentation (vesiculation) and rhoptry disorganization, which can be seen at higher magnification in the inset. (**B**) Tachyzoites treated with 1.5 µM merimepodib also presented large vacuoles containing membranous material (asterisks). (**C**) Fluorescence microscopy analysis of tachyzoites treated with 1.5 µM merimepodib for 24 h. Parasites were labeled with anti-ARO for rhoptries (green), anti-SAG1 for parasite plasma membrane (red), and DAPI for DNA (blue). Images represent the projection of different Z focal planes. (**D**) Quantitative analysis of the number of PVs presenting parasites with rhoptry- altered morphology (arrowheads in (**C**)). Results are the mean ± SD of two independent experiments. * $p < 0.05$. M—mitochondria; N—nucleus; GC—Golgi complex; Rp—rhoptries. Bars = 2 µm.

Treatment with 250 nM mycophenolic acid affected the tachyzoites' division process (Figure 9A,B). Parasites in the division process presenting multiple lobules (arrowheads in Figure 9A) or mitotic nuclei (horseshoe shape) without the construction of new daughter cells were observed (arrow in Figure 9A). Even daughter cells presented a mitotic nucleus before completing the division process (N in Figure 9B). Tachyzoites treated with 125 and 250 nM salinomycin showed an extensive vacuolization process (asterisks), suggesting an advanced cell death process (Figure 10A,B, respectively). PVs containing lysed parasites (arrow in Figure 10B) were also observed. Analysis by fluorescence microscopy after treatment with 125 nM salinomycin and labeling the parasite plasma membrane (red) and host cell lysosomes (green) showed that differently from untreated cells that had PVs with parasites' typical morphology and organized in rosettes (Figure 10C), treated parasites showed vacuoles containing fragmented parasites (Figure 10C′). PVs with lysed parasites did not show fusion with lysosomes.

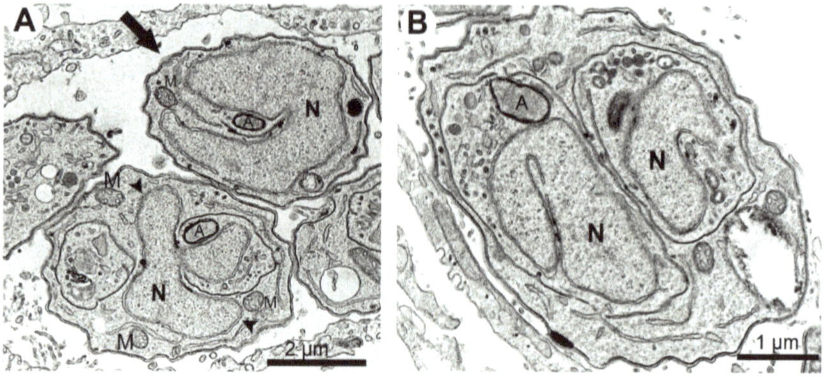

Figure 9. Transmission electron microscopy of *T. gondii* tachyzoites after treatment with 250 nM mycophenolic acid for 48 h. (**A**) Tachyzoites in the division process present multiple lobules (arrowheads) or mitotic nuclei (horseshoe shape) without the construction of new daughter cells (arrow). (**B**) Daughter cells without the completion of the division process with mitotic nuclei. A—apicoplast; M—mitochondria; N—nucleus.

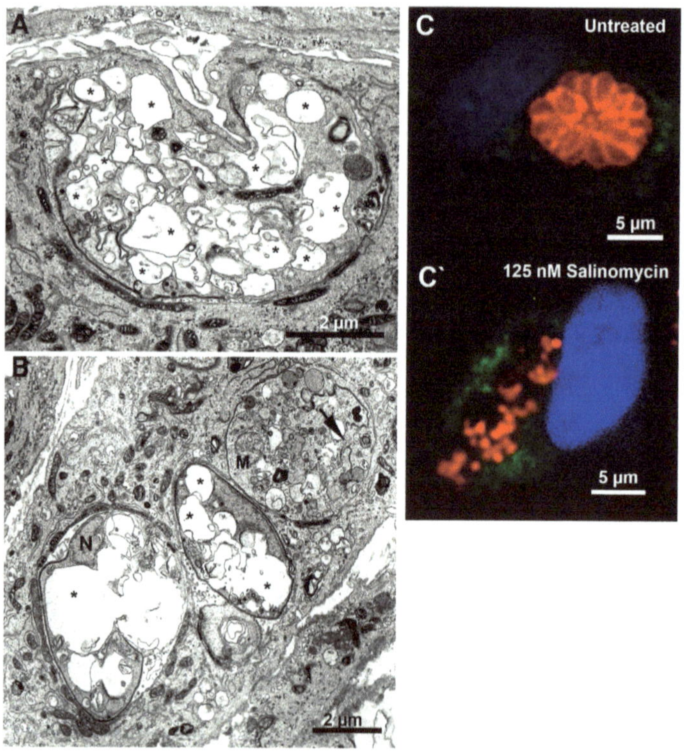

Figure 10. Morphological analysis of *T. gondii* tachyzoites after treatment with salinomycin for 24 h. (**A**) Treatment with 125 nM caused an extensive vacuolization process (asterisks) on the parasite. (**B**) Tachyzoites treated with 250 nM showed an extensive vacuolization process (asterisks) and cell lysis (arrow). (**C**,**C′**) Fluorescence microscopy analysis of tachyzoites treated with 1.5 µM merimepodib for 24 h. Parasites were labeled with anti-LAMP1 for host cell lysosomes (green), anti-SAG1 for parasite plasma membrane (red), and DAPI for DNA (blue).

4. Discussion

In vitro tests were performed to evaluate the potential of applying the drugs and compounds present in the COVID-Box against *T. gondii* infection and to demonstrate the possibility of the antiproliferative effect of the drugs and compounds in the COVID-Box. After these, we selected 23 drugs and compounds that could inhibit the proliferation of the parasite by more than 70%. The discovery of new uses of drugs previously used for other pharmaceutical purposes is a cheaper solution for treating neglected diseases. The antiproliferative analysis showed that the selected compounds inhibited *T. gondii* proliferation with values of IC_{50}s ranging from 0.02 µM to 0.74 µM, and ten showed IC_{50} lower than 100 nM (Table 1). Cytotoxicity analysis by the MTS assay also showed that most compounds were highly selective for *T. gondii* (Table 1). Of the 23 drugs, 11 were recently reported in a study of the anti-*T. gondii* effect of COVID-Box compounds, but two (apilimod and midostaurin) have been reported here for the first time. However, this is the first work to demonstrate the ultrastructural alterations caused by cycloheximide, bortezomib, (-)-anisomycin, ivermectin, almitrine, merimepodib, midostaurin, and salinomycin in tachyzoites through TEM analysis.

Cycloheximide, (-)-anisomycin, and bortezomib are the most potent drugs against *T. gondii* tachyzoites contained in the COVID-Box, inhibiting parasite proliferation with IC_{50}s in the range of 20–30 nM. This finding is in line with the study by Fichera, Bhopale, and Ross (1995) [43], which found an IC_{50} value of 0.01 µM for (-)-anisomycin against *T. gondii* after 48 h of treatment. In addition, in silico analyses demonstrated that cycloheximide and (-)-anisomycin show the predictors of good oral bioavailability according to Lipinski's rule of five (RO5) and Veber, and the potential to cross the BBB (Tables 2 and S2 and Figure S6). In silico analysis also showed that cycloheximide and (-)-anisomycin are non-P-gp substrates. Published in vivo pharmacokinetic studies (Table S3) showed that cycloheximide and (-)-anisomycin have gastrointestinal absorption and reach plasma concentrations higher than the IC_{50} obtained for *T. gondii*. The brain availability for (-)-anisomycin was also confirmed in vivo. Bortezomib violated only one predictor (Table 2) and showed a prediction for high intestinal permeability (Table S2). A published pharmacokinetic study in humans showed that bortezomib is absorbed and reaches a plasma concentration of 147.6 nM, which is higher than the IC_{50} found for *T. gondii* (Table S3).

Cycloheximide and (-)-anisomycin are bacterial antibiotics isolated from *Streptomyces* species [50]. Although cycloheximide and (-)-anisomycin are protein synthesis inhibitors in eukaryotes, acting during the protein elongation state, their targets on the ribosomes are different. Cycloheximide acts by binding to the E-site and inhibits the mRNA–tRNA translocation. (-)-Anisomycin binds to the A-site, which inhibits the protein synthesis by impairing peptide bond formation, preventing the elongation [50–55]. MET analyses showed an increase in the endoplasmic reticulum area of *T. gondii* after treatment with 62.5 nM cycloheximide (Figure 2C) and 100 nM (-)-anisomycin (Figure 4A,B). The endoplasmic reticulum synthesizes essential lipids to maintain the plasma membrane [56,57]. Interestingly, similar ultrastructural changes were observed when *T. gondii* was treated with the antifungal drugs itraconazole, eberconazole, and thiolactomycin analogs [21,29]. The treatment of *T. gondii* with this last drug affected the acylglycerol synthesis by the endoplasmic reticulum [58]. TEM analysis also showed that (-)-anisomycin affected *T. gondii* endodyogeny, confirmed by immunofluorescence microscopy analysis (Figure 5A–C). Interestingly, treatment with 62.5 nM (-)-anisomycin affected the division causing the tethered daughter cell phenotype (Figure 5A), which is typically caused by inhibitors that target the apicoplast pathways, including inhibitors that target the organelle protein synthesis [28,57–59]. Treatment with 125 nM (-)-anisomycin disrupted completely parasite division and daughter cell construction (Figure 6B). The increase in the (-)-anisomycin concentration possibly affected other targets, as besides protein synthesis, this drug is also known to inhibit DNA synthesis [60] and to activate stress-activated protein kinases and mitogen-activated protein kinases [50].

A previous study involving bortezomib in tachyzoites of *T. gondii* found an IC_{50} value of 0.10 μM after 72 h of treatment [34]. Cajazeiro et al. (2022) [24] found a different EC_{50} value of 0.22 μM after 72 h of treatment, which differs slightly from the IC_{50} value found in this study after treatment with bortezomib (0.03 μM). This difference is possible due to the longer treatment time used by us in this study, which suggests a time-dependent effect of this drug. Bortezomib is a known potent, selective, and reversible inhibitor of the proteasome, an organelle responsible for the degradation of defective proteins in the cell and crucial for the stability of regulatory proteins. Proteasome inhibitors are known to cause cell death [61,62]. Other authors have reported the inhibition of catalytic subunits of the proteasome in *Plasmodium falciparum* [63,64]. TEM and immunofluorescence microscopy analysis showed that the treatment with 62.5 nM bortezomib significantly affected the parasite division process (Figure 2E,F). As we observed in *T. gondii*, a study that evaluated the effect of bortezomib in a mantle cell lymphoma cell line showed that this drug affected the cell cycle through the G2/M phase arrest [65]. This result is in line with ours, as we also observed parasites with a large undivided nucleus, typical of arrestment of the G2/M phase during the cell division cycle. Therefore, the sum of the results obtained here supports that cycloheximide, (-)-anisomycin, and bortezomib can be potential drugs for treating the acute phase of toxoplasmosis.

This is the first work to study the effect of midostaurin against *T. gondii.* This drug is a potent inhibitor of protein kinase C and several class III receptor tyrosine kinases involved in hematopoiesis and leukemia. It was approved for leukemia treatment and has an oral bioavailability estimated at >90% [66]. Midostaurin showed anti-*T. gondii* IC50 of 80 nM, which is fifteen times lower than its plasma concentration in humans.

TEM and immunofluorescence analysis showed that around 80% of the vacuoles treated with 250 nM of this drug had drastic morphological alterations, such as an aberrant division process and the induction of cell death (Figures 6 and 7). A similar effect was observed in HMC1 cells (neoplastic human mast cells) after treatment with 500 nM midostaurin for 24 h [67].

Ivermectin was another drug with anti-*T. gondii* activity and good pharmacokinetic prediction (Tables 1, 2, S2 and S3). The potential activity of ivermectin against *T. gondii* and other protozoa, such as *Giardia lamblia*, *Trypanosoma cruzi*, *Leishmania infantum*, and *Trypanosoma evansi*, has also been reported in the literature [37,68–71]. However, this is the first study exploring the mode of action of this drug in a parasite from the Apicomplexa phylum. Using TEM, we observed that the treatment with 1 μM ivermectin for 48 h caused multi-membrane structures and large cytoplasmic vacuoles containing membranous material (Figure 4C,D), suggesting induction of cell death by autophagy [72]. Ivermectin also significantly affected the parasite division (Figure 5D,E). A similar effect was observed after treating glioma cells with ivermectin [73]. Treatment of *T. gondii* tachyzoites with almitrine also caused an intense vacuolization and formation of multi-membrane structures (Figure 6A,B) and aberrant cell division (Figure 7A,C), which are suggestive of the induction of an autophagic process [72]. Almitrine is already used in clinics to treat diseases that affect the respiratory system and is a good predictor of oral and brain bioavailability (Tables 2 and S2). Pharmacokinetic studies on humans showed that it could reach plasma concentrations up to 599 nM (Table S3), which is higher than the IC_{50} obtained for *T. gondii* in this study. The in vitro and in vivo effects against *T. gondii* have been recently reported [23,24], with IC_{50} values of 0.42 μM and 0.32 μM after 72 h of treatment, which are close to what we found in this study. The in vivo administration reduced the number of cysts in a murine model of chronic toxoplasmosis.

Mycophenolic acid and merimepodib are antiviral inhibitors of inositol monophosphate dehydrogenase, affecting DNA and RNA synthesis [74]. These drugs affected *T. gondii* proliferation with IC_{50} of 0.07 μM and 0.48 μM, respectively. Previous studies showed IC_{50}s of 211 μM (mycophenolic acid) and 0.78 μM (merimepodib) for *T. gondii* tachyzoites after 24 and 72 h of treatment, respectively [23]. Merimepodib also showed a high selective index for *T. gondii* compared with HFF cells (human foreskin fibroblasts) [23]. Analyses by TEM showed that merimepodib caused *T. gondii* Golgi complex fragmentation, rhoptry disorganization, and intense vacuolation (Figure 8A,B). The effect against rhoptries was analyzed by immunofluorescence, confirming that this is a significant alteration on the parasite (Figure 8C). Similar results were observed in tachyzoites after depletion of the vacuolar protein sorting nine (*Tg*Vps9), Vps11, and a membrane inositol phospholipid binding protein [75–77]. These results suggest that this drug could affect *T. gondii*, interfering with its secretory pathway. Although merimepodib and mycophenolic acid target the same enzyme, tachyzoites treated with the latter showed a different mode of action, as cell division alteration was the main observed effect after treatment with mycophenolic acid. A previous study also showed rounded tachyzoites with multiple nucleus profiles after treatment with mycophenolic acid [39]. This drug is also widely used in studies of molecular manipulation of *T. gondii* for selecting mutants that express the selectable marker HXGPRT [78].

We also investigated the mode of action of the H^+/K^+ ionophore salinomycin on *T. gondii*, a known anticoccidial drug commonly used for poultry and cattle [79,80]. TEM analysis showed that salinomycin induces parasite death, causing its lysis. Immunofluorescence analysis confirmed that cell death is directly caused by this drug and not by a secondary effect due to the fusion of lysosomes with the PVs.

Developing a new infectious disease treatment is complex because medicines must be absorbed, reach adequate plasma concentrations, and be distributed to tissues and cellular compartments where the infection is present in the body. In the case of toxoplasmosis, this is even more critical, as one of the main sites of infection is the CNS. Based on an anti-*T. gondii* activity assay and according to Lipinski's and Veber's predictors' analysis, the 23 drugs and compounds identified in this work are good candidates to become oral drugs since they inhibit *T. gondii* proliferation at a submicromolar range and comply with RO5, showing no more than one violation. We can highlight that the compounds cycloheximide, bortezomib, anisomycin, almitrine, midostaurin, and mycophenolic acid presented an IC_{50} range lower than 0.10 μM and did not violate Lipinski's rules and the predictors of Veber. Ivermectin and merimepodib are good candidates, inhibiting *T. gondii* with an IC_{50} < 0.5 μM. In addition, these drugs also demonstrated desirable predictors for oral absorption (Tables 2, S2 and S3). However, we should not disregard the potential of drugs that did not comply with RO5 or presented values lower than expected for oral absorption or BBB permeability since AZT, a drug already commercialized, presented a value lower than expected (−0.211 log Papp at 10^{-6} cm/s) and intestinal absorption (human = 45.808%), but despite AZT not having good intestinal absorption, this drug is used to treat toxoplasmosis [9].

5. Conclusions

After COVID-Box screening, we identified two new drugs with anti-*T. gondii* activity, making this the first study to report their effectiveness against this parasite. In total, 23 drugs were found to be promising candidates for further pre-clinical studies on toxoplasmosis. The discovery of these new drug candidates for the treatment of toxoplasmosis is of great relevance and should be further explored for in vivo analysis in the future. The results presented here have shown that drug repurposing is a potential alternative for treating infectious and neglected diseases and that the boxes provided by MMV are crucial for solving the problems involved in treating these diseases.

Supplementary Materials: The following supporting information can be downloaded at https://www.mdpi.com/article/10.3390/microorganisms12122602/s1, Table S1: Plate position and trivial name of each compound of COVID-Box; Table S2: Pharmacokinetic properties of drugs and compounds of COVID-Box according to pkCSM; Table S3: In vivo bioavailability parameters of the best COVID-Box molecules; Figure S1: Preliminary evaluation of the effectiveness of 160 drugs and compounds from COVID-Box against *T. gondii* tachyzoites; Figure S2: Post-treatment recovery assay of *T. gondii* RH strain tachyzoites after treatment with the 23 drugs and compounds of the COVID-Box at a concentration of 1 μM; Figure S3: Chemical structure of 23 drugs and compounds the COVID

Box; Figure S4: Antiproliferative effect of 23 drugs and compounds the COVID-Box after seven days treatment. After obtaining the NDHF cell monolayer, the cells were infected with 600 *T. gondii* RH strain tachyzoites; Figure S5: Cell viability of host cells infected with *T. gondii* tachyzoites after treatment with the 23 drugs and compounds selected from the COVID-Box after seven days of incubation; Figure S6: Boiled-egg graph obtained through the SwissADME platform.

Author Contributions: A.L.O.C.: investigation, methodology, formal analysis, writing—original draft, visualization; M.d.S.: investigation, methodology, formal analysis, visualization, writing—original draft; G.C.D.-V.: investigation, formal analysis; R.E.N.L.: investigation; R.C.V.: investigation, writing—review and editing, resources; É.S.M.-D.: performed experiments, analyzed the results, supervision, conceptualization, investigation, methodology, formal analysis, funding acquisition, visualization, writing—original draft, project administration, resources. All authors have read and agreed to the published version of the manuscript.

Funding: This work was supported by the Coordenação de Aperfeiçoamento de Pessoal de Nível Superior (CAPES), Conselho Nacional de Desenvolvimento Científico e Tecnológico (CNPq)—408964/2018-9, and Pró-Reitoria de Pesquisa-Universidade Federal de Minas Gerais—ADRC 09/2019.

Data Availability Statement: The original contributions presented in this study are included in the paper. Further inquiries can be directed to the corresponding author.

Acknowledgments: We would like to thank MMV for supplying the COVID-Box. TEM analyses were carried out in the Center of Microscopy at the Universidade Federal de Minas Gerais, Belo Horizonte, and in the Centro Nacional de Biologia Estrutural e Bioimagem, Rio de Janeiro, Brazil.

Conflicts of Interest: The authors declare no conflicts of interest.

References

1. Bigna, J.J.; Tochie, J.N.; Tounouga, D.N.; Bekolo, A.O.; Ymele, N.S.; Youda, E.L.; Sime, P.S.; Nansseu, J.R. Global, Regional, and Country Seroprevalence of *Toxoplasma gondii* in Pregnant Women: A Systematic Review, Modelling and Meta-Analysis. *Sci. Rep.* **2020**, *10*, 12102. [CrossRef] [PubMed]
2. Robert-Gangneux, F.; Dardé, M.-L. Epidemiology of and Diagnostic Strategies for Toxoplasmosis. *Clin. Microbiol. Rev.* **2012**, *25*, 264–296. [CrossRef] [PubMed]
3. Zangerle, R.; Allerberger, F.; Pohl, P.; Fritsch, P.; Dierich, M.P. High Risk of Developing Toxoplasmic Encephalitis in AIDS Patients Seropositive to *Toxoplasma gondii*. *Med. Microbiol. Immunol.* **1991**, *180*, 59–66. [CrossRef] [PubMed]
4. Vasconcelos-Santos, D.V.; Machado Azevedo, D.O.; Campos, W.R.; Oréfice, F.; Queiroz-Andrade, G.M.; Carellos, É.V.M.; Castro Romanelli, R.M.; Januário, J.N.; Resende, L.M.; Martins-Filho, O.A. Congenital Toxoplasmosis in Southeastern Brazil: Results of Early Ophthalmologic Examination of a Large Cohort of Neonates. *Ophthalmology* **2009**, *116*, 2199–2205.e1. [CrossRef] [PubMed]
5. Gilbert, R.E. Is Ocular Toxoplasmosis Caused by Prenatal or Postnatal Infection? *Br. J. Ophthalmol.* **2000**, *84*, 224–226. [CrossRef]
6. Gilbert, R.E.; Freeman, K.; Lago, E.G.; Bahia-Oliveira, L.M.G.; Tan, H.K.; Wallon, M.; Buffolano, W.; Stanford, M.R.; Petersen, E. For the European Multicentre Study on Congenital Toxoplasmosis (EMSCOT) Ocular Sequelae of Congenital Toxoplasmosis in Brazil Compared with Europe. *PLoS Negl. Trop. Dis.* **2008**, *2*, e277. [CrossRef]
7. Jones, J.L.; Lopez, A.; Wilson, M.; Schulkin, J.; Gibbs, R. Congenital Toxoplasmosis: A Review. *Obstet. Gynecol. Surv.* **2001**, *56*, 296–305. [CrossRef]
8. Diesel, A.A.; Zachia, S.D.A.; Müller, A.L.L.; Perez, A.V.; Uberti, F.A.D.F.; Magalhães, J.A.D.A. Follow-up of Toxoplasmosis during Pregnancy: Ten-Year Experience in a University Hospital in Southern Brazil. *Rev. Bras. Ginecol. Obs.* **2019**, *41*, 539–547. [CrossRef]
9. Dunay, I.R.; Gajurel, K.; Dhakal, R.; Liesenfeld, O.; Montoya, J.G. Treatment of Toxoplasmosis: Historical Perspective, Animal Models, and Current Clinical Practice. *Clin. Microbiol. Rev.* **2018**, *31*, e00057-17. [CrossRef]
10. Neville, A.J.; Zach, S.J.; Wang, X.; Larson, J.J.; Judge, A.K.; Davis, L.A.; Vennerstrom, J.L.; Davis, P.H. Clinically Available Medicines Demonstrating Anti-Toxoplasma Activity. *Antimicrob. Agents Chemother.* **2015**, *59*, 7161–7169. [CrossRef]
11. Silva, L.A.; Reis-Cunha, J.L.; Bartholomeu, D.C.; Vítor, R.W.A. Genetic Polymorphisms and Phenotypic Profiles of Sulfadiazine-Resistant and Sensitive *Toxoplasma gondii* Isolates Obtained from Newborns with Congenital Toxoplasmosis in Minas Gerais, Brazil. *PLoS ONE* **2017**, *12*, e0170689. [CrossRef] [PubMed]
12. Silva, L.A.; Fernandes, M.D.; Machado, A.S.; Reis-Cunha, J.L.; Bartholomeu, D.C.; Almeida Vitor, R.W. Efficacy of Sulfadiazine and Pyrimetamine for Treatment of Experimental Toxoplasmosis with Strains Obtained from Human Cases of Congenital Disease in Brazil. *Exp. Parasitol.* **2019**, *202*, 7–14. [CrossRef] [PubMed]
13. De Lima Bessa, G.; Vitor, R.W.D.A.; Lobo, L.M.S.; Rêgo, W.M.F.; De Souza, G.C.A.; Lopes, R.E.N.; Costa, J.G.L.; Martins-Duarte, E.S. In Vitro and in Vivo Susceptibility to Sulfadiazine and Pyrimethamine of *Toxoplasma gondii* Strains Isolated from Brazilian Free Wild Birds. *Sci. Rep.* **2023**, *13*, 7359. [CrossRef] [PubMed]
14. Xue, H.; Li, J.; Xie, H.; Wang, Y. Review of Drug Repositioning Approaches and Resources. *Int. J. Biol. Sci.* **2018**, *14*, 1232–1244. [CrossRef]

15. Jourdan, J.-P.; Bureau, R.; Rochais, C.; Dallemagne, P. Drug Repositioning: A Brief Overview. *J. Pharm. Pharmacol.* **2020**, *72*, 1145–1151. [CrossRef]
16. Boyom, F.F.; Fokou, P.V.T.; Tchokouaha, L.R.Y.; Spangenberg, T.; Mfopa, A.N.; Kouipou, R.M.T.; Mbouna, C.J.; Donfack, V.F.D.; Zollo, P.H.A. Repurposing the Open Access Malaria Box To Discover Potent Inhibitors of *Toxoplasma gondii* and *Entamoeba histolytica*. *Antimicrob. Agents Chemother.* **2014**, *58*, 5848–5854. [CrossRef]
17. Subramanian, G.; Belekar, M.A.; Shukla, A.; Tong, J.X.; Sinha, A.; Chu, T.T.T.; Kulkarni, A.S.; Preiser, P.R.; Reddy, D.S.; Tan, K.S.W.; et al. Targeted Phenotypic Screening in *Plasmodium falciparum* and *Toxoplasma gondii* Reveals Novel Modes of Action of Medicines for Malaria Venture Malaria Box Molecules. *mSphere* **2018**, *3*, e00534-17. [CrossRef]
18. Varberg, J.M.; LaFavers, K.A.; Arrizabalaga, G.; Sullivan, W.J. Characterization of Plasmodium Atg3-Atg8 Interaction Inhibitors Identifies Novel Alternative Mechanisms of Action in *Toxoplasma gondii*. *Antimicrob. Agents Chemother.* **2018**, *62*, e01489-17. [CrossRef]
19. Spalenka, J.; Escotte-Binet, S.; Bakiri, A.; Hubert, J.; Renault, J.-H.; Velard, F.; Duchateau, S.; Aubert, D.; Huguenin, A.; Villena, I. Discovery of New Inhibitors of *Toxoplasma gondii* via the Pathogen Box. *Antimicrob. Agents Chemother.* **2018**, *62*, e01640-17. [CrossRef]
20. Radke, J.B.; Burrows, J.N.; Goldberg, D.E.; Sibley, L.D. Evaluation of Current and Emerging Antimalarial Medicines for Inhibition of *Toxoplasma gondii* Growth in Vitro. *ACS Infect. Dis.* **2018**, *4*, 1264–1274. [CrossRef]
21. Dos Santos, M.; Oliveira Costa, A.L.; De Souza Vaz, G.H.; De Souza , G.C.A.; De Almeida Vitor , R.W.; Martins-Duarte, É.S. Medicines for Malaria Venture Pandemic Box In Vitro Screening Identifies Compounds Highly Active against the Tachyzoite Stage of Toxoplasma gondii. *Trop. Med. Infect. Dis.* **2023**, *8*, 510. [CrossRef]
22. Dos Santos, B.R.; Ramos, A.B.D.S.B.; De Menezes, R.P.B.; Scotti, M.T.; Colombo, F.A.; Marques, M.J.; Reimão, J.Q. Anti- *Toxoplasma gondii* Screening of MMV Pandemic Response Box and Evaluation of RWJ-67657 Efficacy in Chronically Infected Mice. *Parasitology* **2023**, *150*, 1226–1235. [CrossRef] [PubMed]
23. Barltrop, J.A.; Owen, T.C.; Cory, A.H.; Cory, J.G. 5-(3-Carboxymethoxyphenyl)-2-(4,5-Dimethylthiazolyl)-3-(4-Sulfophenyl)Tetrazolium, Inner Salt (MTS) and Related Analogs of 3-(4,5-Dimethylthiazolyl)-2,5-Diphenyltetrazolium Bromide (MTT) Reducing to Purple Water-Soluble Formazans As Cell-Viability Indicators. *Bioorganic Med. Chem. Lett.* **1991**, *1*, 611–614. [CrossRef]
24. Martins-Duarte, E.S.; de Araujo Portes, J.; de Araujo Portes, J.; da Silva, R.B.; Pires, H.S.; Garden, S.J.; de Souza, W. In Vitro Activity of N-Phenyl-1,10-Phenanthroline-2-Amines against Tachyzoites and Bradyzoites of *Toxoplasma gondii*. *Bioorganic Med. Chem.* **2021**, *50*, 116467. [CrossRef]
25. De Araújo Jorge, T.C.; de Souza, W. Effect of Carbohydrates, Periodate and Enzymes in the Process of Endocytosis of Trypanosoma Cruzi by Macrophages. *Acta Trop.* **1984**, *41*, 17–28.
26. Martins-Duarte, E.S.; Dubar, F.; Lawton, P.; França Da Silva, C.; Soeiro, M.D.N.C.; De Souza, W.; Biot, C.; Vommaro, R.C. Ciprofloxacin Derivatives Affect Parasite Cell Division and Increase the Survival of Mice Infected with *Toxoplasma gondii*. *PLoS ONE* **2015**, *10*, e0125705. [CrossRef]
27. Cajazeiro, D.C.; Toledo, P.P.M.; De Sousa, N.F.; Scotti, M.T.; Reimão, J.Q. Drug Repurposing Based on Protozoan Proteome: In Vitro Evaluation of In Silico Screened Compounds against *Toxoplasma gondii*. *Pharmaceutics* **2022**, *14*, 1634. [CrossRef]
28. Martins-Duarte, É.D.S.; De Souza, W.; Vommaro, R.C. Itraconazole Affects *Toxoplasma gondii* Endodyogeny. *FEMS Microbiol. Lett.* **2008**, *282*, 290–298. [CrossRef]
29. Martins-Duarte, É.S.; Lemgruber, L.; De Souza, W.; Vommaro, R.C. *Toxoplasma gondii*: Fluconazole and Itraconazole Activity against Toxoplasmosis in a Murine Model. *Exp. Parasitol.* **2010**, *124*, 466–469. [CrossRef]
30. Chang, H.R.; Comte, R.; Pechère, J.C. In Vitro and in Vivo Effects of Doxycycline on *Toxoplasma gondii*. *Antimicrob. Agents Chemother.* **1990**, *34*, 775–780. [CrossRef]
31. McCabe, R.E.; Luft, B.J.; Remington, J.S. The Effects of Vyclosporine on Toxoplasma gondii In Vivo and In Vitro. *Transplantation* **1986**, *41*, 611–615. [CrossRef] [PubMed]
32. Dittmar, A.J.; Drozda, A.A.; Blader, I.J. Drug Repurposing Screening Identifies Novel Compounds That Effectively Inhibit *Toxoplasma gondii* Growth. *mSphere* **2016**, *1*, e00042-15. [CrossRef]
33. Adeyemi, O.S.; Atolani, O.; Awakan, O.J.; Olaolu, D.; Nwonuma, C.O.; Alejolowo, O.; Otohinoyi, D.A.; Rotimi, D.; Owolabi, A.; Batiha, G.E.-S. In Vitro Screening to Identify Anti-*Toxoplasma* Compounds and In Silico Modeling for Bioactivities and Toxicity. *Yale J. Biol. Med.* **2019**, *92*, 369–383. [PubMed]
34. Dos Santos, B.R.; Ramos, A.B.D.S.B.; De Menezes, R.P.B.; Scotti, M.T.; Colombo, F.A.; Marques, M.J.; Reimão, J.Q. Repurposing the Medicines for Malaria Venture's COVID Box to Discover Potent Inhibitors of *Toxoplasma gondii*, and In Vivo Efficacy Evaluation of Almitrine Bismesylate (MMV1804175) in Chronically Infected Mice. *PLoS ONE* **2023**, *18*, e0288335. [CrossRef]
35. Zhang, J.L.; Si, H.F.; Shang, X.F.; Zhang, X.K.; Li, B.; Zhou, X.Z.; Zhang, J.Y. New Life for an Old Drug: In Vitro and in Vivo Effects of the Anthelmintic Drug Niclosamide against *Toxoplasma gondii* RH Strain. *Int. J. Parasitol. Drugs Drug Resist.* **2019**, *9*, 27–34. [CrossRef] [PubMed]
36. Gurnett, A.M.; Dulski, P.M.; Darkin-Rattray, S.J.; Carrington, M.J.; Schmatz, D.M. Selective Labeling of Intracellular Parasite Proteins by Using Ricin. *Proc. Natl. Acad. Sci. USA* **1995**, *92*, 2388–2392. [CrossRef]
37. Bilgin, M.; Yildirim, T.; Hokelek, M. In Vitro Effects ofIvermectin and Sulphadiazine on *Toxoplasma gondii*. *Balk. Med. J.* **2013**, *30*, 19–22. [CrossRef]

38. Shang, F.-F.; Wang, M.-Y.; Ai, J.-P.; Shen, Q.-K.; Guo, H.-Y.; Jin, C.-M.; Chen, F.-E.; Quan, Z.-S.; Jin, L.; Zhang, C. Synthesis and Evaluation of Mycophenolic Acid Derivatives as Potential Anti-*Toxoplasma gondii* Agents. *Med. Chem. Res.* **2021**, *30*, 2228–2239. [CrossRef]
39. Castro-Elizalde, K.N.; Hernández-Contreras, P.; Ramírez-Flores, C.J.; González-Pozos, S.; Gómez De León, C.T.; Mondragón-Castelán, M.; Mondragón-Flores, R. Mycophenolic Acid Induces Differentiation of *Toxoplasma gondii* RH Strain Tachyzoites into Bradyzoites and Formation of Cyst-like Structure in Vitro. *Parasitol. Res.* **2018**, *117*, 547–563. [CrossRef]
40. Garrison, E.M.; Arrizabalaga, G. Disruption of a Mitochondrial MutS DNA Repair Enzyme Homologue Confers Drug Resistance in the Parasite *Toxoplasma gondii*. *Mol. Microbiol.* **2009**, *72*, 425–441. [CrossRef]
41. Egawa, Y.; Oshima, S.; Umezawa, S. Studies on Cycloheximide-Related Compounds. I Esters of Cycloheximide and Their Antitoxoplasmic Activity. *J. Antibiot. Ser. A* **1965**, *18*, 171–174.
42. Beckers, C.J.; Roos, D.S.; Donald, R.G.; Luft, B.J.; Schwab, J.C.; Cao, Y.; Joiner, K.A. Inhibition of Cytoplasmic and Organellar Protein Synthesis in *Toxoplasma gondii*. Implications for the Target of Macrolide Antibiotics. *J. Clin. Investig.* **1995**, *95*, 367–376. [CrossRef] [PubMed]
43. Fichera, M.E.; Bhopale, M.K.; Roos, D.S. In Vitro Assays Elucidate Peculiar Kinetics of Clindamycin Action against *Toxoplasma gondii*. *Antimicrob. Agents Chemother.* **1995**, *39*, 1530–1537. [CrossRef] [PubMed]
44. Daina, A.; Michielin, O.; Zoete, V. SwissADME: A Free Web Tool to Evaluate Pharmacokinetics, Drug-Likeness and Medicinal Chemistry Friendliness of Small Molecules. *Sci. Rep.* **2017**, *7*, 42717. [CrossRef]
45. Lipinski, C.A.; Lombardo, F.; Dominy, B.W.; Feeney, P.J. Experimental and Computational Approaches to Estimate Solubility and Permeability in Drug Discovery and Development Settings. *Adv. Drug Deliv. Rev.* **1997**, *23*, 3–25. [CrossRef]
46. Veber, D.F.; Johnson, S.R.; Cheng, H.-Y.; Smith, B.R.; Ward, K.W.; Kopple, K.D. Molecular Properties That Influence the Oral Bioavailability of Drug Candidates. *J. Med. Chem.* **2002**, *45*, 2615–2623. [CrossRef]
47. Pires, D.E.V.; Blundell, T.L.; Ascher, D.B. pkCSM: Predicting Small-Molecule Pharmacokinetic and Toxicity Properties Using Graph-Based Signatures. *J. Med. Chem.* **2015**, *58*, 4066–4072. [CrossRef]
48. Lynch, T.; Price, A. The Effect of Cytochrome P450 Metabolism on Drug Response, Interactions, and Adverse Effects. *Am. Fam. Physician.* **2007**, *76*, 391–396.
49. Sanchez-Covarrubias, L.; Slosky, L.; Thompson, B.; Davis, T.; Ronaldson, P. Transporters at CNS Barrier Sites: Obstacles or Opportunities for Drug Delivery? *Curr. Pharm. Des.* **2014**, *20*, 1422–1449. [CrossRef]
50. Macias-Silva, M.; Vazquez-Victorio, G.; Hernandez-Damian, J. Anisomycin Is a Multifunctional Drug: More than Just a Tool to Inhibit Protein Synthesis. *Curr. Chem. Biol.* **2010**, *4*, 124–132. [CrossRef]
51. Osborn, C.D.; Holloway, F.A. Can Commonly Used Antibiotics Disrupt Formation of New Memories? *Bull. Psychon. Soc.* **1984**, *22*, 356–358. [CrossRef]
52. Myasnikov, A.G.; Kundhavai Natchiar, S.; Nebout, M.; Hazemann, I.; Imbert, V.; Khatter, H.; Peyron, J.-F.; Klaholz, B.P. Structure–Function Insights Reveal the Human Ribosome as a Cancer Target for Antibiotics. *Nat. Commun.* **2016**, *7*, 12856. [CrossRef] [PubMed]
53. Schneider-Poetsch, T.; Ju, J.; Eyler, D.E.; Dang, Y.; Bhat, S.; Merrick, W.C.; Green, R.; Shen, B.; Liu, J.O. Inhibition of Eukaryotic Translation Elongation by Cycloheximide and Lactimidomycin. *Nat. Chem. Biol.* **2010**, *6*, 209–217. [CrossRef] [PubMed]
54. Ch'Ih, J.J.; Duhl, D.M.; Faulkner, L.S.; Devlin, T.M. Regulation of mammalian protein synthesis In Vivo. Protein synthesis in rat liver and kidney after the administration of sublethal doses of cyclohyximide. *Biochem. J.* **1976**, *178*, 151–157.
55. Garreau De Loubresse, N.; Prokhorova, I.; Holtkamp, W.; Rodnina, M.V.; Yusupova, G.; Yusupov, M. Structural Basis for the Inhibition of the Eukaryotic Ribosome. *Nature* **2014**, *513*, 517–522. [CrossRef]
56. Ramakrishnan, S.; Serricchio, M.; Striepen, B.; Bütikofer, P. Lipid Synthesis in Protozoan Parasites: A Comparison between Kinetoplastids and Apicomplexans. *Prog. Lipid Res.* **2013**, *52*, 488–512. [CrossRef]
57. Martins-Duarte, É.S.; Carias, M.; Vommaro, R.; Surolia, N.; De Souza, W. Apicoplast Fatty Acid Synthesis Is Essential for Pellicle Formation at the End of Cytokinesis in *Toxoplasma gondii*. *J. Cell Sci.* **2016**, *129*, 3320–3331. [CrossRef]
58. Martins-Duarte, E.S.; Jones, S.M.; Gilbert, I.H.; Atella, G.C.; De Souza, W.; Vommaro, R.C. Thiolactomycin Analogues as Potential Anti-*Toxoplasma gondii* Agents. *Parasitol. Int.* **2009**, *58*, 411–415. [CrossRef]
59. Camps, M.; Arrizabalaga, G.; Boothroyd, J. An rRNA Mutation Identifies the Apicoplast as the Target for Clindamycin in *Toxoplasma gondii*. *Mol. Microbiol.* **2002**, *43*, 1309–1318. [CrossRef]
60. Grollman, A.P. Inhibitors of Protein Biosynthesis. *J. Biol. Chem.* **1967**, *242*, 3226–3233. [CrossRef]
61. Kisselev, A.F.; Callard, A.; Goldberg, A.L. Importance of the Different Proteolytic Sites of the Proteasome and the Efficacy of Inhibitors Varies with the Protein Substrate. *J. Biol. Chem.* **2006**, *281*, 8582–8590. [CrossRef] [PubMed]
62. Groll, M.; Huber, R.; Moroder, L. The Persisting Challenge of Selective and Specific Proteasome Inhibition. *J. Pept. Sci.* **2009**, *15*, 58–66. [CrossRef] [PubMed]
63. Sridhar, S.; Bhat, G.; Guruprasad, K. Analysis of Bortezomib Inhibitor Docked within the Catalytic Subunits of the *Plasmodium falciparum* 20S Proteasome. *SpringerPlus* **2013**, *2*, 566. [CrossRef] [PubMed]
64. Kreidenweiss, A.; Kremsner, P.G.; Mordmüller, B. Comprehensive Study of Proteasome Inhibitors against *Plasmodium falciparum* Laboratory Strains and Field Isolates from Gabon. *Malar. J.* **2008**, *7*, 187. [CrossRef] [PubMed]

65. Hutter, G.; Rieken, M.; Pastore, A.; Weigert, O.; Zimmermann, Y.; Weinkauf, M.; Hiddemann, W.; Dreyling, M. The Proteasome Inhibitor Bortezomib Targets Cell Cycle and Apoptosis and Acts Synergistically in a Sequence-Dependent Way with Chemotherapeutic Agents in Mantle Cell Lymphoma. *Ann. Hematol.* **2012**, *91*, 847–856. [CrossRef]
66. Yin, O.Q.P.; Wang, Y.; Schran, H. A Mechanism-Based Population Pharmacokinetic Model for Characterizing Time-Dependent Pharmacokinetics of Midostaurin and Its Metabolites in Human Subjects. *Clin. Pharmacokinet.* **2008**, *47*, 807–816. [CrossRef]
67. Gleixner, K.V.; Mayerhofer, M.; Aichberger, K.J.; Derdak, S.; Sonneck, K.; Böhm, A.; Gruze, A.; Samorapoompichit, P.; Manley, P.W.; Fabbro, D.; et al. PKC412 Inhibits In Vitro Growth of Neoplastic Human Mast Cells Expressing the D816V-Mutated Variant of KIT: Comparison with AMN107, Imatinib, and Cladribine (2CdA) and Evaluation of Cooperative Drug Effects. *Blood* **2006**, *107*, 752–759. [CrossRef]
68. Mayol, G.F.; Revuelta, M.V.; Salusso, A.; Touz, M.C.; Rópolo, A.S. Evidence of Nuclear Transport Mechanisms in the Protozoan Parasite *Giardia lamblia*. *Biochim. Biophys. Acta (BBA) Mol. Cell Res.* **2020**, *1867*, 118566. [CrossRef]
69. Gupta, S.; Vohra, S.; Sethi, K.; Gupta, S.; Bera, B.C.; Kumar, S.; Kumar, R. In Vitro Anti-Trypanosomal Effect of Ivermectin on *Trypanosoma evansi* by Targeting Multiple Metabolic Pathways. *Trop. Anim. Health Prod.* **2022**, *54*, 240. [CrossRef]
70. Fraccaroli, L.; Ruiz, M.D.; Perdomo, V.G.; Clausi, A.N.; Balcazar, D.E.; Larocca, L.; Carrillo, C. Broadening the Spectrum of Ivermectin: Its Effect on *Trypanosoma cruzi* and Related Trypanosomatids. *Front. Cell. Infect. Microbiol.* **2022**, *12*, 885268. [CrossRef]
71. Reis, T.A.R.; Oliveira-da-Silva, J.A.; Tavares, G.S.V.; Mendonça, D.V.C.; Freitas, C.S.; Costa, R.R.; Lage, D.P.; Martins, V.T.; Machado, A.S.; Ramos, F.F.; et al. Ivermectin Presents Effective and Selective Antileishmanial Activity In Vitro and in Vivo against *Leishmania infantum* and Is Therapeutic against Visceral Leishmaniasis. *Exp. Parasitol.* **2021**, *221*, 108059. [CrossRef] [PubMed]
72. Smith, D.; Kannan, G.; Coppens, I.; Wang, F.; Nguyen, H.M.; Cerutti, A.; Olafsson, E.B.; Rimple, P.A.; Schultz, T.L.; Mercado Soto, N.M.; et al. *Toxoplasma* TgATG9 Is Critical for Autophagy and Long-Term Persistence in Tissue Cysts. *eLife* **2021**, *10*, e59384. [CrossRef] [PubMed]
73. Liu, J.; Liang, H.; Chen, C.; Wang, X.; Qu, F.; Wang, H.; Yang, K.; Wang, Q.; Zhao, N.; Meng, J.; et al. Ivermectin Induces Autophagy-Mediated Cell Death through the AKT/mTOR Signaling Pathway in Glioma Cells. *Biosci. Rep.* **2019**, *39*, BSR20192489. [CrossRef] [PubMed]
74. Sintchak, M.D.; Nimmesgern, E. The Structure of Inosine 5′-Monophosphate Dehydrogenase and the Design of Novel Inhibitors. *Immunopharmacology* **2000**, *47*, 163–184. [CrossRef]
75. Morlon-Guyot, J.; Pastore, S.; Berry, L.; Lebrun, M.; Daher, W. *Toxoplasma gondii* Vps11, a Subunit of HOPS and CORVET Tethering Complexes, Is Essential for the Biogenesis of Secretory Organelles. *Cell Microbiol.* **2015**, *17*, 1157–1178. [CrossRef]
76. Sakura, T.; Sindikubwabo, F.; Oesterlin, L.K.; Bousquet, H.; Slomianny, C.; Hakimi, M.-A.; Langsley, G.; Tomavo, S. A Critical Role for *Toxoplasma gondii* Vacuolar Protein Sorting VPS9 in Secretory Organelle Biogenesis and Host Infection. *Sci. Rep.* **2016**, *6*, 38842. [CrossRef]
77. Houngue, R.; Sangaré, L.O.; Alayi, T.D.; Dieng, A.; Bitard-Feildel, T.; Boulogne, C.; Slomianny, C.; Atindehou, C.M.; Fanou, L.A.; Hathout, Y.; et al. *Toxoplasma* Membrane Inositol Phospholipid Binding Protein TgREMIND Is Essential for Secretory Organelle Function and Host Infection. *Cell Rep.* **2024**, *43*, 113601. [CrossRef]
78. Donald, R.G.K.; Carter, D.; Ullman, B.; Roos, D.S. Insertional Tagging, Cloning, and Expression of the Hypoxanthine-Xanthine-Guanine Phosphoribosyltransferase Gene. *J. Biol. Chem.* **1996**, *271*, 14010–14019. [CrossRef]
79. Van Dongen, K.C.W.; De Lange, E.; Van Asseldonk, L.L.M.; Zoet, L.; Van Der Fels-Klerx, H.J. Safety and Transfer of Veterinary Drugs from Substrate to Black Soldier Fly Larvae. *Animal* **2024**, *18*, 101214. [CrossRef]
80. Bolder, N.; Wagenaar, J.; Putirulan, F.; Veldman, K.; Sommer, M. The Effect of Flavophospholipol (Flavomycin) and Salinomycin Sodium (Sacox) on the Excretion of *Clostridium perfringens*, *Salmonella enteritidis*, and *Campylobacter jejuni* in Broilers after Experimental Infection. *Poult. Sci.* **1999**, *78*, 1681–1689. [CrossRef]

microorganisms

MDPI

Article

Determinants of Anemia in Schoolchildren in the Highland Bolivia

Washington R. Cuna [1], Ivonne Contreras [1], Armando Rodriguez [1], Roberto Passera [2] and Celeste Rodriguez [1,*]

[1] Unidad de Inmunología Parasitaria, Facultad de Medicina, Universidad Mayor de San Andrés, La Paz 10077, Bolivia; washingtoncuna@gmail.com (W.R.C.); civcont@gmail.com (I.C.); arjoroze@gmail.com (A.R.)

[2] Department of Medical Sciences, University of Torino, 10126 Torino, Italy; passera.roberto@gmail.com

* Correspondence: celesterodriguez6@gmail.com

Abstract: Anemia is a health problem of concern among schoolchildren in underprivileged rural regions, where recurrent parasitic infections are common. A cross-sectional study was conducted in 229 schoolchildren in rural highland Bolivia in the department of La Paz, an area with a high prevalence of protozoan and helminth infections, to determine the types and mechanisms of anemia. A substantial proportion of children (40.2%) were found to be anemic based on hemoglobin measurements. No associations were found between low hemoglobin levels and helminth or protozoan infections when evaluating infectious causes of anemia, nor with *Giardia lamblia* or *Blastocystis hominis*, which are associated with iron deficiency and nutrient malabsorption and were highly prevalent in this study. The significant association between anemia and hypochromia suggests iron deficiency, aligned with low hemoglobin levels. A total of 39 out of 150 children (26%) had markers consistent with iron deficiency anemia (IDA), 26 out of 127 children (20%) met the criteria for anemia of inflammation (AI). Furthermore, 12 of the 127 tested children (9.4%) met the criteria for mixed AI with IDA according to the soluble transferrin receptor (sTfR)/log ferritin levels, which increased significantly due to overall infections by *Hymenolepis nana* and *Ascaris lumbricoides* helminths. The findings highlight the need for integrated public health interventions to address iron nutrition and parasitic infections to effectively prevent anemia in this vulnerable population.

Keywords: children; anemia; protozoa; helminths; iron deficiency anemia; anemia of inflammation

1. Introduction

Anemia is a global public health problem that affects approximately one third of the world's population, mainly in developing countries [1], although pregnant women and preschool children are the populations most vulnerable to anemia due to reduced oxygen supply at a time when metabolic needs are high, anemia in schoolchildren is also of concern due to its impact on their physical and cognitive development [2,3], predisposing them to a higher frequency of morbidity [4–6]. It is generally recognized that anemia is one of the consequences of iron deficiency due to low dietary intake and iron-deficiency anemia (IDA) has been associated with significantly lower scores of on cognitive tests and significantly lower achievements than in non-anemic children [7]. In addition to IDA, another type of anemia, includes anemia of chronic disease, also termed anemia of inflammation (AI); these are the two most prevalent forms of anemia [8]. AI ranks second; its actual prevalence is difficult to assess as it often coexists with IDA. Iron-restricted erythropoiesis is part of the pathophysiology of both AI and IDA, making differential diagnosis difficult when both diseases coexist [9,10]. IDA and AI often affect the same individuals, especially in low-income countries with a high burden of parasitic diseases, where chronic blood loss and iron deficiency emerge due to helminth infestations (e.g., hookworms, *Trichuris trichiura*) [11–13]. The combination of serum transferrin receptor (sTfR) and ferritin measurements for the calculation of the sTfR/Ferritin index has been shown to accurately determine whether the occurrence of IDA, AI, or a combination of these mechanisms is involved in anemia, particularly in anemia with active inflammation [8]. Polyparasitism of concurrent helminth and protozoan parasites and a strongly increased risk of anemia are observed in schoolchildren from rural Bolivian altiplano (highland) regions, where clean water, adequate housing and sewage systems are not available [14]. This study builds on our previous research by examining the most likely mechanisms and biomarkers associated with anemia in the highland regions. Therefore, the aim of this study was to use more comprehensive laboratory testing in the differential diagnosis of various types of anemia among schoolchildren. To our knowledge, there are no studies in the literature evaluating the sTfR-Ferritin index which should be considered in the management of regional childhood anemia in this underprivileged area of Bolivia, in which recurrent gastrointestinal protozoa and helminth infections are endemic.

2. Materials and Methods

2.1. Study Area, Population, and Assessment

Cross-sectional data were collected from schoolchildren of the Bolivian highland, where helminths and protozoan gastrointestinal parasites are known to be endemic. Surveys were carried out in different school units located at an altitude between 3.810 and 4.050 m above sea level, in which health and basic service conditions are regular to inappropriate; 55% have a water pipe supply, barely 30% have a flush toilet or a latrine, and most (99%) of the households lack piped sewers. A total of 229 children aged 5–13 years who participated in this study had blood drawn for anemia testing. Anthropometric measurements (i.e., height, weight) to assess the growth of children, were taken by trained local nurses following standardized procedures. The study received approval from the ethics committee of the Universidad Mayor de San Andres.

2.2. Diagnosis of Gastrointestinal Parasites

Stool samples were analyzed microscopically, samples under high suspicion for amebic dysentery (that is, bloody diarrhea with mucus) were carefully examined in fresh and in smears fixed and stained with Wright's stain, a portion of the other samples was analyzed through Ritchie's formol-ether concentration technique after Wright staining. Samples with eggs, larvae, or cysts, were considered positive for a species of parasite.

2.3. Testing for Anemia

Venous whole blood (5 mL) was collected via venipuncture and divided: 1 mL in a tube containing EDTA for complete blood count and 4 mL in a tube for serum preparation. The serum was frozen and used later for ferritin and sTfR measurements. In the highlands, cases of anemia with hemoglobin (Hb) levels < 14.4 g/dL corresponded to World Health Organization levels [15]; these levels together with serum ferritin (SF) < 30 ng/mL defined IDA. Anemia of inflammation was identified in children with SF levels 30–100 ng/mL and an sTfR/log SF ratio of less than 1. Cases of combined AI with IDA included samples with SF 30–100 ng/mL, and an sTfR/log ferritin ratio > 2 [8]. Ferritin and sTfR were assayed using commercial enzyme immunoassays (Monobind Inc., Lake Forest, CA, USA). Initially, all participants underwent hemoglobin measurements on-site by point of care testing using a HemoCue analyzer (Angelholm, Sweden).

2.4. Statistical Data Analysis

Categorical variables were described as absolute and relative frequencies, while continuous ones as the median and inter quartile range (IQR). Non-parametric inferential analysis for categorical and continuous covariates was performed using Fisher's exact test and the Mann–Whitney and Kruskal–Wallis ones, respectively. All p values were obtained via the two-sided exact method at the conventional 5% significance level. Data were analyzed in October 2024 by R 4.4.1 (R Foundation for Statistical Computing, Vienna, Austria. https://www.r-project.org).

3. Results

3.1. Study Population

Demographic characteristics and prevalence of helminth and protozoan parasite infections for 229 participants are shown in Table 1. There was a fairly balanced gender distribution with a slight majority of female participants (55%) compared to males (45%). Over 70% and 20% of children were infected with protozoan and helminth parasites, respectively. A substantial proportion of participants (40.2%) were anemic (90/229), indicating a significant public health problem. The percentages of malnutrition indicators such as stunting (4.8%), wasting (3.5%), and malnourished children (3.0%) were relatively low, suggesting that severe malnutrition is not widespread in this population. In general, there were no significant differences in the prevalence of sex, age, infection, malnutrition, stunting, wasting, or anemia as detected through point-of-care testing during the surveys.

Table 1. Demographic characteristics of participants and parasitic infections for the anemia study in the highlands of Bolivia.

Characteristic	Highlands (N = 229)
Sex (%)	
Female	126 (55)
Male	103 (45)
Age (%)	
5–7	106 (46.3)
8–10	76 (33.2)
11–13	47 (20.5)
Helminths (%)	57 (24.9)
Protozoa (%)	170 (74.2)
Helminths-protozoa (%)	180 (78.6)
Anemic * (%)	90 (40.2)
Malnourished † (%)	7 (3.0)
Stunted ‡ (%)	11 (4.8)
Wasted § (%)	8 (3.5)

* Hemoglobin < 14.4 g/dL. † Weight for hight z score (WAZ) < −2. ‡ Height for age z score (HAZ) < −2. § Body mass index (BMI) for age z score (BAZ) < −2.

3.2. Prevalence of Parasitic Infections

Figure 1 shows the prevalence of parasitic, protozoa, and helminth infections, and coinfections (protozoa and helminths) among children grouped by three age categories, 5–7, 8–10, and 11–13 years. Consistent with an earlier set up, the division of age into three age groups was defined in our previous study [14]. The data on parasitic infections highlighted a high prevalence of protozoan infections (74.2%), particularly *Entamoeba coli* (48.9%) and *Blastocystis hominis* (40.2%). *Entamoeba coli* mature cysts present up to eight nuclei and *E. histolytica/dispar* four. Microscopic identification of *E. histolytica* hematophagous trophozoites was accomplished using fresh and fixed stool samples. As previously reported, hematophagy is considered a distinguishing microscopic criterion for identifying *E. histolytica* [16–18]. Specific parasitic and helminth infections increased significantly with age. Across the age groups of 5–7-, 8–10-, and 11–13-years, the prevalence of each infection and polyparasitism increased significantly. There was a progressive increase in *H. nana* (99% CI, 0.031 to 0.041, $p = 0.036$), *A. lumbricoides* ($p < 0.001$), *B. hominis* (99% CI, 0.038 to 0.048, $p = 0.043$), and helminths ($p < 0.001$). There was a statistically significant association in the *E. coli* ($p < 0.001$), protozoan ($p = 0.004$), and coinfection ($p < 0.001$) infection rates, across age groups, particularly showing higher positivity in children over the age of 7 years.

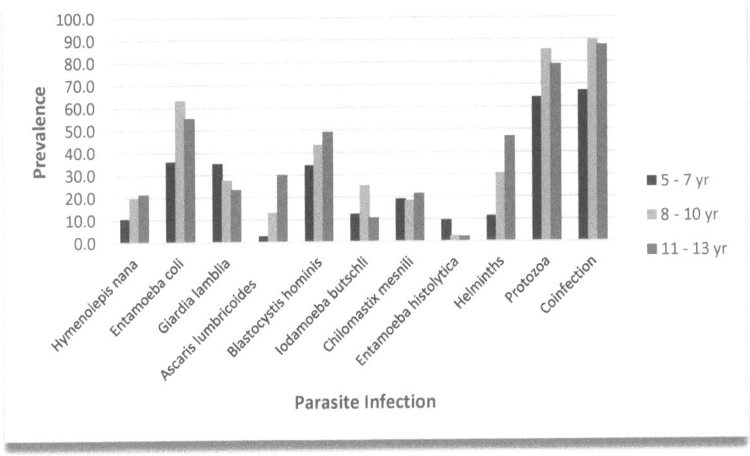

Figure 1. Parasite prevalence by age group. Vertical bars indicate the prevalence of specific parasite infections, helminth, protozoa, and coinfections. Percentages are shown for the 5–7, 8–10, and 11–13 year subgroups.

3.3. *Types of Anemia and Iron Biomarkers Across Age Groups*

Table 2 details data on hemoglobin levels, anemia status, and iron biomarkers among children classified as 5–7, 8–10, and 11–13 years of age. Hemoglobin measurements performed in venous blood revealed that 40.2% of children were anemic (hemoglobin < 14.4 g/dL). The percentage of children who were anemic was quite high across all age groups, with a slight decrease in 8–10-year-olds (36.0%) compared to 5–7 (43.1%) and 11–13-year olds (40.4%) (Table 2).

Table 2. Anemia and anemia biomarkers by age group among the highland children of Bolivia.

Variable	5- to 7-Year-Olds	8- to 10-Year-Olds	11- to 13-Year-Olds	All Children
Hemoglobin in g/dL, median (IQR)	14.5 (13.8–15.4)	14.6 (13.5–14.8)	14.3 (14.0–15.2)	14.6 (13.8–15.2)
Percent anemic	43.1	36.0	40.4	40.2
MCV in fL median (IQR)	84.2 (80.4–90.1)	84.5 (78.8–90.3)	84.2 (75.6–88.1)	84.2 (80.3–90.1)
Percent macrocytosis	43.1	41.3	29.8	40.0
MCHC in g/dL median (IQR)	32.7 (32.0–33.5)	32.6 (31.8–33.3)	32.4 (31.6–33.0)	32.6 (31.9–33.3)
Percent hypochromasia	25.5	29.0	36.2	27.7
Serum ferritin in ng/mL, median (IQR)	17.8 (10.3–30.8)	16.0 (9.0–24.5)	18.8 (11.2–30.8)	17.3 (10.2–30.3)
sTfR mg/L, median (IQR)	1.26 (0.93–1.48)	2.35 (1.27–5.03)	2.18 (0.90–3.53)	1.39 (0.93–2.96) †
sTfR/log ferritin, median (IQR)	0.98 (0.77–1.31)	2.17 (0.67–5.47)	0.70 (0.55–2.23)	1.01 (0.71–2.30)
Anemic children meeting IDA criteria (%)	21/78 (26.9)	13/51 (25.5)	5/21 (23.8)	39/150 (26.0) §
Anemic children meeting AI criteria (%)	15/72 (20.8)	7/38 (18.4)	4/17 (23.5)	26/127 (20.5)
Anemic children meeting criteria for mixed AI with IDA (%)	5/72 (6.9)	4/38 (10.5)	3/17 (17.6)	12/127 (9.4)

MCV = Mean corpuscular volume; MCHC = Mean corpuscular hemoglobin concentration; IDA = iron-deficiency anemia, AI = anemia of inflammation; IQR = inter quartile range; Significant differences among age groups by x^2 testing † $p < 0.001$; Overall significant association between anemia and IDA regardless of age, by x^2 testing § $p < 0.001$.

Median serum ferritin levels varied slightly between age groups, with values ranging from 16.0 to 18.82 ng/mL and an overall median level of 17.27 ng/mL for all children. Median MCV measurements to determine the underlying cause of anemia (for example, nutritional deficiencies), fell within normal ranges (84.2 to 84.5 fL) in 120 of 225 (53%) cases, after accounting for four missing values; however, red cell macrocytosis (elevated MCV) was observed in 89 of 225 (39%) children. MCHC levels showed a small decrease in the median levels with increasing age, from 32.7 g/dL in the youngest group to 32.4 g/dL in the oldest group. The prevalence of hypochromasia (lower MCHC) increased with age, from 25.5% in children aged 5–7 years to 36.2% in those aged 11–13 years, being significantly higher ($p = 0.006$) among anemic children (36.7%) compared to those with normal hemoglobin status (21.6%). The 8–10-year group showed a significant increase ($p < 0.001$) in the median level of sTfR (2.35 mg/L) compared to the other groups, which had median values of 1.26 mg/L (5–7 years) and 2.18 mg/L (11–13 years). This is reflected in the higher sTfR/log ferritin ratio in the 8–10 years age group, with a median of 2.17 compared to the other groups (0.98, 0.70); however, the observed median differences were not statistically significant ($p = 0.09$). Detailed percentages of children classified as iron-deficient and anemic (26%), meeting the AI criteria (20.5%), and a smaller percentage (9.4%) characterized as mixed AI with IDA, are shown (Table 2). Reflecting deficient iron levels in our data, we observed a significant association between anemia and IDA, characterized by low levels of hemoglobin and ferritin ($p > 0.001$). Figure 2 presents box plots that compare the sTfR/log ferritin ratio in different types of infection (helminth, protozoa, and helminth–protozoa), with each category divided into negative and positive infection statuses. The positive groups show a wider range of sTfR/log ferritin ratio, suggesting greater variability in iron status or inflammatory response resulting from infection. In particular, children who are positive for helminth infections show a much higher median sTfR/log ferritin ratio compared to those who are negative. Children infected with *H. nana* and *A. lumbrioides* exhibited a significantly higher median sTfR/log SF value of 3.18 (range 2.04–5.93), compared to 0.97 (range 0.69–1.81) in uninfected children ($p < 0.001$). Mixed AI with IDA is a relatively rare condition in our study, observed only in 9.4% of cases, after accounting for missing data (102 cases, or 44.5%).

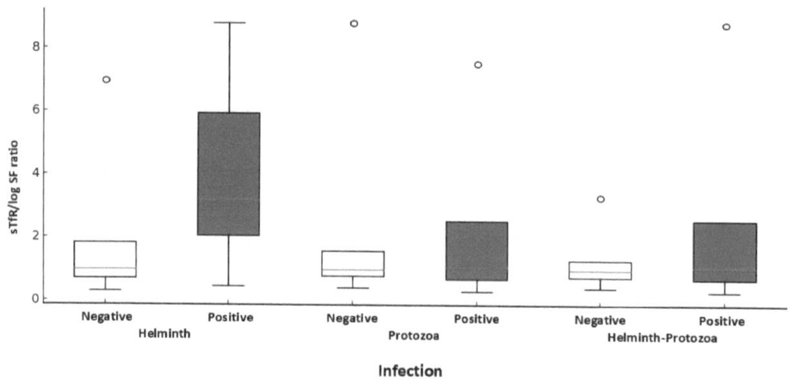

Figure 2. Boxplot comparing the soluble transferrin receptor (sTfR) to log ferritin ratios, defining combined anemia of inflammation with iron-deficiency anemia among study children with and without detectable infections. The distributions of sTfR/log SF ratios are presented for sTfR-tested children who either have (N = 12) or do not have (N = 127) helminth infections. The *p*-value for group differences is 0.001, determined via the Mann–Whitney U test.

4. Discussion

The findings of this study conducted in the highlands of Bolivia indicate a substantial prevalence of anemia (40%) among schoolchildren in a rural area where protozoan infections predominate, followed by helminth infections [14,19]. While evaluating potential infectious causes of anemia, no statistically significant associations were found with helminth, or protozoan infections, despite a notable, progressive increase in the prevalence of these infections with age. Furthermore, we did not observe a statistically significant correlation between lower hemoglobin levels (<14.4 g/dL) and the presence of *G. lamblia* or *B. hominis,* protozoa known to be associated with iron and nutrient malabsorption, malnutrition, and highly prevalent in this survey [20–22]. Specifically, the capacity of *G. lamblia* to lower hemoglobin levels may be associated with its tendency to cause vitamin B12 and folic acid deficiencies, which could explain the macrocytosis (40%) observed in our study in the context of anemia [23,24]. Whether these metabolic alterations persist in our setting remains to be confirmed. The occurrence of macrocytosis and hypochromasia (27.7%) suggests overlapping causes for our observations in addition to deficiencies. The significant association between anemia and hypochromasia could be an indicator of iron deficiency reflecting lower hemoglobin levels.

The lower rates of malnutrition observed suggested that anemia was driven by factors other than malnutrition, likely, micronutrient deficiencies. The availability of iron rich food, such as meat, is limited, and children's diet consists mainly of carbohydrates, vegetables, and fruits. The absence of heme iron from meat, a primary source of iron that is more easily absorbed by the body compared to non-heme iron from plant-based sources, along with increased vulnerability to inhibitors such as phytates and polyphenols, further hinders iron absorption [25]. Consequently, a significant proportion of anemia cases may be due to nutritional deficiencies, particularly a lack of dietary iron. Our findings indicate that 26% of anemic children had markers compatible with iron deficiency, including low hemoglobin and ferritin levels. This suggests that IDA together with true iron deficiency, are the main contributors to childhood anemia in the highlands of Bolivia. This assumption is further supported by a significant association between anemia and IDA. At the same time, 20% of anemic children with normal or moderately elevated ferritin levels, met the criteria for anemia of inflammation, suggesting that both altered iron status and coexisting inflammatory processes play a role in the development of anemia in our study. In the context of chronic inflammation, persistent parasitic infections have been found to impair the body's ability to absorb dietary iron by promoting ferroportin degradation, thus reducing the use of existing iron stores for erythropoiesis [26–28]. Ferroportin internalization and degradation are mediated by the regulatory protein hepcidin which is up-regulated by interleukin (IL) 6 during inflammation [29,30]. This process decreases blood iron levels and leads to AI.

Significantly higher levels of sTfR in relation to serum ferritin, in our study, underscore the impact of IDA and AI as additional causes of anemia in 9.4% of highland children, and the effects of combined infections by helminths of *H. nana* and *A. lumbricoides* lead to mixed types of anemia, likely due to a variety of symptoms, including nutritional malabsorption, malnutrition, gastrointestinal bleeding, and iron deficiency attributed to these helminths [31–34].

There were strengths and weaknesses in our study. The use of comprehensive laboratory tests enabled the differentiation between IDA, AI, and mixed forms of anemia for this region with chronic infections, where overlapped symptoms could complicate diagnosis. However, due to budget limitations, we were unable to measure ferritin and sTfR levels for all children, which may have hindered an accurate representation of the impact of combined types of anemia in our setting. Analyzing only a single stool sample per child may lead to an underestimation of the true prevalence of parasitic infections. Furthermore, there is a lack of data on the intensity of helminth infections since health is intensity-related. Moreover, data on socioeconomic status, such as family size, income, and education level, which may contribute to inadequate nutrition linked to anemia and IDA, is lacking.

The high prevalence of anemia among children in highland areas is a significant public health concern, given its impact on childhood cognitive development and physical fitness. By identifying the underlying causes and the most impactful parasitic infections, this study provides actionable insights to help reduce the burden of anemia in this vulnerable population and in similar settings across different regions, particularly in high-altitude, low-income areas. The specific parasite patterns and the diagnostic approach such as the use of sTfR/log ferritin ratios provide unique information into anemia in the Bolivian highlands which may differ in some aspects from the more acute or geographically distinct forms of anemia seen in malaria or schistosomiasis [27,35,36]. Therefore, this study enhances the global understanding of how various parasites contribute to anemia and provides insights into the regional epidemiology of the condition.

Our findings highlight key challenges that need to be addressed to improve childhood health conditions. These include the need for integrated public health interventions, such as enhanced regular deworming and sanitation to prevent recurrent infections, as well as improved iron nutrition to effectively combat anemia in this vulnerable population.

Author Contributions: Conceptualization, W.R.C. and C.R.; data curation, W.R.C.; formal analysis, R.P.; funding acquisition, C.R.; investigation, W.R.C. and C.R.; methodology, W.R.C., I.C., A.R. and C.R.; project administration, C.R.; resources, C.R.; supervision, C.R.; writing-original draft, W.R.C. and C.R.; writing-review and editing R.P. and C.R. All authors have read and agreed to the published version of the manuscript.

Funding: This work was supported by the "Impuesto Directo a los Hidrocarburos (IDH)", the Swedish Research Council, and the Swedish International Developing Cooperation Agency (SIDA) through the Program SIDA-Universidad Mayor de San Andrés.

Institutional Review Board Statement: The study was conducted in accordance with the Declaration of Helsinki, and approved by the Research Ethics of the Universidad Mayor de San Andrés (UMSA) CEI-UMSA; Approval Code, None; Approval Date, 14 September 2011.

Informed Consent Statement: Prior to participation, informed consent was obtained from the parents or legal guardians of each child.

Data Availability Statement: The data supporting the findings of this study are provided within the article.

Acknowledgments: We thank the school teachers, parents, and health authorities of each study location for their support and authorization to conduct this research.

Conflicts of Interest: The authors declare no conflicts of interest.

References

1. Kassebaum, N.J.; Jasrasaria, R.; Naghavi, M.; Wulf, S.K.; Johns, N.; Lozano, R.; Regan, M.; Weatherall, D.; Chou, D.P.; Eisele, T.P.; et al. A systematic analysis of global anemia burden from 1990 to 2010. *Blood* **2014**, *123*, 615–624. [CrossRef] [PubMed]
2. Grantham-McGregor, S.; Ani, C. A review of studies on the effect of iron deficiency on cognitive development in children. *J. Nutr.* **2001**, *131*, 649S–668S. [CrossRef] [PubMed]
3. Nelson, M. Anemia in adolescent girls: Effects on cognitive function and activity. *Proc. Nutr. Soc.* **1996**, *55*, 359–367. [CrossRef]
4. Azab, S.F.A.; Abdelsalam, S.M.; Saleh, S.H.A.; Elbehedy, R.M.; Lotfy, S.M.; Esh, A.M.H.; Srea, M.A.; Aziz, K.A. Iron deficiency anemia as a risk factor for cerebrovascular events in early childhood: A case-control study. *Ann. Hematol.* **2014**, *93*, 571–576. [CrossRef]
5. Ramakrishnan, K.; Borade, A. Anemia as a risk factor for childhood Asthma. *Lung India* **2010**, *27*, 51–53. [CrossRef]
6. Opoka, R.O.; Hamre, K.E.S.; Brand, N.; Bangirana, P.; Idro, R.; John, C.C. High postdischarge morbidity in Ugandan children with severe malarial anemia or cerebral malaria. *J. Pediatr. Infect. Dis. Soc.* **2017**, *6*, e41–e48. [CrossRef]
7. Lozoff, B.; Beard, J.; Connor, J.; Barbara, F.; Georgieff, M.; Schallert, T. Long lasting neural and behavioral effects of iron deficiency in infancy. *Nutr. Rev.* **2006**, *64*, S34–S43. [CrossRef]
8. Weiss, G.; Goodnough, L.T. Anemia of chronic disease. *N. Engl. J. Med.* **2005**, *352*, 1011–1023. [CrossRef]
9. Goodnough, L.T.; Nemeth, E.; Ganz, T. Detection, evaluation, and management of iron-restricted erythropoiesis. *Blood* **2010**, *116*, 4754–4761. [CrossRef]
10. Weiss, G. Anemia of chronic disorders: New diagnostic tools and new treatment strategies. *Semin. Hematol.* **2015**, *52*, 313–320. [CrossRef]
11. Brooker, S.; Jardim-Botelho, A.; Quinnell, R.J.; Geiger, S.M.; Caldas, I.R.; Fleming, F.; Hotez, P.J.; Correa-Oliveira, R.; Rodrigues, L.C.; Bethony, J.M. Age-related changes in hookworm infection, anaemia and iron deficiency in an area of high Necator americanus hookworm transmission in south-eastern Brazil. *Trans. R. Soc. Trop. Med. Hyg.* **2007**, *101*, 146–154. [CrossRef] [PubMed]
12. Degarege, A.; Animut, A.; Medhin, G.; Legesse, M.; Erko, B. The association between multiple intestinal helminth infections and blood group, anaemia and nutritional status in human populations from Dore Bafeno, southern Ethiopia. *J. Helminthol.* **2014**, *88*, 152–159. [CrossRef] [PubMed]
13. Ulukanligil, M.; Seyrek, A. Anthropometric status, anaemia and intestinal helminthic infections in shantytown and apartment schoolchildren in the Sanliurfa province of Turkey. *Eur. J. Clin. Nutr.* **2004**, *58*, 1056–1061. [CrossRef] [PubMed]
14. Apaza, C.; Cuna, W.; Brañez, F.; Passera, R.; Rodriguez, C. Frequency of gastrointestinal parasites, anemia, and nutritional status among children from different geographical regions of Bolivia. *J. Trop. Med.* **2023**, *2023*, 5020490. [CrossRef]

15. WHO. Hemoglobin Concentrations for the Diagnosis of Anemia and Assessment of Severity. 2011. Available online: https://apps.who.int/iris/handle/10665/85839 (accessed on 18 August 2022).
16. Gonzalez-Ruiz, A.; Haque, R.; Aguirre, A.; Castañón, G.; Hall, A.; Guhl, F.; Ruiz-Palacios, G.; Miles, M.A.; Warhurst, D.C. Value of microscopy in the diagnosis of dysentery associated with invasive *Entamoeba histolytica*. *J. Clin. Pathol.* **1994**, *47*, 236–239. [CrossRef]
17. Haque, R.; Hall, A.; Tzipori, S. Zymodemes of *Entamoeba histolytica* in Dhaka, Bangladesh. *Ann. Trop. Med. Parasitol.* **1990**, *84*, 629–632.18. [CrossRef]
18. Van Den Broucke, S.; Verschueren, J.; Van Esbroeck, M.; Bottieau, E.; Van den Ende, J. Clinical and microscopic predictors of *Entamoeba histolytica* intestinal infection in travelers and migrants diagnosed with *Entamoeba histolytica*/dispar infection. *PLoS Negl. Trop. Dis.* **2018**, *12*, e0006892. [CrossRef]
19. Aruni Chura, J.; Macchioni, F.; Furzi, F.; Balboa, V.; Mercado, E.; Gómez, J.; Rojas Gonzales, P.; Poma, V.; Loup, A.; Roselli, M.; et al. Cross-sectional study on intestinal parasite infections in different ecological zones of the Department of La Paz, Bolivia. *One Health* **2021**, *13*, 100271. [CrossRef]
20. Martins, A.S.; Santos, S.A.; Lisboa, C.A.D.S.; Barros, T.F.; Ribeiro, T.C.M.; Da Costa-Ribeiro, H.; Mattos, Â.P.; Almeida Mendes, P.S.; Mendes, C.M.C.; Souza, E.L.; et al. Infectious etiology and indicators of malabsorption or intestinal injury in childhood diarrhea. *Biomedica* **2024**, *44*, 80–91. [CrossRef]
21. Solaymani-Mohammadi, S. Mucosal defense against *Giardia* at the intestinal epithelial cell interface. *Front. Immunol.* **2022**, *13*, 817468. [CrossRef]
22. Zaman Wahid, B.; Haque, M.A.; Gazi, M.A.; Fahim, S.M.; Faruque, A.S.G.; Mahfuz, M.; Ahmed, T. Site-specific incidence rate of *Blastocystis hominis* and its association with childhood malnutrition: Findings from a multi-country birth cohort study. *Am. J. Trop. Med. Hyg.* **2023**, *108*, 887–894. [CrossRef] [PubMed]
23. Olivares, J.L.; Fernández, R.; Fleta, J.; Ruiz, M.Y.; Clavel, A. Vitamin B12 and folic acid in children with intestinal parasitic infection. *J. Am. Coll. Nutr.* **2002**, *21*, 109–113. [CrossRef] [PubMed]
24. Aslinia, F.; Mazza, J.J.; Yale, S.H. Megaloblastic anemia and other causes of macrocytosis. *Clin. Med. Res.* **2006**, *4*, 236–241. [CrossRef]
25. Petry, N.; Egli, I.; Zeder, C.; Walczyk, T.; Hurrell, R. Polyphenols and phytic acid contribute to the low iron bioavailability from common beans in young women. *J. Nutr.* **2010**, *140*, 1977–1982. [CrossRef]
26. Glinz, D.; Hurrell, R.F.; Righetti, A.A.; Zeder, C.; Adiossan, L.G.; Tjalsma, H.; Utzinger, J.; Zimmermann, M.B.; N'Goran, E.K.; Wegmuller, R. In Ivorian school-age children, infection with hookworm does not reduce dietary iron absorption or systemic iron utilization, whereas afebrile *Plasmodium falciparum* infection reduces iron absorption by half. *Am. J. Clin. Nutr.* **2015**, *101*, 462–470. [CrossRef]
27. Leenstra, T.; Coutinho, H.M.; Acosta, L.P.; Langdon, G.C.; Su, L.; Olveda, R.M.; McGarvey, S.T.; Kurtis, J.D.; Friedman, J.F. *Schistosoma japonicum* reinfection after praziquantel treatment causes anemia associated with inflammation. *Infect. Immun.* **2006**, *74*, 6398–6407. [CrossRef]
28. Zhang, D.L.; Wu, J.; Shah, B.N.; Greutelaers, K.C.; Ghosh, M.C.; Ollivierre, H.; Su, X.Z.; Thuma, P.E.; Bedu-Addo, G.; Mockenhaupt, F.P.; et al. Erythrocytic ferroportin reduces intracellular iron accumulation, hemolysis, and malaria risk. *Science* **2018**, *359*, 1520–1523. [CrossRef]
29. Nemeth, E.; Rivera, S.; Gabayan, V.; Keller, C.; Taudorf, S.; Pedersen, B.K.; Ganz, T. IL-6 mediates hypoferremia of inflammation by inducing the synthesis of the iron regulatory hormone hepcidin. *J. Clin. Investig.* **2004**, *113*, 1271–1276. [CrossRef]
30. Nemeth, E.; Tuttle, M.S.; Powelson, J.; Vaughn, M.B.; Donovan, A.; Ward, D.M.; Ganz, T.; Kaplan, J. Hepcidin regulates cellular iron efflux by binding to ferroportin and inducing its internalization. *Science* **2004**, *306*, 2090–2093. [CrossRef]
31. Barakat, M.; Shebani, A.; Rehman, A. Acute Massive Gastrointestinal Bleeding Caused by *Ascaris lumbricoides* Infection: A Case Report. *Cureus* **2022**, *14*, e30935. [CrossRef]
32. Tripathy, K.; Duque, E.; Bolaños, O.; Lotero, H.; Mayoral, L.G. Malabsorption syndrome in ascariasis. *Am. J. Clin. Nutr.* **1972**, *25*, 1276–1281. [CrossRef] [PubMed]
33. Ngui, R.; Lim, Y.A.; Chong Kin, L.; Sek Chuen, C.; Jaffar, S. Association between anaemia, iron deficiency anaemia, neglected parasitic infections and socioeconomic factors in rural children of West Malaysia. *PLoS Negl. Trop. Dis.* **2012**, *6*, e1550. [CrossRef]
34. Ikumapayi, U.N.; Sanyang, C.; Pereira, D.I. A case report of an intestinal helminth infection of human hymenolepiasis in Rural Gambia. *Clin. Med. Rev. Case Rep.* **2019**, *6*, 251. [PubMed]

35. Righetti, A.A.; Wegmuller, R.; Glinz, D.; Ouattara, M.; Adiossan, L.G.; N'Goran, E.K.; Utzinger, J.; Hurrell, R.F. Effects of inflammation and *Plasmodium falciparum* infection on soluble transferrin receptor and plasma ferritin concentration in different age groups: A prospective longitudinal study in Cote d'Ivoire. *Am. J. Clin. Nutr.* **2013**, *97*, 1364–1374. [CrossRef] [PubMed]
36. Chang Cojulun, A.; Bustinduy, A.L.; Sutherland, L.J.; Mungai, P.L.; Mutuku, F.; Muchiri, E.; Kitron, U.; King, C.H. Anemia among children exposed to polyparasitism in Coastal Kenya. *Am. J. Trop. Med. Hyg.* **2015**, *93*, 1099–1105. [CrossRef]

microorganisms

MDPI

Review

Cryptosporidium Species Infections Detected from Fecal Samples of Animal and Human Hosts in South Africa: Systematic Review and Meta-Analysis

Mpho Tawana [1,2], ThankGod E. Onyiche [1,3], Tsepo Ramatla [1,4,*], Sebolelo Jane Nkhebenyane [4], Dennis J. Grab [5,6] and Oriel Thekisoe [1]

1 Unit for Environmental Sciences and Management, North-West University, Potchefstroom 2531, South Africa; 2027028865@ufs4life.ac.za (M.T.); et.onyiche@unimaid.edu.ng (T.E.O.); oriel.thekisoe@nwu.ac.za (O.T.)
2 Department of Zoology and Entomology, University of the Free State, Phuthaditjhaba 9866, South Africa
3 Department of Veterinary Parasitology and Entomology, University of Maiduguri, P.M.B. 1069, Maiduguri 600230, Nigeria
4 Department of Life Sciences, Central University of Technology, Bloemfontein 9300, South Africa; snkheben@cut.ac.za
5 Department of Pathology, Uniformed Services University of the Health Sciences, Bethesda, MD 20814, USA; dennis.grab@usuhs.edu
6 Department of Tropical Medicine, Medical Microbiology and Pharmacology, University of Hawaii at Manoa, Honolulu, HI 96822, USA
* Correspondence: tramatla@cut.ac.za

Abstract: This study presents a systematic review and meta-analysis approach of *Cryptosporidium* species prevalence studies in animal and human hosts published between 1980 and 2020 in South Africa. Extensive searches were conducted on three electronic databases including PubMed, ScienceDirect and Google Scholar. The findings indicated an overall pooled prevalence estimate (PPE) of *Cryptosporidium* spp. infections in animals and humans at 21.5% and 18.1%, respectively. The PCR–RFLP appeared to be the most sensitive diagnostic method with a PPE of 77.8% for the detection of *Cryptosporidium* spp. infections followed by ELISA (66.7%); LAMP (45.4%); PCR (25.3%); qPCR (20.7%); microscopy (10.1%); IFAT (8.4%); and RDT (7.9%). In animal hosts, *C. parvum* had the highest PPE of 3.7%, followed by *C. andersoni* (1.5%), *C. ubiquitum* (1.4%) and *C. bovis* (1.0%), while in humans, *C. parvum* also had the highest PPE of 18.3% followed by *C. meleagridis* at 0.4%. The data generated in this study indicated that *Cryptosporidium* spp. infections were highly prevalent in both animals and humans in South Africa, especially in the KwaZulu-Natal and North West provinces. However, we further observed that there was a lack of prevalence studies for both animals and humans in some of the provinces. This study highlights the necessity for a "One Health" strategic approach promoting public hygiene, animal husbandry and regular screening for *Cryptosporidium* spp. infections in both animals and humans.

Keywords: *Cryptosporidium* species; prevalence; South Africa

1. Introduction

South Africa is a developing country with over 59 million of human population and ranks number 13th in the global list for countries living in the poverty line [1,2]. Accounting for about 17% of the world's HIV infections, South Africa's population is already vulnerable to secondary opportunistic infections which include parasitic diseases such as cryptosporidiosis, giardiasis and toxoplasmosis [3]. Cryptosporidiosis is a zoonotic disease caused by protozoan parasites of the genus *Cryptosporidium*. Of the *Cryptosporidium* species causing disease in humans and animals, *C. hominis* and *C. parvum* are known to cause gastroenteritis among the general public [4,5]. While the disease is often self-limiting, cryptosporidiosis infection can be life-threatening to humans living with HIV/AIDS as well as children and young animals [6–9]. The main symptoms and signs for both animals and humans are watery diarrhea, weight loss, nausea, vomiting, fatigue and low-grade fever [10–12]. In southern Africa and Asia, about 2.9 million and 4.7 million cases of *Cryptosporidium* spp. infections have been reported among children less than 2 years old, respectively [13]. The first *Cryptosporidium* spp. infection cases to be recorded in South Africa were of four children in Durban in 1987, and since then, there has been an increase in the studies on *Cryptosporidium* parasitic prevalence in the country [14].

Cryptosporidium species can spread through the fecal–oral route either primarily (direct contact) or secondarily through the consumption of contaminated food or water with human or animal feces [15–17]. The annual quantity of excreted *Cryptosporidium* spp. oocysts by domestic animals globally has been estimated to be approximately 3.2×10^{23} [18]. Animals are important contributing factors of environmental contamination of *Cryptosporidium* oocyst distribution [19,20], spreading via water [21] or food [22]. *Cryptosporidium* spp. distribution and infection are exacerbated by their resistance to normal water treatment, including chlorination [23].

Diagnostic techniques for *Cryptosporidium* spp. infections includes microscopy, polymerase chain reaction (PCR) and enzyme-linked immunosorbent assay (ELISA) with reported sensitivities of 56.0% to 75.4%; 100%; and 89.7% to 100%, respectively [24].

Omolabi et al. [25] conducted a meta-analysis on *Cryptosporidium* species in humans from southern Africa. However, there is limited information on comprehensive data available to estimate the prevalence of *Cryptosporidium* spp. in humans and animals in South Africa. Hence, the premise of this review is founded around the wildlife–domestic–human interface, highlighting the shared, interconnected links between the health of humans, wildlife and domestic animals. Our meta-analytical approach allowed for the identification of study gaps, examination of the pooled prevalence for animal and human *Cryptosporidium* spp. in South Africa and further investigated the influence of risk factors such as age, sex, fecal consistency, HIV status and diagnostic techniques on the spread of the *Cryptosporidium* parasite.

2. Materials and Methods

2.1. Study Design

This systematic review and meta-analysis were conducted on published articles reporting *Cryptosporidium* spp. infections in South Africa, confirmed by examining feces of animals and humans for the presence of *Cryptosporidium* spp. oocysts using microscopy, immunological and molecular techniques.

2.2. Search Strategy and Criteria

Literature searches were conducted in PubMed, ScienceDirect and Google Scholar for articles published in English between 1980 and 2020 on the prevalence or epidemiology of *Cryptosporidium* spp. infections across South Africa in animals and humans. The search keywords were "prevalence", "*Cryptosporidium*" and "South Africa". Keywords used in the search were entered individually or in combination with the "AND" and/or "OR" operators. None of the authors of original studies were contacted for additional information and no attempt was made to retrieve unpublished articles. Titles and abstracts were scanned, and relevant full-text articles were downloaded and obtained through library resources and online platforms.

2.3. Inclusion and Exclusion Criteria

Articles were included only if they fulfilled the following inclusion criteria: cross section (prevalence study) conducted within South Africa, vertebrate host (humans or animals) used, study conducted on fecal samples, exact total numbers and positive cases clearly provided, sample size ($\geq$25 for enabling statistical calculations) and written in English. Studies without these characteristics were all excluded such as review studies,

studies on water, case report studies, ones with a lower sample size and ones not written in English.

2.4. Data Quality Control Measures

To confirm the methodological soundness of the research articles selected for quantitative synthesis, two authors independently used the Joanna Briggs Institute (JBI) Critical Appraisal Tools Checklist 2017 review guideline for prevalence studies. Studies that achieved a score of five or higher for the evaluation criteria were included.

2.5. Data Extraction

The data extraction protocol consisted of the name of the author and region, hosts, total sample size, number of positive cases, estimated prevalence, species of intestinal parasites, and diagnostic technique. Moreover, studies that were conducted in more than one province and those that had both animal and human studies simultaneously were separated accordingly.

2.6. Statistical/Meta-Analytic Procedures

The meta-analysis was performed using the Comprehensive Meta-Analysis (CMA) program [26]. The random effects model was used to estimate the pooled prevalence and corresponding 95% confidence interval (CI). Statistical heterogeneity between studies was estimated with Cochran's Q statistic and I-square (I^2) test (values of 25%, 50% and 75% were considered to represent low, medium and high heterogeneities, respectively). The funnel plot and Begg's rank correlation test were used to evaluate the possibility of publication bias, where $p < 0.05$ was considered as indicative of statistically significant publication bias [27,28].

3. Results

3.1. Literature Search and Eligible Studies

A total of 8028 studies were identified from three electronic databases, namely PubMed (95), ScienceDirect (2223) and Google Scholar (5710). After the removal of duplicates and a subsequent review of study titles and abstracts, 4167 studies were excluded and 43 studies were found to be eligible and subjected to full text evaluation for inclusion. Furthermore, seven studies were excluded for the following reasons: (i) no clear focus on the sample of choice (n = 2); (ii) secondary data (n = 1); and (iii) carried out in other countries in sub-Saharan Africa outside of South Africa (n = 4). Finally, 36 studies that assessed the prevalence of *Cryptosporidium* spp. in animal and human feces were included for quantitative synthesis (Figure 1).

3.2. Characteristics of Eligible Studies

With respect to the animal studies, 10 studies in total were included in the meta-analysis. These studies were published between the years 2008 and 2014 on the prevalence of *Cryptosporidium* spp. in various animals, including buffaloes, cats, cattle, dogs, elephants, goats, impala and sheep in South Africa (Table 1). Almost all studies were from the northern region of South Africa, which included four provinces, namely, Gauteng (n = 1), Limpopo (n = 4), Mpumalanga (n = 4) and North West (n = 2) (Table 1). Individually, the prevalence ranged from 0.00% to 80.0% across the various provinces (Table 1), with the highest prevalence recorded from the North West province and the lowest from the Mpumalanga province.

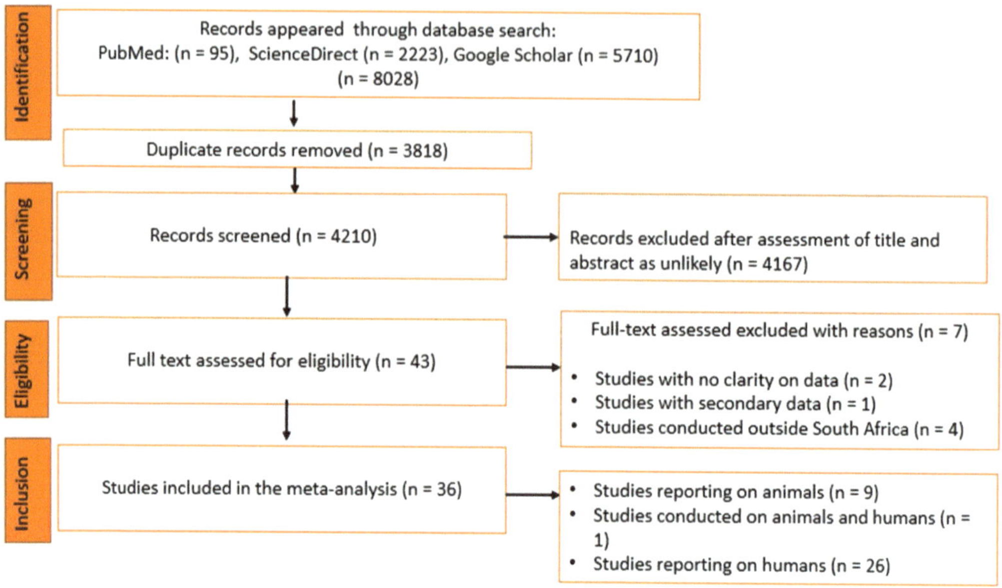

Figure 1. Flow chart of included studies according to PRISMA guidelines.

Table 1. List and characteristics of eligible studies included in the meta-analysis with respect to animal study prevalence by different provinces in South Africa.

Study Authors	Animal Host (n)	Study Area (Province)	Sample Size	No. of Positives	Prevalence (%)
Bakheit et al. [29]	Cattle (n = 107) Horse (n = 78) Sheep (n = 85)	Free State	270	79	29.26
Hlungwani [30]	Cattle (n = 52) Goat (n = 33)	Limpopo	85	48	56.47
Lukášová et al. [31]	Cat (n = 1)	Gauteng	1	0	0.00
Lukášová et al. [31]	African civet (n = 2) African wild cat (n = 1) Banded mongoose (n = 3) Black-backed jackal (n = 1) Caracal (n = 2) Cat (n = 1) Cheetah (n = 4) Dog (n = 11) Lion (n = 13) Serval (n = 2) Spotted hyena (n = 3) Striped polecat (n = 2)	Limpopo	45	7	15.56
Lukášová et al. [31]	Bat-eared fox (n = 1) Black-backed jackal (n = 6) Caracal (n = 1)	North West	8	0	0.00
Samie et al. [32]	Cat (n = 25) Dog (n = 25)	Limpopo	50	19	38.00

Table 1. *Cont.*

Study Authors	Animal Host (n)	Study Area (Province)	Sample Size	No. of Positives	Prevalence (%)
Samie et al. [33]	Chicken (n = 28) Goat (n = 93) Sheep (n = 4) Cattle (n = 187)	Limpopo	312	98	31.41
Samra et al. [34]	Buffalo (n = 141) Elephant (n = 144) Impala (n = 161)	Mpumalanga	446	29	6.50
Samra et al. [35]	Buffalo (n = 71) Elephant (n = 72) Impala (n = 71)	Mpumalanga	214	79	36.92
Samra et al. [36]	Calf (Bovine) (n = 352)	Mpumalanga	352	2	7.95
Syakalima et al. [37]	Pig (n = 90)	North West	90	72	80.00
Vink [38]	Calf (Bovine) (n = 345)	Mpumalanga	345	2	0.58

Additionally, 27 studies focusing on humans published between 1986 and 2020 were included in the meta-analysis on the prevalence of *Cryptosporidium* spp. in humans. These studies were from both the northern and southern regions of South Africa. Studies from the northern region included the Gauteng (n = 7), Limpopo (n = 6), Mpumalanga (n = 1), North West (n = 1) provinces, while the southern region included the Eastern Cape (n = 6) and KwaZulu-Natal (n = 8) provinces (Table 2). *Cryptosporidium* spp. prevalence for all the different provinces ranged from 2.89% to 72.94%, respectively (Table 2), with the highest pooled prevalence from the KwaZulu-Natal province and the lowest from the Gauteng province.

Table 2. List and characteristics of eligible studies included in the meta-analysis with respect to human study prevalence by different provinces in South Africa.

Study Authors	Study Area (Province)	Sample Size	No. of Positives	Prevalence (%)
Abebe et al. [39]	Limpopo	84	25	29.76
Bartelt et al. [40]	Limpopo	251	165	65.74
Becker et al. [41]	Eastern Cape	1428	164	11.49
Berkowitz et al. [42]	Gauteng	121	18	14.88
Etinosa [43]	Eastern Cape	180	122	67.78
Fripp et al. [44]	Gauteng	6870	289	4.21
Geyer et al. [45]	Gauteng	78	20	25.64
Hlungwani [30]	Gauteng	362	80	22.10
Hlungwani [30]	Limpopo	218	159	72.94
Htun et al. [46]	Eastern Cape	842	179	21.26
Jarmey-Swan et al. [47]	KwaZulu-Natal	2800	1300	46.43
Leav et al. [48]	KwaZulu-Natal	101	25	24.75
Moodley et al. [49]	KwaZulu-Natal	1229	111	9.03
Msolo et al. [50]	Eastern Cape	53	3	5.66
Müller et al. [51]	Eastern Cape	934	28	2.99

Table 2. *Cont.*

Study Authors	Study Area (Province)	Sample Size	No. of Positives	Prevalence (%)
Omoruyi et al. [52]	Eastern Cape	180	47	26.11
Samie et al. [53]	Limpopo	244	44	18.03
Samie et al. [54]	Limpopo	255	46	18.04
Samie et al. [55]	Limpopo	528	143	27.08
Samie et al. [56]	Limpopo	322	42	13.04
Samie et al. [57]	Limpopo	151	15	9.93
Samra et al. [58]	Gauteng	141	25	17.73
Samra et al. [58]	Mpumalanga	128	11	8.59
Samra et al. [58]	North West	147	14	9.52
Samra et al. [58]	KwaZulu-Natal	26	4	15.39
Smith and Van den Ende [59]	KwaZulu-Natal	259	31	11.97
Steele et al. [60]	Gauteng	1316	38	2.89
Steele et al. [14]	Gauteng	3186	129	4.05
Trönnberg et al. [61]	KwaZulu-Natal	120	25	20.83
Walters [62]	KwaZulu-Natal	91	53	58.24
Witienberg et al. [63]	KwaZulu-Natal	194	30	15.46

3.3. Pooling, Heterogeneity and Subgroup Analysis

3.3.1. Prevalence in Animals Based on Hosts, Study Years and *Cryptosporidium* Species

Studies examining the prevalence of *Cryptosporidium* in animals found high heterogeneity based on the host, year of study and *Cryptosporidium* species (Table 3). In total, 2579 samples were screened, of which 374 tested positive to various species of *Cryptosporidium* spp. with a pooled prevalence estimate (PPE) of 21.5% (95%CI: 10.5–39.2%; Q = 391.34; I^2 = 97.70; Q–p = 0.0003) (Table 3).

Table 3. Pooled prevalence estimates and risk factors associated with *Cryptosporidium* species infection in animals.

Risk Factor	No. of Studies	Pooled Prevalence Estimates			Measure of Heterogeneity		Q–P	Publication Bias
		Sample Size	No. of Positives	Prevalence 95%CI (%)	Q	I^2		Begg and Mazumdar Rank p-Value
Overall animals	10	2579	374	21.5 (10.5–39.2)	391.34	97.70	0.000	0.123
Study year								
2001–2010	4	1015	102	11.7 (4.4–27.5)	63.94	95.31	0.000	0.500
2011–2020	4	872	124	11.3 (1.1–58.8)	134.15	97.76	0.000	0.248
Cryptosporidium species								
C. andersoni	2	299	4	1.5 (0.6–3.9)	0.86	0.00	0.352	–
C. bovis	2	566	5	1.0 (0.4–2.3)	0.98	0.00	0.320	
C. parvum	3	882	32	3.7 (1.1–12.0)	17.58	88.62	0.000	0.301
C. ubiquitum	1	214	3	1.4				–

Table 3. *Cont.*

Risk Factor	No. of Studies	Pooled Prevalence Estimates			Measure of Heterogeneity		*Q–P*	Publication Bias
		Sample Size	No. of Positives	Prevalence 95%CI (%)	*Q*	I^2		Begg and Mazumdar Rank *p*-Value
Animal species								
Buffalo	2	212	10	4.9 (2.7–8.9)	0.82	0.00	0.364	–
Cattle	5	749	100	11.4 (4.7–25.1)	49.31	91.89	0.000	0.025
Dog	2	38	13	30.4 (9.7–64.1)	2.84	64.84	0.092	–
Elephant	2	216	37	5.9 (0.1–73.8)	7.48	86.62	0.006	–
Goat	2	126	48	31.3 (11.2–62.0)	6.94	85.60	0.008	–
Impala	2	232	9	3.9 (2.1–7.4)	0.31	0.00	0.581	–
Sheep	2	89	28	31.5 (22.7–41.9)	0.64	0.00	0.425	–

Animal studies conducted during the 2001–2010 duration had a slightly higher PPE of [11.7% (95%CI: 4.4–27.5); $Q = 63.94$; $I^2 = 95.31$; $Q–p = 0.0001$] than those of the 2011–2020 duration [11.3% (95%CI: 1.1–58.8%); $Q = 134.15$, $I^2 = 97.76$; $Q–p = 0.0001$] (Figure 2).

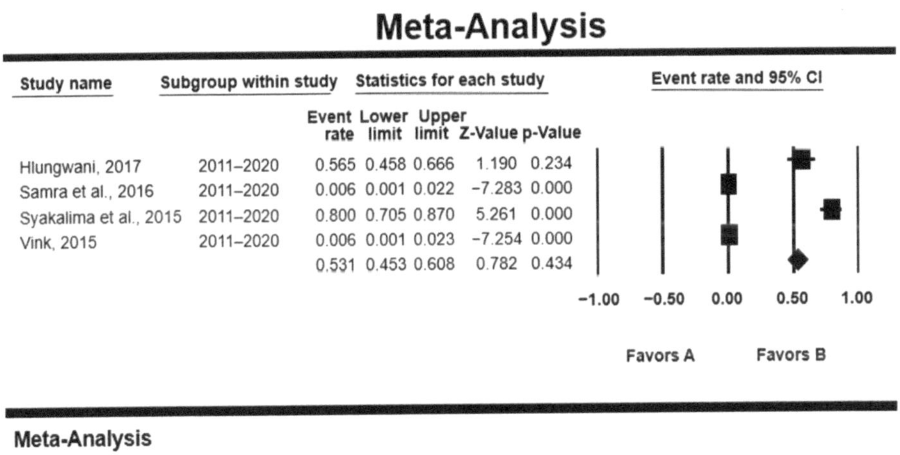

Figure 2. Forest plot of the prevalence of *Cryptosporidium* spp. from animal studies conducted during 2001–2010. The squares demonstrate the individual point estimates. The diamond at the base indicates the pooled estimate from the overall studies [30,36–38].

With references to species, *C. parvum* had the highest PPE of [3.7% (95%CI: 1.1–12.0%); $Q = 17.58$; $I^2 = 88.62$, $Q–p = 0.000$], followed by *C. andersoni* [1.5% (95CI: 0.6–3.9%); $Q = 0.86$; $I^2 = 0.00$; $Q–p = 0.352$] and *C. ubiquitum* 1.4%, and *C. bovis* had the lowest PPE [1.0% (95%CI: 0.4–2.3%); $Q = 0.98$; $I^2 = 0.00$; $Q–p = 0.320$] (Table 3). The prevalence of *Cryptosporidium* spp. by animal host varied as the following: sheep had the highest PPE [31.5% (95%CI: 22.7–41.9%); $Q = 0.64$; $I^2 = 0.00$; $Q–p = 0.425$], followed by goats [31.3% (95%CI: 11.2–62.0%); $Q = 6.94$; $I^2 = 85.60$; $Q–p = 0.008$], dogs [30.4% (95%CI: 9.7–64.1%); $Q = 30.4$; $I^2 = 64.84$; $Q–p = 0.092$], cattle [11.4% (95%CI 4.7–25.1%); $Q = 49.31$; $I^2 = 91.89$; $Q–p < 0.000$], elephants [5.9% (95%CI: 0.1–73.8%); $Q = 7.48$; $I^2 = 86.62$; $Q–p = 0.006$] and buffaloes [4.9% (95%CI; 2.7–8.9%); $Q = 0.82$; $I^2 = 0.00$; $Q–p = 0.36$], and the lowest was for impala with a PPE of [3.9% (95%CI: 2.1–7.4%); $Q = 0.31$; $I^2 = 0.00$; $Q–p = 0.581$] (Table 3).

3.3.2. Assessment of Publication Bias in Animals

A funnel plot of standard error by logit event rate was used to ascertain the presence of publication bias in the eligible studies. No significant bias was observed overall in the animal studies using the Begg and Mazumdar rank correlation test except for the subgroup analysis, where with respect to the prevalence in cattle, significant bias was observed as evident by the asymmetry of the plot with a *p*-value of 0.0432 (Table 3; Figure 3).

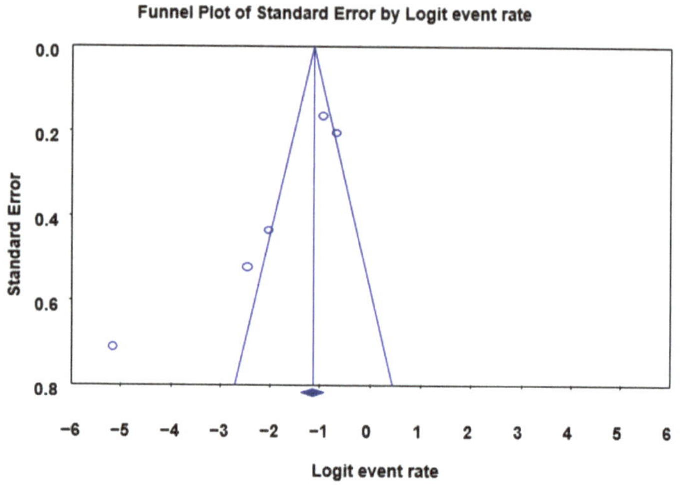

Figure 3. Funnel plot with 95% confidence limits of *Cryptosporidium* spp. pooled prevalence estimates of cattle subgroup studies that tested positive for *Cryptosporidium* species. The diamond at the base indicates the pooled estimate from the studies overall.

3.3.3. Prevalence in Humans Based on Study Years, Areas, Ages, HIV Statuses and Diagnostic Techniques

High heterogeneity was observed in studies looking at the prevalence of *Cryptosporidium* in humans depending on factors like age, HIV status, area, year of study and diagnostic method (Table 4). The 27 eligible studies for the evaluation of the prevalence of *Cryptosporidium* spp. in humans was conducted with data of studies published from 1983 to 2018. A total of 22,994 human fecal samples were examined, of which 3589 samples tested positive for *Cryptosporidium* spp., with a PPE of 18.1% (95%CI: 11.8–26.6). Substantial heterogeneity was observed [Q = 3655.54; I^2 = 99.23; Q–p = 0.000] (Table 4).

Table 4. Pooled prevalence estimates and risk factors associated with *Cryptosporidium* species infection in humans.

Risk Factor	No. of Studies	Pooled Prevalence Estimates			Measure of Heterogeneity		Q–P	Publication Bias
		Sample Size	No. of Positives	Prevalence 95%CI (%)	Q	I^2		Begg and Mazumdar Rank (*p*-Value)
Overall humans	27	22,787	3510	18.1 (11.8–26.6)	3655.54	99.23	0.000	0.411
Study region								
Northern region	15	13,825	1175	16.9 (8.7–30.3)	1578.25	99.11	0.000	0.200
Southern region	14	8437	2122	19.8 (11.8–31.4)	1190.73	98.91	0.000	0.274

Table 4. *Cont.*

Risk Factor	No. of Studies	Pooled Prevalence Estimates			Measure of Heterogeneity		Q–P	Publication Bias
		Sample Size	No. of Positives	Prevalence 95%CI (%)	Q	I^2		Begg and Mazumdar Rank (p-Value)
Sex								
Female	5	2067	785	41.1 (19.5–66.7)	187.25	97.86	0.000	0.500
Male	5	1786	615	38.1 (20.5–59.6)	66.58	93.99	0.000	0.500
Age								
<6 months–25 years	19	3636	1214	28.7 (23.3–34.7)	112.60	84.01	0.000	0.376
26–45 years	8	369	130	30.0 (14.1–52.9)	91.33	92.34	0.000	0.310
>45 years	8	153	43	24.2 (9.1–50.5)	40.44	82.69	0.000	0.310
Diagnostic technique								
Microscopy	21	25,475	1570	10.1 (6.1–16.2)	1761.79	98.87	0.000	0.359
ELISA	4	1081	598	66.7 (46.4–82.3)	101.98	97.06	0.000	0.087
IFAT	2	546	36	8.4 (0.7–53.2)	44.89	97.77	0.000	–
LAMP	2	237	93	45.4 (26.6–56.6)	3.91	74.39	0.048	–
RDT	4	3257	331	7.9 (3.2–18.0)	143.71	97.91	0.000	0.249
qPCR	3	717	139	20.7 (11.1–35.4)	30.50	93.44	0.000	0.059
PCR	9	991	252	25.3 (11.5–46.9)	203.65	96.07	0.000	0.500
PCR-RFLP	2	64	50	77.8 (65.9–86.4)	0.60	0.00	0.438	–
Study year								
1981–1990	8	13,557	768	9.2 (4.9–16.4)	479.10	98.54	0.000	0.042
1991–2000	2	2901	1325	35.4 (17.5–58.6)	17.19	94.18	0.000	–
2001–2010	5	1223	502	42.7 (24.4–63.2)	167.58	97.61	0.000	0.500
2011–2020	7	4422	675	11.2 (5.7–21.0)	366.02	98.36	0.000	0.410
Cryptosporidium species								
C. hominis	3	1165	46	4.0 (3.0–5.3)	1.43	0.000	0.489	0.301
C. meleagridis	2	585	2	0.4 (0.1–1.6)	0.64	0.00	0.424	–
C. muris	1	580	1	–	–	–	–	–
C. parvum	4	1636	403	18.3 (5.3–47.0)	223.28	98.66	0.000	0.500
HIV status								
HIV+	3	259	180	59.3 (19.8–89.6)	32.72	93.89	0.000	0.301
HIV−	3	396	149	39.8 (12.3–75.8)	83.34	97.60	0.000	0.059
Fecal consistency								
Diarrhea	3	691	243	24.4 (9.4–50.3)	70.81	97.18	0.000	0.301
Non-diarrhea	3	565	148	21.7 (8.7–44.8)	45.30	95.59	0.000	0.301

ELISA, enzyme-linked immunosorbent assay; HIV, human immunodeficiency virus; IFAT, immunofluorescence antibody test; qPCR, quantitative polymerase chain reaction; PCR, polymerase chain reaction; LAMP, loop-mediated isothermal amplification; PCR-RFLP, polymerase chain reaction–restriction fragment length polymorphism.

The southern region had the highest PPE [19.8% (95%CI: 11.8–31.9%); Q = 1190.73, I^2 = 98.91, Q–p = 0.000] compared to the northern region [16.9% (95%CI: 8.7–30.3%), Q = 1578.25, I^2 = 99.11, Q–p = 0.000], with the highest PPE from the KwaZulu-Natal province (Table 4). Also, studies conducted during the 2001–2010 duration had the highest PPE, while studies conducted between 2011 and 2020 had the lowest [11.2% (95%CI: 5.7–21.0%); Q = 366.02; I^2 = 98.36; Q–p = 0.000]. Despite the 1981–1990 period having had the highest number of studies and sample size, we observed a low PPE of 9.2% [95%CI: 4.9–16.4%; Q = 479.10; I^2 = 98.54; Q–p = 0.000] (Table 4). With reference to species, *C. parvum* had the highest PPE of 18.3% [95%CI: 5.3–47.0%; Q = 223.28; I^2 = 98.66; Q–p = 0.000] while *C. meleagridis* had the lowest PPE of 0.4% [95%CI: 0.1–1.6%; Q = 0.64; I^2 = 0.00; Q–p = 0.424] (Table 4).

The age interval of 26–45 years had the highest PPE at 30.0% [Q = 91.33; 95%CI 14.1–52.9; I^2 = 92.34; Q–p = 0.000], while the lowest was in the >45 yrs age interval at 24.2% [95%CI: 9.1–50.5%; Q = 40.44; I^2 = 82.69; Q–p = 0.000] (Table 4).

In all studies, the PPE was higher in females at 41.1% [95%CI: 19.5–66.7%; Q = 187.25; I^2 = 97.86; Q–p = 0.000], than 38.1% [95%CI: 20.5–59.6%; Q = 66.58; I^2 = 93.99; Q–p = 0.000] in male participants (Table 4). Figure 4 shows a forest plot of individual point estimates for the combined prevalence estimates of males (A) and females (B).

With regards to HIV infection, the HIV-positive (HIV+) population had a comparatively higher PPE at 59.3% [95%CI: 19.8–89.6%; Q = 32.72; I^2 = 93.89; Q–p = 0.000] as compared to 39.8% (95%CI: 12.3–75.8%); Q = 83.34; I^2 = 97.60; Q–p = 0.000] in the HIV-negative (HIV–) population (Table 4).

Cryptosporidium spp. infections was high in diarrheal patients with a PPE of 24.4% [95%CI: 9.4–50.3); Q = 70.81; I^2 = 97.18; Q–p = 0.000], as compared to non-diarrheal patients at 21.7% [95%CI: 8.7–44.8%); Q = 45.30; I^2 = 95.59; Q–p = 0.000] (Table 4). With respect to diagnostic techniques, our analyses showed that polymerase chain reaction–restriction fragment length polymorphism (PCR-RFLP) had the highest *Cryptosporidium* spp. detection sensitivity with a PPE at [77.8% (95%CI: 65.9–86.4%); Q = 0.60; I^2 = 0.00; Q–p = 0.438], followed by ELISA [66.7% (95%CI 46.4–82.3%); Q = 101.98; I^2 = 97.06; Q–p = 0.000], loop-mediated isothermal amplification (LAMP) [45.4% (95%CI: 26.6–56.6%); Q = 3.91; I^2 = 74.39; Q–p = 0.048], PCR [25.3% (95%CI: 11.5–46.9%); Q = 203.65; I^2 = 96.07; Q–p = 0.000], quantitative polymerase chain reaction (qPCR) [20.7% (95%CI: 11.1–35.4%); Q = 30.50; I^2 = 93.44; Q–p = 0.000], microscopy [10.1% (95%CI: 6.1–16.2%); Q = 1761.79; I^2 = 98.87; Q–p = 0.000] and immunofluorescence antibody test (IFAT) [8.4% (95%CI: 0.7–53.2%); Q = 44.89; I^2 = 97.77; Q–p = 0.000], and rapid diagnostic test (RDT) had the lowest PPE [7.9% (95%CI: 3.2–18.0%); Q = 143.71; I^2 = 97.91; Q–p = 0.000] (Table 4).

Meta-Analysis

Study name	Subgroup within study	Event rate	Lower limit	Upper limit	Z-Value	p-Value
Barteit et al., 2013	M	0.737	0.576	0.852	2.795	0.005
Jarmey-Swan et al., 2001	M	0.344	0.321	0.368	-12.022	0.000
Omoruyi et al., 2014	M	0.711	0.564	0.824	2.739	0.006
Samie et al., 2006	M	0.183	0.122	0.267	-6.033	0.000
Samie et al., 2014	M	0.093	0.039	0.204	-4.862	0.000
		0.347	0.325	0.370	-12.381	0.000

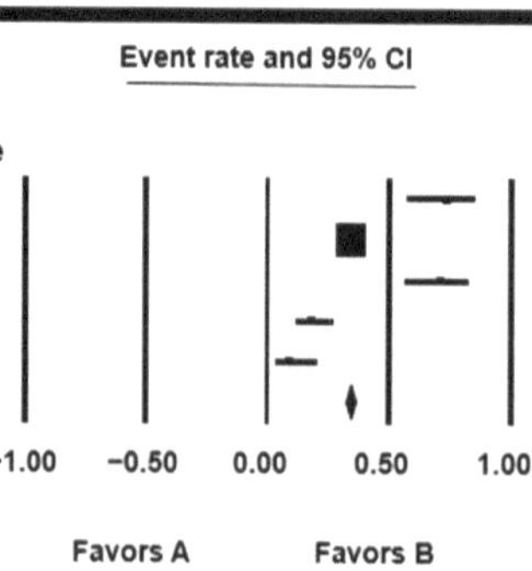

Meta-Analysis

(A)

Figure 4. *Cont.*

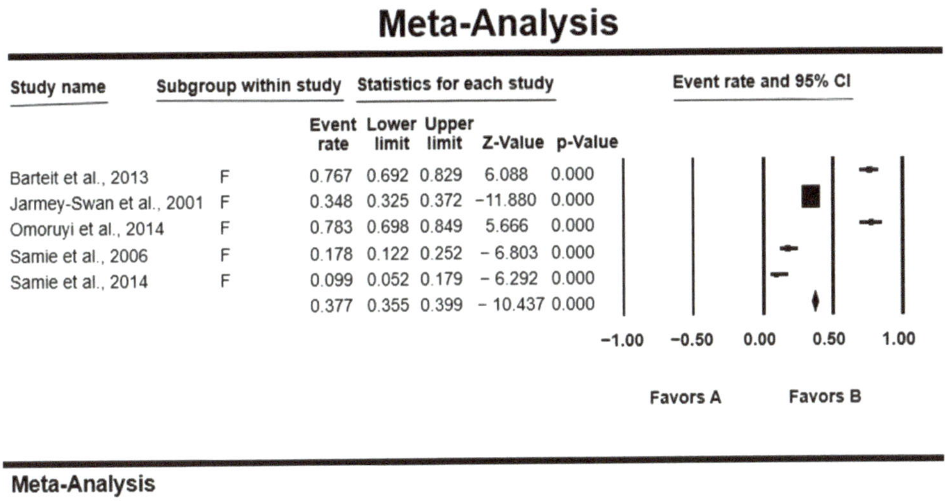

(**B**)

Figure 4. Forest plot showing the pooled estimates of *Cryptosporidium* spp. from studies conducted on (**A**) males and (**B**) females. The squares demonstrate the individual point estimates. The diamonds at the base indicate the pooled estimates from the overall studies [40,47,52,53,57].

3.4. Publication Bias Assessment in Human Studies

The Begg and Mazumdar rank correlation test revealed no significant publication bias for almost all the parameters except for the study year period 1981–1990, where significant bias was observed of both the asymmetry of the funnel plots and *p*-value of 0.042 (Table 4; Figure 5).

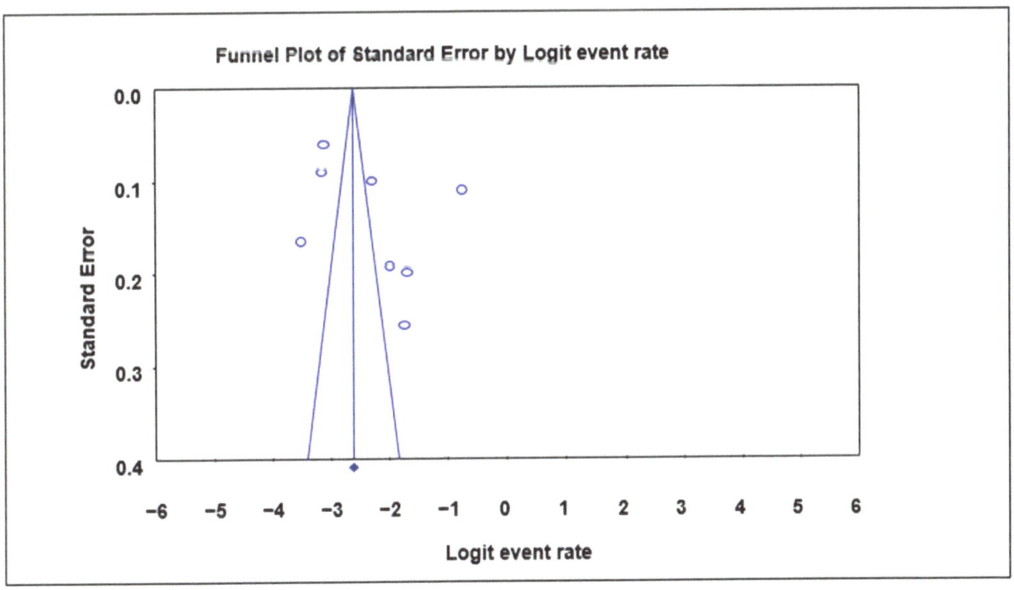

Figure 5. Funnel plot with 95% confidence limits of *Cryptosporidium* spp. pooled prevalence estimates of 1981–1990 interval subgroup studies that tested positive for *Cryptosporidium* spp. in humans. The diamond at the base indicates the pooled estimate from the studies overall.

4. Discussion

This study recorded an overall PPE of 21.5% for *Cryptosporidium* spp. infection in animals. Similar findings have also been reported in Australia (22.3%) and Tunisia (18.9%) [5,64]. On the other hand, higher prevalence above 50.0% was reported in an animal study from China and in a global review of *Cryptosporidium* spp. [65,66]. Most of the animal studies included in this meta-analysis were focused on the northern region of South Africa, where our analysis recorded a PPE of 13.7% of *Cryptosporidium* spp. infections. Data of animal studies were scarce in the southern regions; this paucity of studies may be due to a low research interest, as cryptosporidiosis is possibly not regarded as problematic for livestock. Furthermore, the results indicated a declining trend of *Cryptosporidium* spp. infection prevalence overtime, whereby the 2011–2020 period had a slightly lower pooled estimated prevalence as compared to the 2001–2010 period of study. This observation could be associated with the use of proper sanitary toilets, medication and improved animal husbandry practice. These observations are in accordance with the World Health Organization (WHO) and Global Roadmap 2012 mandate to decrease the prevalence of zoonotic diseases by 2020 through the improvement in veterinary public health practice and a focus on the supply of safe and palatable water, good sanitary infrastructures and proper hygiene practice [67]. Moreover, treatment, sanitation and proper hygiene practices have been proven to assist in reducing the prevalence of *Cryptosporidium* spp. infections in livestock [68–70].

Among all *Cryptosporidium* spp. observed in this study, *C. parvum* (3.7%) had the highest PPE, followed by *C. bovis* (1.0%). Similar findings have been reported in Greece and Peru, whereby *C. parvum* (64.3%) was more prevalent in comparison to both *C. andersoni* + *C. bovis* (7.1%) prevalence in animals [71,72]. In contrast, Ref. [73] reported different findings, whereby *C. bovis* (57.0%) was the most common *Cryptosporidium* spp. of health concern and *C. parvum* (7.0%), the species of least concern in veterinary medicine.

Animal hosts facilitate the spread of *Cryptosporidium* spp. differently according to their level of relationship with humans. This meta-analysis of pooled data indicated that there were more cattle studies as compared to other animal hosts. Sheep had the highest PPE (31.5%) as compared to cattle (11.4%). Similar results have been reported in India, where sheep had a 35.0% prevalence compared to cattle, of 5.0% prevalence [74]. Additionally, Odenrian and Ademola [75] observed that cattle (26.1%) in Nigeria appeared to be more exposed to *Cryptosporidium* spp. infections as compared to other domestic animals.

We observed an overall PPE of *Cryptosporidium* spp. of 18.1% in our present human study. Similarly, this result has been reported in studies from the southern region (20.0%), the Oromia (18.0%) province in Ethiopia and in humans from southern Africa (6.8%) [76,77]. Urban areas are known to have access to potable water with better sanitation practices, which can lower the spread of cryptosporidiosis [78]. In the current study, the northern region of South Africa had a lower prevalence (16.9%) compared to the southern region (19.8%), which we believe is due to better sanitary practices in the north than the resource-poor southern region. This is in accordance with the findings reported by Kalantari et al. [79]. Our analysis indicated an increase in the prevalence of *Cryptosporidium* spp. with decreasing sample size. This accounted for the increase in prevalence from 1981 to 2010 and then a decline in the 2011–2020 period, which could have been due to an increase in the population sample size, which agrees with the findings in Ethiopia [80].

All characterized *Cryptosporidium* spp. were detected by serological and molecular techniques and our findings showed *C. parvum* to be the most abundant species infecting humans. These findings are in agreement with the findings obtained in Iran, whereby *C. parvum* (84.4%) was the most prevalent species followed by *C. hominis* (13.4%) [79]. However, other studies have reported *C. hominis* as the most common *Cryptosporidium* spp. that infected humans followed by *C. parvum* in Malawi and India [81,82]. The present study observed a higher PPE for the population group aged 26–45 years, followed by <6 months–25 years and the lowest in the >45 years age group. Similar results were obtained whereby the 1–25 years (3.0%; 15.4%) group had high *Cryptosporidium* spp. prevalence as compared to the >45 years group (0.4%; 4.0%) in studies from Iran and Scotland [83,84]. Interestingly, the literature generally indicated that younger children were more susceptible to contracting *Cryptosporidium* spp. infections as compared to other older age groups [85]. This is because children tend to ignore hygiene while playing outside in grounds that might harbor zoonotic microorganisms of fecal origin [86]. Moreover, higher prevalence among children may reflect a lack of immunity as compared to older groups who acquire it due to exposure to *Cryptosporidium* spp. infection during their lifetime due to activities such as farming and swimming [87,88].

Our results revealed that females (41.1%) had a higher PPE compared to males (38.1%), which is consistent with previous reports from villages around Lake Atitlan, Guatemala, and in Delta State, Nigeria, where prevalences of 42.9% and 3.5% were found in females, respectively, and of 24.1%, and 2.1% in males, respectively [88,89]. However, higher prevalence was observed in males (13.0% and 13.3%) compared to females (6.1% and 7.1%) as recorded in Zambia and Pakistan, respectively [90,91]. This high prevalence in females could be associated with a lack of access to clean water, participation in day care, conducting house chores including cleaning and washing clothes and sometimes bad sanitary activities due to socioeconomic conditions [92]. Historically, it is rare for males to consult health practitioners whenever they are ill, and they will either try home remedies, while females consult practitioners for most health complications happening in their bodies, and hence, they appear to have higher records of infection [93,94].

Undoubtably, available data have shown that the *Cryptosporidium* parasite is an opportunistic infection, particularly in immunosuppressed individuals [95]. Our results indicated that HIV+ individuals (59.3%) were more exposed to infection with *Cryptosporidium* spp. as compared to HIV− individuals (39.8%). The peak occurrence of *Cryptosporidium* spp. in HIV+ individuals was consistent with previous observations from Uganda, which had 73.6% and 5.9% HIV+ and HIV− individuals, respectively [96]. Our findings appeared to be higher as compared to those reported in Nigeria [97] and also fell within the epidemiological range of the world's rate (0–78.1%) for *Cryptosporidium* spp. infections [98]. This relatively high PPE of *Cryptosporidium* spp. infection could be linked to poor hygiene practices, water scarcity, close contact with animals and high rate of immunocompromised individuals [24,99].

Numerous diagnostic methods can be used to detect *Cryptosporidium* infection in humans and animals worldwide including histology, immunology, microscopy and molecular techniques [100]. The findings from this study suggest that the frequently applied diagnostic method for *Cryptosporidium* species in South Africa was microscopy, followed by PCR, RDT, ELISA, qPCR, IFAT and LAMP, and the least used was PCR-RFLP. This agrees with documented reports by Kalantari et al. [80] in Iran, where they reported microscopy as the most employed diagnostic approach for the detection of *Cryptosporidium* spp. infection. Mohebali et al. [81] detected similar *Cryptosporidium* prevalence (10%) in Ethiopia using a modified Ziehl–Neelsen staining diagnostic technique. This increase in the detection of *Cryptosporidium* spp. prevalence might be linked to the use of serological and molecular techniques, which are more sensitive and less time-consuming.

This study confirmed the prevalence of *Cryptosporidium* spp. in animals, humans and the environment (soil and water). Our results highlight the importance of "One Health" because *Cryptosporidium* spp. has been proven to exist in humans, animals and the environment. Future researchers should be encouraged to use the "One Health" approach to developing methods that explicitly examine the relationships between human–animal–environment frameworks, with a particular focus on *Cryptosporidium* infections.

5. Highlights and Limitations

This systematic review and meta-analysis study used good-quality studies to present a summary of unbiased results of both animal and human *Cryptosporidium* prevalence in South Africa and revealed that there are some provinces where *Cryptosporidium* spp. infections have not yet been studied. With respect to humans, there are no studies published in the Northern Cape, Western Cape and Free State provinces, while for animals, there are no published studies in the KwaZulu-Natal, Northern Cape, Western Cape and Eastern Cape provinces. Additionally, this study demonstrated the impact of the domestic–wildlife–human interface on the prevalence of *Cryptosporidium* spp. infection and distribution, which emphasizes the need for studies focusing on a "One Health" approach to produce multi-data covering animal and human hosts as well as the environment, such as contaminated water and soil.

It must be noted that this systematic review and meta-analysis had some limitations. The study by Lukasova et al. [34] had a very small number of samples in the Gauteng (n = 1) and North West (n = 8) provinces, which were examined for the presence of *Cryptosporidium* spp., and there was also lack of similar studies in some provinces (e.g., the Northern Cape province with zero publications/representation in both animal and human studies). Additionally, the majority of studies included in this meta-analysis were conducted using microscopic diagnostic techniques, which have a lesser diagnostic sensitivity as compared to molecular and immunological techniques. Moreover, there was no repeated fecal sample examination conducted, which might have resulted in possible false-positive or -negative results. This means that the reported prevalence might have been underestimated.

Due to the small number of studies on some subgroups such as (i) various *Cryptosporidium* species such as *C. andersoni, C. bovis, C. ubiquitum, C. hominis* and *C. muris;* (ii) various hosts such as buffaloes, dogs, elephants, goats, sheep and impala; (iii) the wide use of various diagnostic methods such as IFAT, LAMP and PCR-RFLP; and (iv) the lack of studies in the period 1991–2000, the identified formal assessment of publication bias using funnel plots and Begg's rank and Mazumdar test was not possible. However, meta-analyses that include fewer than 10 studies or have a high degree of heterogeneity between studies may lead to misleading results from these assessment tools. When there is a high level of heterogeneity, it is very difficult to evaluate the actual results of statistically significant publication bias tests. Because there is high heterogeneity across analyses, readers should exercise caution when interpreting pooled analyses and subgroups.

6. Conclusions

The data generated in this study indicated that the prevalence of *Cryptosporidium* spp. infections was slightly higher in animal than human hosts in South Africa. However, we further observed that there was a lack of *Cryptosporidium* spp. prevalence studies for both animals and humans in some of the provinces. Furthermore, human infections were prevalent in HIV+ and immunocompromised patients, emphasizing that they were a high-risk group for opportunistic diseases such as cryptosporidiosis. However, the results of the included studies varied greatly, between their sampling methods, sample sizes, study locations and diagnostic techniques used, and this needs to be taken into account and may explain some of the inconsistencies. The occurrence and prevalence of *Cryptosporidium* spp. infections in animals is of public health importance, hence, more studies involving both domestic and wild animals are required. The findings of this study suggest the necessity for a "One Health" strategy to promote public hygiene, animal husbandry and regular screening for *Cryptosporidium* spp. infections of animals, humans and the environment (soil and water) in all nine provinces of South Africa.

Author Contributions: M.T., T.E.O. and O.T. conceptualized the study. M.T. and T.R. collected data and conducted data analyses. T.E.O., T.R., D.J.G., S.J.N. and O.T. edited the draft manuscript. All authors have read and agreed to the published version of the manuscript.

Funding: This research work did not receive any specific grant from funding agencies.

Data Availability Statement: The datasets generated and/or analyzed during the current study are available from the corresponding author on reasonable request.

Acknowledgments: The first author was financially supported by the National Research Foundation (NRF) of South Africa block grant (UID number: 129080) and the North-West University (NWU) Postgraduate student study bursaries.

Conflicts of Interest: The authors declare no conflict of interest.

References

1. Burns, J.K. The mental health gap in South Africa: A human rights issue. *Equal. Rights Rev.* **2011**, *6*, 99–113.
2. Stiegler, N.; Bouchard, J.P. South Africa: Challenges and successes of the COVID-19 lockdown. *Ann. Médico-Psychol.* **2020**, *178*, 695–698. [CrossRef] [PubMed]
3. Hammond-Aryee, K.; Esser, M.; Van Helden, P.D. Toxoplasma gondii sero-prevalence studies on humans and animals in Africa. *S. Afr. Fam. Pract.* **2014**, *56*, 119–124. [CrossRef]
4. Abubakar, I.; Aliyu, S.H.; Arumugam, C.; Usman, N.K.; Hunter, P.R. Treatment of cryptosporidiosis in immunocompromised individuals: Systematic review and meta-analysis. *Br. J. Clin. Pharmacol.* **2007**, *63*, 387–393. [CrossRef] [PubMed]
5. Hatam-Nahavandi, K.; Ahmadpour, E.; Carmena, D.; Spotin, A.; Bangoura, B.; Xiao, L. *Cryptosporidium* infections in terrestrial ungulates with focus on livestock: A systematic review and meta-analysis. *Parasit. Vectors* **2019**, *12*, 453. [CrossRef]
6. Brook, E.; Hart, C.A.; French, N.; Christley, R. Prevalence and risk factors for *Cryptosporidium* spp. infection in young calves. *Vet. Parasitol.* **2008**, *152*, 46–52. [CrossRef]
7. Gambhir, I.S.; Jaiswal, J.P.; Nath, G. Significance of *Cryptosporidium* as an aetiology of acute infectious diarrhoea in elderly Indians. *Trop. Med. Int. Health* **2003**, *8*, 415–419. [CrossRef]
8. Hawker, J.; White, J.; Catchpole, M. Quarterly communicable disease review April to June 2000. *J. Public Health* **2000**, *22*, 546–550. [CrossRef]
9. Khalil, I.A.; Troeger, C.; Rao, P.C.; Blacker, B.F.; Brown, A.; Brewer, T.G.; Colombara, D.V.; De Hostos, E.L.; Engmann, C.; Guerrant, R.L.; et al. Morbidity, mortality, and long-term consequences associated with diarrhoea from *Cryptosporidium* infection in children younger than 5 years: A meta-analyses study. *Lancet Glob. Health* **2018**, *6*, e758–e768. [CrossRef]
10. Bouzid, M.; Steverding, D.; Tyler, K.M. Detection and surveillance of waterborne protozoan parasites. *Curr. Opin. Biotechnol.* **2008**, *19*, 302–306. [CrossRef]
11. Chalmers, R.M.; Campbell, B.; Crouch, N.; Davies, A.P. Clinical laboratory practices for detection and reporting of *Cryptosporidium* in community cases of diarrhoea in the United Kingdom, 2008. *Eurosurveillance* **2010**, *15*, 19731. [CrossRef] [PubMed]
12. Naciri, M.; Lefay, M.P.; Mancassola, R.; Poirier, P.; Chermette, R. Role of *Cryptosporidium* parvum as a pathogen in neonatal diarrhoea complex in suckling and dairy calves in France. *Vet. Parasitol.* **1999**, *85*, 245–257. [CrossRef] [PubMed]
13. Sow, S.O.; Muhsen, K.; Nasrin, D.; Blackwelder, W.C.; Wu, Y.; Farag, T.H.; Saha, D. The burden of *Cryptosporidium* diarrheal disease among children< 24 months of age in moderate/high mortality regions of sub-Saharan Africa and South Asia, utilizing data from the Global Enteric Multicenter Study (GEMS). *PLoS Negl. Trop. Dis.* **2016**, *10*, e0004729. [CrossRef]
14. Steele, A.D.; Gove, E.; Meewes, P.J. Cryptosporidiosis in white patients in South Africa. *J. Infect.* **1989**, *19*, 281–285. [CrossRef] [PubMed]
15. Hunter, P.R.; Thompson, R.A. The zoonotic transmission of Giardia and *Cryptosporidium*. *Int. J. Parasitol.* **2005**, *35*, 1181–1190. [CrossRef]
16. Xiao, L.; Fayer, R.; Ryan, U.; Upton, S.J. *Cryptosporidium* taxonomy: Recent advances and implications for public health. *Clin. Microbiol. Rev.* **2004**, *17*, 72–97. [CrossRef]
17. Wang, R.; Wang, H.; Sun, Y.; Zhang, L.; Jian, F.; Qi, M.; Xiao, L. Characteristics of *Cryptosporidium* transmission in pre-weaned dairy cattle in Henan, China. *J. Clin. Microbiol.* **2011**, *49*, 1077–1082. [CrossRef]
18. Vermeulen, L.C.; Benders, J.; Medema, G.; Hofstra, N. Global *Cryptosporidium* loads from livestock manure. *Environ. Sci. Technol.* **2017**, *51*, 8663–8671. [CrossRef]
19. Oates, S.C.; Miller, M.A.; Hardin, D.; Conrad, P.A.; Melli, A.; Jessup, D.A.; Miller, W.A. Prevalence, environmental loading, and molecular characterization of *Cryptosporidium* and Giardia isolates from domestic and wild animals along the Central California Coast. *Appl. Environ. Microbiol.* **2012**, *78*, 8762–8772. [CrossRef]
20. Silverlås, C.; Bosaeus-Reineck, H.; Näslund, K.; Björkman, C. Is there a need for improved *Cryptosporidium* diagnostics in Swedish calves? *Int. J. Parasitol.* **2013**, *43*, 155–161. [CrossRef]
21. Efstratiou, A.; Ongerth, J.; Karanis, P. Evolution of monitoring for Giardia and *Cryptosporidium* in water. *Water Res.* **2017**, *123*, 96–112. [CrossRef]
22. Ryan, U.; Hijjawi, N.; Xiao, L. Foodborne cryptosporidiosis. *Inter. J. Parasitol.* **2018**, *48*, 1–12. [CrossRef] [PubMed]
23. Havelaar, A.H.; Melse, J.M. *Quantifying Public Health Risk in the WHO Guidelines for Drinking-Water Quality: A Burden of Disease Approach*; National Institute of Public Health and the Environment RIVM report 734301022/2003; RIVM: Bilthoven, The Netherlands, 2003.
24. Shaposhnik, E.G.; Abozaid, S.; Grossman, T.; Marva, E.; On, A.; Azrad, M.; Peretz, A. The prevalence of *Cryptosporidium* among children hospitalized because of gastrointestinal symptoms and the efficiency of diagnostic methods for *Cryptosporidium*. *Am. J. Trop. Med. Hyg.* **2019**, *101*, 160–163. [CrossRef]

25. Omolabi, K.F.; Odeniran, P.O.; Soliman, M.E. A meta-analysis of *Cryptosporidium* species in humans from southern Africa (2000–2020). *J. Parasit. Dis.* **2022**, *46*, 304–316. [CrossRef]
26. Ramatla, T.; Tawana, M.; Lekota, K.E.; Thekisoe, O. Antimicrobial resistance genes of Escherichia coli, a bacterium of "One Health" importance in South Africa: Systematic review and meta-analysis. *AIMS Microbiol.* **2023**, *9*, 75. [CrossRef] [PubMed]
27. Tawana, M.; Onyiche, T.E.; Ramatla, T.; Thekisoe, O. A "One Health" perspective of Africa-wide distribution and prevalence of Giardia species in humans, animals and waterbodies: A systematic review and meta-analysis. *Parasitology* **2023**, *150*, 769–780. [CrossRef] [PubMed]
28. Ramatla, T.; Tawana, M.; Mphuthi, M.B.; Onyiche, T.E.; Lekota, K.E.; Monyam, M.C.; Ndou, R.; Bezuidenhout, C.; Thekisoe, O. Prevalence and antimicrobial resistance profiles of Campylobacter species in South Africa: A "One Health" approach using systematic review and meta-analysis. *Int. J. Infect. Dis.* **2022**, *125*, 294–304. [CrossRef]
29. Bakheit, M.A.; Torra, D.; Palomino, L.A.; Thekisoe, O.M.; Mbati, P.A.; Ongerth, J.; Karanis, P. Sensitive and specific detection of *Cryptosporidium* species in PCR-negative samples by loop-mediated isothermal DNA amplification and confirmation of generated LAMP products by sequencing. *Vet. Parasitol.* **2008**, *158*, 11–22. [CrossRef]
30. Hlungwani, H.A. Molecular Detection and Identification of *Cryptosporidium* Species Isolated from Human and Animal Sources in Limpopo and Gauteng Provinces. Doctoral Dissertation, University of Venda, Thohoyandou, South Africa, 2017. Available online: https://univendspace.univen.ac.za/handle/11602/917 (accessed on 13 February 2022).
31. Lukášová, R.; Halajian, A.; Bártová, E.; Kobédová, K.; Swanepoel, L.H.; O'Riain, M.J. The occurrence of some nonblood protozoan parasites in wild and domestic mammals in South Africa. *J. Wildl. Dis.* **2018**, *54*, 392–396. [CrossRef]
32. Samie, A.; Tsipa, M.A.; Bessong, P. The epidemiology of *Cryptosporidium* in cats and dogs in the Thohoyandou region, South Africa. *Afr. J. Microbiol. Res.* **2013**, *7*, 2510–2518.
33. Samie, A.; Hlungwani, A.H.; Mbati, P.A. Prevalence and risk factors of *Cryptosporidium* species among domestic animals in rural communities in Northern South Africa. *Trop. Biomed.* **2017**, *34*, 636–647.
34. Samra, N.A.; Jori, F.; Samie, A.; Thompson, P. The prevalence of *Cryptosporidium* spp. oocysts in wild mammals in the Kruger National Park, South Africa. *Vet. Parasitol.* **2011**, *175*, 155–159. [CrossRef] [PubMed]
35. Samra, N.A.; Thompson, P.N.; Jori, F.; Frean, J.; Poonsamy, B.; Du Plessis, D.; Mogoye, B.; Xiao, L. Genetic characterization of *Cryptosporidium* spp. in diarrhoeic children from four provinces in South Africa. *Zoonoses Public Health* **2013**, *60*, 154–159. [CrossRef] [PubMed]
36. Samra, N.A.; Jori, F.; Cacciò, S.M.; Frean, J.; Poonsamy, B.; Thompson, P.N. *Cryptosporidium* genotypes in children and calves living at the wildlife or livestock interface of the Kruger National Park, South Africa. *Onderstepoort J. Vet. Res.* **2016**, *83*, a1024. [CrossRef] [PubMed]
37. Syakalima, M.; Noinyane, M.I.; Ramaili, T.; Motsei, L.; Nyirenda, M. A coprological assessment of cryptosporidiosis and giardiosis in pigs of Mafikeng villages, North West province of South Africa. *Indian J. Anim. Res.* **2015**, *49*, 132–135. [CrossRef]
38. Vink, J.J.W.G. Cryptosporidiosis in Pre-Weaned Calves at the Wildlife/Livestock Interface of the Kruger National Park, South Africa. Master's Thesis, Utrecht University VET, Utrecht, The Netherlands, 2015. Available online: https://studenttheses.uu.nl/bitstream/handle/20.500.12932/19531/Onderzoek%20Cryptosporidiosis%20South%20Africa.docx?sequence=2&isAllowed=y (accessed on 19 March 2022).
39. Abebe, L.S.; Smith, J.A.; Narkiewicz, S.; Oyanedel-Craver, V.; Conaway, M.; Singo, A.; Amidou, S.; Mojapelo, P.; Brant, J.; Dillingham, R. Ceramic water filters impregnated with silver nanoparticles as a point-of-use water-treatment intervention for HIV-positive individuals in Limpopo Province, South Africa: A pilot study of technological performance and human health benefits. *J. Water Health* **2014**, *12*, 288–300. [CrossRef]
40. Bartelt, L.A.; Sevilleja, J.E.; Barrett, L.J.; Warren, C.A.; Guerrant, R.L.; Bessong, P.O.; Dillingham, R.; Samie, A. High anti-*Cryptosporidium* parvum IgG seroprevalence in HIV-infected adults in Limpopo, South Africa. *Am. J. Trop. Med. Hyg.* **2013**, *89*, 531–534. [CrossRef]
41. Becker, S.L.; Müller, I.; Mertens, P.; Herrmann, M.; Zondie, L.; Beyleveld, L.; Gerber, M.; du Randt, R.; Pühse, U.; Walter, C.; et al. PCR-based verification of positive rapid diagnostic tests for intestinal protozoa infections with variable test band intensity. *Acta Trop.* **2017**, *174*, 49–55. [CrossRef]
42. Berkowitz, F.E.; Vallabh, W.; Buqwana, A.; Heney, C. Cryptosporidiosis in black South African children. *S. Afr. Med. J.* **1988**, *74*, 272. [CrossRef]
43. Etinosa, O.B. Immunological and Molecular Characterization of *Cryptosporidium* Species in HIV-Positive and HIV-Negative Diarrhoea Patients in the Nkonkobe Municipality of the Eastern Cape Province of South Africa: A Pilot Study. Doctoral Dissertation, University of Fort Hare, Alice, South Africa, 2010. Available online: https://core.ac.uk/reader/145049160 (accessed on 22 June 2022).
44. Fripp, P.J.; Bothma, M.T.; Crewe-Brown, H.H. Four years of cryptosporidiosis at GaRankuwa Hospital. *J. Infect.* **1991**, *23*, 93–100. [CrossRef]
45. Geyer, A.; Crewe-Brown, H.H.; Greeff, A.S.; Fripp, P.J.; Steele, A.D.; Van, T.S.; Clay, C.G. The microbial aetiology of summer paediatric gastroenteritis at Ga-Rankuwa Hospital in South Africa. *East. Afr. Med. J.* **1993**, *70*, 78–81.
46. Htun, N.S.N.; Odermatt, P.; Müller, I.; Yap, P.; Steinmann, P.; Schindler, C.; Gerber, M.; Du Randt, R.; Walter, C.; Pühse, U.; et al. Association between gastrointestinal tract infections and glycated haemoglobin in school children of poor neighbourhoods in Port Elizabeth, South Africa. *PLoS Negl. Trop. Dis.* **2018**, *12*, e0006332. [CrossRef] [PubMed]

47. Jarmey-Swan, C.; Bailey, I.W.; Howgrave-Graham, A.R. Ubiquity of the water-borne pathogens, *Cryptosporidium* and Giardia, in KwaZulu-Natal populations. *Water S. Afr.* **2001**, *27*, 57–64. [CrossRef]
48. Leav, B.A.; Mackay, M.R.; Anyanwu, A.; O'Connor, R.M.; Cevallos, A.M.; Kindra, G.; Rollins, N.C.; Bennish, M.L.; Nelson, R.G.; Ward, H.D. Analysis of sequence diversity at the highly polymorphic Cpgp40/15 locus among *Cryptosporidium* isolates from human immunodeficiency virus-infected children in South Africa. *Infect. Immun.* **2002**, *70*, 3881–3890. [CrossRef] [PubMed]
49. Moodley, D.; Jackson, T.F.H.G.; Gathiram, V.; Van Den Ende, J. *Cryptosporidium* infections in children in Durban Seasonal variation, age distribution and disease status. *S. Afr. Med. J.* **1991**, *79*, 296–297.
50. Msolo, L.; Iweriebor, B.C.; Okoh, A.I. Rotavirus and *Cryptosporidium* pathogens as etiological proxies of gastroenteritis in some pastoral communities of the Amathole District Municipality, Eastern Cape, South Africa. *BMC Res. Notes* **2020**, *13*, 1–6. [CrossRef]
51. Müller, I.; Yap, P.; Steinmann, P.; Damons, B.P.; Schindler, C.; Seelig, H.; Htun, N.S.; Probst-Hensch, N.; Gerber, M.; du Randt, R.; et al. Intestinal parasites, growth and physical fitness of schoolchildren in poor neighbourhoods of Port Elizabeth, South Africa: A cross-sectional survey. *Parasit. Vectors* **2016**, *9*, 488. [CrossRef]
52. Omoruyi, B.E.; Nwodo, U.U.; Udem, C.S.; Okonkwo, F.O. Comparative diagnostic techniques for *Cryptosporidium* infection. *Molecules* **2014**, *19*, 2674–2683. [CrossRef]
53. Samie, A.; Bessong, P.O.; Obi, C.L.; Sevilleja, J.E.A.D.; Stroup, S.; Houpt, E.; Guerrant, R.L. *Cryptosporidium* species: Preliminary descriptions of the prevalence and genotype distribution among school children and hospital patients in the Venda region, Limpopo Province, South Africa. *Exp. Parasitol.* **2006**, *114*, 314–322. [CrossRef]
54. Samie, A.; Obi, C.L.; Tzipori, S.; Weiss, L.M.; Guerrant, R. Microsporidiosis in South Africa: PCR detection in stool samples of HIV-positive and HIV-negative individuals and school children in Vhembe district, Limpopo Province. *Trans. R. Soc. Trop. Med. Hyg.* **2007**, *101*, 547–554. [CrossRef]
55. Samie, A.; Guerrant, R.L.; Barrett, L.; Bessong, P.O.; Igumbor, E.O.; Obi, C.L. Prevalence of intestinal parasitic and bacterial pathogens in diarrhoeal and non-diarroeal human stools from Vhembe district, South Africa. *J. Health Popul. Nutr.* **2009**, *27*, 739. [CrossRef] [PubMed]
56. Samie, A.; Bessong, P.O.; Obi, C.L.; Dillingham, R.; Guerrant, R.L. Bacterial and Parasitic Agents of Infectious Diarrhoea in the Era of HIV and AIDS-The Case of a Semi Rural Community in South Africa. In *Microbes, Viruses and Parasites in AIDS Process*; Ar Zajae, V., Ed.; IntechOpen: Rijeka, Croatia, 2011; ISBN 978-938-307-601-0.
57. Samie, A.; Makuwa, S.; Mtshali, S.; Potgieter, N.; Thekisoe, O.; Mbati, P.; Bessong, P.O. Parasitic infection among HIV/AIDS patients at Bela-Bela clinic, Limpopo province, South Africa with special reference to *Cryptosporidium*. *Southeast Asian J. Trop. Med. Public Health* **2014**, *45*, 783. [PubMed]
58. Samra, N.A.; Jori, F.; Xiao, L.; Rikhotso, O.; Thompson, P.N. Molecular characterization of *Cryptosporidium* species at the wildlife/livestock interface of the Kruger National Park, South Africa. *Comp. Immunol. Microbiol. Infect. Dis.* **2013**, *36*, 295–302. [CrossRef] [PubMed]
59. Smith, G.; Van den Ende, J. Cryptosporidiosis among black children in hospital in South Africa. *J. Infect.* **1986**, *13*, 25–30. [CrossRef]
60. Steele, A.D.; Geyer, A.; Alexander, J.J.; Crewe-Brown, H.H.; Fripp, P.J. Enteropathogens isolated from children with gastro-enteritis at Ga-Rankuwa Hospital, South Africa. *Ann. Trop. Paediatr.* **1988**, *8*, 262–267. [CrossRef]
61. Trönnberg, L.; Hawksworth, D.; Hansen, A.; Archer, C.; Stenström, T.A. Household-based prevalence of helminths and parasitic protozoa in rural KwaZulu-Natal, South Africa, assessed from faecal vault sampling. *Trans. R. Soc. Trop. Med. Hyg.* **2010**, *104*, 646–652. [CrossRef]
62. Walters, I.N.; Miller, N.M.; Van den Ende, J.; Dees, G.C.; Taylor, L.A.; Taynton, L.F.; Bennett, K.J. Outbreak of cryptosporidiosis among young children attending a day-care centre in Durban. *S. Afr. Med. J.* **1988**, *74*, 496–499. [PubMed]
63. Wittenberg, D.F.; Smith, E.G.; Den Ende, J.V.; Becker, P.J. *Cryptosporidium*-associated diarrhoea in children. *Ann. Trop. Paediatr.* **1987**, *7*, 113–117. [CrossRef] [PubMed]
64. Ng, J.; Yang, R.; McCarthy, S.; Gordon, C.; Hijjawi, N.; Ryan, U. Molecular characterization of *Cryptosporidium* and Giardia in pre-weaned calves in Western Australia and New South Wales. *Vet. Parasitol.* **2011**, *176*, 145–150. [CrossRef]
65. Zambrano, L.D.; Levy, K.; Menezes, N.P.; Freeman, M.C. Human diarrhoea infections associated with domestic animal husbandry: A systematic review and meta-analysis. *Trans. R. Soc. Trop. Med. Hyg.* **2014**, *108*, 313–325. [CrossRef]
66. Zhang, W.; Wang, R.; Yang, F.; Zhang, L.; Cao, J.; Zhang, X.; Ling, H.; Liu, A.; Shen, Y. Distribution and genetic characterizations of *Cryptosporidium* spp. in pre-weaned dairy calves in North-eastern China's Heilongjiang Province. *PLoS ONE* **2013**, *8*, e54857. [CrossRef]
67. Matilla, F.; Velleman, Y.; Harrison, W.; Nevel, M. Animal influence on water, sanitation and hygiene measures for zoonosis control at the household level: A systematic literature review. *PLoS Negl. Trop. Dis.* **2018**, *12*, e0006619. [CrossRef] [PubMed]
68. Bjorkman, C. Disinfection with hydrated lime may help manage cryptosporidiosis in calves. *Vet. Parasitol.* **2018**, *264*, 58–63. [CrossRef] [PubMed]
69. Grinberg, A. Controlling the onset of natural cryptosporidiosis in calves with paromomycin sulphate. *Vet. Rec. J.* **2002**, *151*, 606–608. [CrossRef] [PubMed]
70. Kay, D. Effectiveness of best management practices for attenuating the transport of livestock derived pathogens within catchments. In *Animal Waste Water Quality and Human Health*; Dufour, A., Bartram, J., Eds.; IWA Publishing: London, UK, 2012; pp. 195–255.

71. Ligda, P.; Claerebout, E.; Kostopoulou, D.; Zdragas, A.; Casaert, S.; Robertson, L.J.; Sotiraki, S. *Cryptosporidium* and Giardia in surface water and drinking water: Animal sources and towards the use of a machine-learning approach as a tool for predicting contamination. *Environ. Pollut.* **2020**, *264*, 114766. [CrossRef] [PubMed]
72. Xiao, L.; Bern, C.; Limor, J.; Sulaiman, I.; Roberts, J.; Checkley, W.; Cabrera, L.; Gilman, R.H.; Lal, A.A. Identification of 5 types of *Cryptosporidium* parasites in children in Lima, Peru. *J. Infect Dis.* **2001**, *183*, 492–497. [CrossRef]
73. Krumkamp, R.; Aldrich, C.; Maiga-Ascofare, O.; Mbwana, J.; Rakotozandrindrainy, N.; Borrmann, S.; Caccio, S.M.; Rakotozandrindrainy, R.; Adegnika, A.A.; Lusingu, J.; et al. Transmission of *Cryptosporidium* spp. among human and animal local contact networks in sub-Saharan Africa: A multi-country study. *Clin. Infect. Dis.* **2020**, *in press*. [CrossRef]
74. Daniels, M.E.; Shrivastava, A.; Smith, W.A.; Sahu, P.; Odagiri, M.; Misra, P.R.; Panigrahi, P.; Suar, M.; Clasen, T.; Jenkins, M.W. *Cryptosporidium* and Giardia in humans, domestic animals, and village water sources in rural India. *Am. J. Trop. Med. Hyg.* **2015**, *93*, 596–600. [CrossRef]
75. Odeniran, P.O.; Ademola, I.O. Epidemiology of *Cryptosporidium* infection in different hosts in Nigeria: A meta-analysis. *Parasitol. Int.* **2019**, *71*, 194–206. [CrossRef]
76. Raga, D.K.; Menkir, S.; Getachew, Y. Prevalence of Isospora belli and *Cryptosporidium* parvum Infections among HIV Sero-positive Patients in Asella Hospital, Central Ethiopia. *Res. Rev. A J. Immunol.* **2014**, *4*, 17–23.
77. Assefa, S.; Erko, B.; Medhin, G.; Assefa, Z.; Shimelis, T. Intestinal parasitic infections in relation to HIV/AIDS status, diarrhoea and CD4 T-cell count. *BMC Infect. Dis.* **2009**, *9*, 155. [CrossRef] [PubMed]
78. Berahmat, R.; Spotin, A.; Ahmadpour, E.; Mahami-Oskouei, M.; Rezamand, A.; Aminisani, N.; Ghojazadeh, M.; Ghoyounchi, R.; Mikaeili-Galeh, T. Human cryptosporidiosis in Iran: A systematic review and meta-analysis. *Parasitol. Res.* **2017**, *116*, 1111–1128. [CrossRef] [PubMed]
79. Kalantari, N.; Ghaffari, S.; Bayani, M. *Cryptosporidium* spp. infection in Iranian children and immunosuppressive patients: A systematic review and meta-analysis. *Casp. J. Intern. Med.* **2018**, *9*, 106. [CrossRef]
80. Mohebali, M.; Yimam, Y.; Woreta, A. *Cryptosporidium* infection among people living with HIV/AIDS in Ethiopia: A systematic review and meta-analysis. *Pathog. Glob. Health* **2020**, *114*, 183–193. [CrossRef] [PubMed]
81. Korpe, P.S.; Haque, R.; Gilchrist, C.; Valencia, C.; Niu, F.; Lu, M.; Ma, J.Z.; Petri, S.E.; Reichman, D.; Kabir, M.; et al. Natural history of cryptosporidiosis in a longitudinal study of slum-dwelling Bangladeshi children: Association with severe malnutrition. *PLoS Negl. Trop. Dis.* **2016**, *10*, e0004564. [CrossRef]
82. Kattula, D.; Jeyavelu, N.; Prabhakaran, A.D.; Premkumar, P.S.; Velusamy, V.; Venugopal, S.; Geetha, J.C.; Lazarus, R.P.; Das, P.; Nithyanandhan, K.; et al. Natural history of cryptosporidiosis in a birth cohort in southern India. *Clin. Infect. Dis.* **2017**, *64*, 347–354. [CrossRef]
83. Mirzaei, M. Prevalence of *Cryptosporidium* sp. infection in diarrheic and non-diarrheic humans in Iran. *Korean J. Parasitol.* **2007**, *45*, 133. [CrossRef]
84. Casemore, D.P. Human cryptosporidiosis. In *Recent Advances in Infection*; Reeves, D., Geddes, A., Eds.; Churchill Livingstone: Edinburgh, UK, 1988; pp. 209–236.
85. Tariuwa, H.O.; Ajogi, I.; Ejembi, C.L.; Awah, I.J.; Green, P.A.; Fadipe, E.O.; Odoba, M.B. Incidence of *Cryptosporidium* infection in port-harcourt rivers state Nigeria based on regular contact with domestic animals. *Niger. Vet. J.* **2007**, *28*, 1–5. [CrossRef]
86. Laubach, H.E.; Bentley, C.Z.; Ginter, E.L.; Spalter, J.S.; Jensen, L.A. A study of risk factors associated with the prevalence of *Cryptosporidium* in villages around Lake Atitlan, Guatemala. *Brazil. J. Infect. Dis.* **2004**, *8*, 319–323. [CrossRef]
87. Garvey, P.; McKeown, P. Epidemiology of human cryptosporidiosis in Ireland, 2004-2006: Analysis of national notification data. *Eurosurveillance* **2009**, *14*, 19128. [CrossRef]
88. Siwila, J. Prevalence, Characterization and Transmission of *Cryptosporidium* Species Between Animals and Humans on Dairy Farms in Zambia. Doctoral Dissertation, University of Zambia, Lusaka, Zambia, 2012. Available online: http://dspace.unza.zm/handle/123456789/1754 (accessed on 11 August 2022).
89. Erhabor, O.; Obunge, O.; Awah, I. Cryptosporidiosis among HIV-infected persons in the Niger Delta of Nihgeria. *Niger. J. Med.* **2011**, *20*, 372–375. [PubMed]
90. Mumtaz, S.; Ahmed, J.; Ali, L. Frequency of *Cryptosporidium* infection in children under five years of age having diarrhoea in the North West of Pakistan. *Afr. J. Biotechnol.* **2010**, *9*, 1230–1235.
91. Sinyangwe, N.N.; Siwila, J.; Muma, J.B.; Chola, M.; Michelo, C. Factors associated with *Cryptosporidium* infection among adult HIV positive population in contact with livestock in Namwala District, Zambia. *Front. Public Health* **2020**, *8*, 74. [CrossRef]
92. Raphael, M.; Mathias, A.; Tirah, G.; Mafindi, M. Prevalence of *Cryptosporidium* species in HIV positive and negative patients attending Hong general hospital and Michika general hospital, Adamawa state, Nigeria. *Am. J. Eng. Res.* **2017**, *6*, 25–28.
93. Galdas, P.M.; Cheater, F.; Marshall, P. Men and health help-seeking behaviour: Literature review. *J. Adv. Nurs.* **2005**, *49*, 616–623. [CrossRef]
94. Noone, J.H.; Stephens, C. Men, masculine identities, and health care utilisation. *Sociol. Health Illn.* **2008**, *30*, 711–725. [CrossRef]
95. Laksemi, D.A.; Suwanti, L.T.; Mufasirin, M.; Suastika, K.; Sudarmaja, M. Opportunistic parasitic infections in patients with human immunodeficiency virus/acquired immunodeficiency syndrome: A review. *Vet. World* **2019**, *13*, 716. [CrossRef]
96. Tumwine, J.K.; Kekitiinwa, A.; Bakeera-Kitaka, S.; Ndeezi, G.; Downing, R.; Feng, X.; Akiyoshi, D.E.; Tzipori, S. Cryptosporidiosis and microsporidiosis in Ugandan children with persistent diarrhoea with and without concurrent infection with the human immunodeficiency virus. *Am. J. Trop. Med. Hyg.* **2005**, *73*, 921–925. [CrossRef]

97. Karshima, S.N.; Karshima, M.N. Epidemiology of *Cryptosporidium* Infections among People Living with HIV/AIDS in Nigeria: Results of Systematic Review and Meta-analysis. *Acta Parasitol.* **2020**, *66*, 60–74. [CrossRef]
98. Wang, R.J.; Li, J.Q.; Chen, Y.C.; Zhang, L.X.; Xiao, L.H. Widespread occurrence of *Cryptosporidium* infections in patients with HIV/AIDS: Epidemiology, clinical feature, diagnosis, and therapy. *Acta Trop.* **2018**, *187*, 257–263. [CrossRef]
99. Nsagha, D.S.; Njunda, A.L.; Assob, N.J.C.; Ayima, C.W.; Tanue, E.A.; Kwenti, T.E. Intestinal parasitic infections in relation to CD4+ T cell counts and diarrhoea in HIV/AIDS patients with or without antiretroviral therapy in Cameroon. *BMC Infect. Dis.* **2015**, *16*, 9. [CrossRef] [PubMed]
100. Khurana, S.; Chaudhary, P. Laboratory diagnosis of cryptosporidiosis. *Trop. Parasitol.* **2018**, *8*, 2. [CrossRef] [PubMed]

microorganisms

MDPI

Article

Recombinant SAG2A Protein from *Toxoplasma gondii* Modulates Immune Profile and Induces Metabolic Changes Associated with Reduced Tachyzoite Infection in Peritoneal Exudate Cells from Susceptible C57BL/6 Mice

Thaíse Anne Rocha dos Santos [1], Mário Cézar de Oliveira [2], Edson Mario de Andrade Silva [3], Uener Ribeiro dos Santos [4], Monaliza Macêdo Ferreira [1], Ana Luísa Corrêa Soares [5], Neide Maria Silva [2], Tiago Antônio de Oliveira Mendes [3], Jamilly Azevedo Leal-Sena [1], Jair Pereira da Cunha-Júnior [2], Tiago Wilson Patriarca Mineo [2], José Roberto Mineo [2], Érica Araújo Mendes [6], Jane Lima-Santos [4,*] and Carlos Priminho Pirovani [1]

1 Centro de Biotecnologia e Genética, Universidade Estadual de Santa Cruz, Ilhéus 45662-900, BA, Brazil; tayseesantos@gmail.com (T.A.R.d.S.); monalizamacedo2@gmail.com (M.M.F.); millybio7@gmail.com (J.A.L.-S.); pirovani@uesc.br (C.P.P.)
2 Instituto de Ciências Biomédicas, Universidade Federal de Uberlândia, Uberlândia 38408-100, MG, Brazil; cezar_cle@yahoo.com.br (M.C.d.O.); nmsilva@ufu.br (N.M.S.); jair.cunha.junior@gmail.com (J.P.d.C.-J.); tiago.mineo@ufu.br (T.W.P.M.); jrmineo@ufu.br (J.R.M.)
3 Departamento de Bioquímica e Biologia Molecular, Universidade Federal de Viçosa, Viçosa 36570-900, MG, Brazil; mariodeandradee@gmail.com (E.M.d.A.S.); tiagoaomendes@ufv.br (T.A.d.O.M.)
4 Laboratório de Imunobiologia, Universidade Estadual de Santa Cruz, Ilhéus 45662-900, BA, Brazil; uener_edu@yahoo.com.br
5 Laboratory of Tropical Crop Improvement, Division of Crop Biotechnics, KU Leuven, 3000 Leuven, Belgium; anacorreasoares@outlook.com
6 Departamento de Microbiologia, Universidade de São Paulo, São Paulo 05508-000, SP, Brazil; ericaarmendes@gmail.com
* Correspondence: jlsantos@uesc.com.br

Abstract: Toxoplasmosis is a neglected disease that represents a significant public health problem. The antigenic profile of *T. gondii* is complex, and the immune response can lead to either susceptibility or resistance. Some antigens, such as surface antigen glycoprotein (SAG), are expressed on the surface of tachyzoite stages and interact with the host immune cells. In this study, we investigated the potential of the recombinant SAG2A protein of *T. gondii* to control parasitism and modulate the immune response in the peritoneal exudate cells (PECs) of both susceptible (C57BL/6) and resistant (BALB/c) mice using an in vitro infection model, gene expression, proteomic analysis, and bioinformatic tools. Our results showed that rSAG2A-treated PECs presented a lower parasitism in C57BL/6 mice but not in the PECs from BALB/c mice, and induced a pro-inflammatory cytokine profile in C57BL/6 mice (*iNOS*, *TNF-α*, and *IL-6*). rSAG2A modulated different exclusive proteins in each mouse lineage, with PECs from the C57BL/6 mice being more sensitive to modulation by rSAG2A. Additionally, biological processes crucial to parasite survival and immune response were modulated by rSAG2A in the C57BL/6 PECs, including fatty acid beta-oxidation, reactive oxygen species metabolism, interferon production, and cytokine-mediated signaling pathways. Together, our study indicates that rSAG2A controls *T. gondii* parasitism in susceptible C57BL/6 PECs through the modulation of pro-inflammatory cytokines and enhanced expression of proteins involved in the cytotoxic response.

Keywords: *Toxoplasma gondii*; macrophages; rSAG2A; cytokine; inflammation

1. Introduction

Toxoplasmosis is a neglected zoonotic disease caused by *Toxoplasma gondii*, which can infect a wide range of warm-blooded vertebrates [1]. Approximately one-third of the world's population has antibodies against *T. gondii*, and in some European countries, around 80% of the population have been exposed to the parasite at some point in their lives. In Brazil, recent years have been marked by several outbreaks of Toxoplasmosis in different states [2–4]. The disease can be fatal in congenital infections and can also cause cranio-cerebral and ocular sequelae in surviving neonates. In fact, infections in immunosuppressed patients and pregnant woman infected during the first trimester of pregnancy present the highest risk [5].

The life cycle of the parasite is heteroxenous, with felines serving as the definitive hosts. When infected, felines shed millions of oocysts per day in their feces. These oocysts sporulate and become infective in the environment. Intermediate hosts likely include all warm-blooded animals, such as most livestock and humans. Hosts become infected by ingesting sporulated oocysts that contaminate crops, soil, and water sources, or by consuming raw or undercooked meat containing cysts, which are the two primary horizontal transmission routes [6]. In the intestinal cells, these forms of the parasite differentiate into tachyzoites, which rapidly replicate in acute-disease-causing forms. As the immune response develops, some tachyzoites escape destruction and develop into bradyzoites, which form cysts in various tissues, including the brain and skeletal muscle. In immunocompetent hosts, these cysts do not cause overt disease and remain undetected as a relatively benign chronic infection. However, in immunocompromised patients, latent infection can reactivate, with bradyzoites converting into rapidly replicating tachyzoites, causing severe, life-threatening disease [6,7]. In pregnant women, acute infections acquired during or shortly after gestation can lead to congenital toxoplasmosis.

T. gondii has a complex antigenic profile that changes over the parasite's life cycle, including tachyzoite, bradyzoite, and sporozoite stages. These antigens include micronem (MIC), apical membrane antigens (AMAs), rhomboid antigens (ROMs), rhoptry antigens (ROPs), dense granule (GRAs) antigens, and surface antigen glycoproteins (SAGs) [8]. However, the immune response that these antigens promote and their immunogenicity vary, making effective vaccine development challenging [8]. The immune response to *T. gondii* can lead to different immunological outcomes, susceptibility, and resistance. Once *T. gondii* cross the mucosal barrier in the hosts, they activate immune cells to produce tumor necrosis factor-alpha (TNF-α) and anti-inflammatory cytokines such as interleukin 10 (IL-10), IL-27, and transforming growth factor-beta (TGF-β) [9]. This downregulates pro-inflammatory cytokines, enhances parasite proliferation, and triggers parasite migration to immune-privileged sites, including the brain, eye, and placenta. Conversely, an immune response that activates cell-mediated immunity and interferon-gamma (IFN-γ) production by natural killer cells and CD8$^+$ T cells controls *T. gondii* infection [9]. Neutrophils, macrophages, and dendritic cells that recognize the parasite evoke a pro-inflammatory response with high levels of IL-1β, IL-12, IL-18, and TNF-α, inhibiting parasite proliferation and infection [9].

Interestingly, different *T. gondii* antigens control the immune response. For instance, IL-22 is produced by CD4$^+$ T_H17 cells in response to the presence of ROP antigens [10,11]. GRAs enhance interferon regulatory factor 8 (IRF8), nuclear factor kappa B (NF-κB), and T-bet, controlling the CD4$^+$ T_H1/T_H2 immune response [12]. Additionally, SAG antigens can induce INF-γ and IL-12 production against *T. gondii* [13].

SAG2A is a protein of 22 kDa that belongs to the SAG2-like sequence family [14]. It differs from the other SAG proteins due to the presence of a disordered structure in its C-terminal region. SAG2A is an immunodominant antigen expressed in the tachyzoite phase of *T. gondii*, which contains a B cell epitope, making it a promising candidate for vaccine and diagnostic purposes [14–16]. SAG2A interacts with the host immune cells during acute infection, as it is abundantly expressed on the surface of tachyzoites [17]. The detection of IgG anti-SAG2A has been proposed for the diagnosis of toxoplasmosis. The use of IgG1 anti-SAG2A [15] and IgG3 anti-recombinant SAG2A [16] has shown promising results and a higher sensitivity in the acute phase of disease. Interestingly, these authors observed that the determination of the IgG3/IgG1 ratio of antibodies specific for rSAG2A associated with classic serum markers could be a tool to distinguish the early acute phase of *T. gondii*-infected patients.

The protein SAG2A is released from *T. gondii* tachyzoites in gliding assays using "freshly harvested" *T. gondii* RH strain tachyzoites [18]. Additionally, murine innate immune cells can be modulated by SAG2A [14]. These results suggest that SAG2A can be released during the beginning of infection, while also being able to interact with compounds from the adaptive immune response, inducing a robust humoral immune response during the initial phases of infection. To understand the role of SAG2A in the parasite immune response in toxoplasmosis, we evaluated the impact of the recombinant SAG2A protein (rSAG2A) during the infection of peritoneal exudate cells (PECs) from BALB/c and C57BL/6 mice in comparison with untreated infected macrophages. Our results showed that rSAG2A-primed PECs presented a lower number of intracellular tachyzoites in susceptible C57BL/6 mice, but not in PECs from resistant BALB/c mice. We characterized the immunological profile in both lineages and analyzed the potential metabolic pathways involved in the reduced parasitism of PECs from C57BL/6 mice infected with *T. gondii* compared to PECs treated with rSAG2A.

2. Materials and Methods

2.1. Ethics Statement

BALB/c (n = 15) and C57BL/6 (n = 15) male mice (5–8 weeks) were purchased from the Federal University of Uberlândia Rodent Bioterium and maintained in standard conditions, with a 12/12 h light/dark cycle under specific pathogen-free conditions (SPF) and with ad libitum water and chow. All the experiments with mice were carried out according to the institutional guidelines and were approved by the ethics institutional care and use committee of the State University of Santa Cruz, under protocol number 004/11.

2.2. Parasites

T. gondii tachyzoites RH strain (2F1 clone), which constitutively expresses cytoplasmic β-galactosidase [19], was kindly donated by Dr. Vern Carruthers of the University of Michigan Medical School (Michigan, MI, USA). The tachyzoites were propagated in human cervix adenocarcinoma (HeLa) cell lines obtained from the American Type Culture Collection (ATCC, Manassas, VA, USA) and cultured in Roswell Park Memorial Institute (RPMI) 1640 medium (Cultilab, Campinas, SP, Brazil) supplemented with L-glutamine, penicillin, streptomycin, and 2% fetal bovine serum (Cultilab) at 37 °C and 5% CO_2.

2.3. Isolation of Peritoenal Exudate Cells (PECs)

The animals received an intraperitoneal injection of 2 mL of 3% thioglycolate medium, and after 72 h, the peritoneal exudate cells (PECs) rich in peritoneal macrophages were collected in 5 mL of PBS after euthanasia, as previously suggested [20]. The cells were washed, counted, and suspended in DMEM medium (Cultilab) supplemented with 10% of fetal bovine serum (FBS) + 40 mg mL^{-1} of gentamicin (Gibco by Invitrogen, Waltham, MA, USA).

2.4. Parasitism and SAG2A

T. gondii parasitism in the PECs was measured using the β-Gal assay, as previously suggested [21]. Two independent experiments were performed using a pool of PECs obtained from n = 4 animals in each experiment. The PECs of BALB/c and C57BL/6 mice were incubated (1 × 10^5 cells/200 μL/well) in 10% DMEM medium in 96-well plates for 24 h at 37 °C and 5% CO_2. After the incubation period, the cells were washed to remove non-adherent cells, infected or not with tachyzoites from 2F1 clone in a 1:1 parasites/cell proportion (multiplicity of infection, MOI = 1). Three hours after the infection, the cell monolayers were rinsed with fresh DMEM to remove non-adherent and extracellular parasites, followed by the addition of 10 μg mL^{-1} of the full-length rSAG2A protein. The full-length rSAG2A was obtained as previously described [22]. The dose concentration (10 μg/mL^{-1}) used presented no toxicity to the PECs from both BALB/c and C57BL/6 mice, as previously described [22]. After 24 h, the parasitism was quantified using chlorophenol red–β-D-galactopyranoside (CPRG; Roche, Mannheim, Germany), as previously described [21]. The β- galactosidase activity was measured at 570 nm in a microplate reader (Versa Max ELISA Microplate Reader, Molecular Devices, Sunnyvale, CA, USA).

2.5. RT-qPCR

The PECs (1×10^6 cells) from BALB/c and C57BL/5 mice were placed in 24-well plates, infected with 2×10^6 tachyzoites, and incubated for 3 h at 37 °C and 5% CO_2. After that, the cells were treated or not with 10 μg/mL^{-1} of rSAG2A protein for 24 h. The samples were collected in 500 μL of TRIzol (Invitrogen, Waltham, MA, USA), followed by RNA extraction according to the manufacturer's protocol. The RNA concentration was determined using a NanoDrop 2000 spectrophotometer (Thermo Fisher Scientific, Waltham, MA, USA) after RNA treatment with DNAse I (Thermo Fisher Scientific), and the RNA integrity was evaluated using 1% agarose gel. cDNA synthesis was performed using the cDNA First Strand kit (Thermo Fisher Scientific), according to the manufacturer's instructions, followed by quantification and sample dilution to achieve 40 ng/1 μL. Amplification was carried out with Maxima SYBR Green/GoTaq qPCR Master Mix (Promega, Madison, WI, USA) using a Stratagene Mx3005P system (Agilent Technologies, Santa Clara, CA, USA). The gene markers included *IL-1β*, *IL-6*, *IL-10*, *TGF-β*, *TNF-α*, *Arginase 1*, and *iNOS*, and the gene *18S* was used for normalization (Table 1). The PCR conditions were as follows: 95 °C for 10 min and 40 cycles of 95 °C for 5 s, 60 °C for 30 s, and 75 °C during 30 s. The reaction products were dissociated to confirm whether the products obtained were unique and specific. The relative expression was measured using the $2^{-\Delta\Delta Ct}$ methodology [23]. The mean values of $2^{-\Delta\Delta Ct}$ for each gene were converted using a Z-score scale and these values were used for a heatmap analysis with clusterization. These analyses were implemented in the R environment (3.6.2) with the Heatmap function, using the ComplexHeatmap package [24,25]. Grouping was performed using the mean Euclidean distance and the complete method.

Table 1. Sequence of primers used in real-time quantitative PCR reactions (RT-qPCR).

Gene	Sequence
IL-1β	Forward: 5′-GCACACCCACCCTGCA-3′ Reverse: 5′-ACCGCTTTTCCATCTTCTTCTT-3′
IL-6	Forward: 5′-TCCAGAAACCGCTATGAAGTTC-3′ Reverse: 5′-CACCAGCATCAGTCCCAAGA-3′
IL-10	Forward: 5′-TGGACAACATACTGCTAACC-3′ Reverse: 5′-GGATCATTTCCGATAAGGCT-3′
TGF-β	Forward: 5′-AAGCTACCAAGTTAGACTTCCCA-3′ Reverse: 5′-TGAAAGTTTAGCATACAGAATCCC-3′
TNF-α	Forward: 5′-TGTTTCGAGGTTGCTTGTCT-3′ Reverse: 5′-GATTGTTCCACCAGCTTGC-3′
Arginase1	Forward: 5′-AAAGCTGGTCTGCTGGAAAA-3′ Reverse: 5′-ACAGACCGTGGGTTCTTCAC-3′
iNOS	Forward: 5′-CAGCTGGGCTGTACAAACCTT-3′ Reverse: 5′-CATTGGAAGTGAAGCGTTTCG-3′

2.6. Quantitative Proteomic

A pool of PECs obtained from n = 7 male BALB/c and C57BL/6 mice were plated in 6-well plates in triplicates using 5×10^6 cells/well. Three conditions were established: a control group with PECs without infection and not treated with rSAG2A, PECs infected with *T. gondii* (MOI = 1), and PECs treated with rSAG2A (10 μg/mL^{-1}). The cells were incubated in a BOD incubator for 24 h at 37 °C and 5% CO_2. Protein extraction was carried out according to the phenol method, as previously described [26]. The cells were treated with 750 μL of extraction buffer (100 mmol L^{-1} Tris-HCL pH 8.3; 5 mmol L^{-1} EDTA; 100 mmol L^{-1} KCL; 1% p/v DTT; 30% sucrose; 1 complete Mini EDTA-free protease inhibitor cocktail tablet (Roche Applied Science, Penzberg, Upper Bavaria, Germany) per 10 mL of buffer. The samples were then vortexed and centrifuged at 12,000 rpm at 4 °C. The phenolic phase was collected, followed by the addition of 750 μL of buffered phenol, vortexing, and centrifugation for five minutes. The phenolic phase was collected once more, added with 100 mmol/L of ammonium acetate in methanol and left overnight at −20 °C to precipitate. After that, the samples were centrifuged for 60 min at 4 °C and 13,000 rpm, the supernatant was discarded and the pellet was kept for 60 min in cold acetone and 0.2% DTT solution at −20 °C and rinsed twice in the same solution. Between the rinsing processes, the samples were centrifuged (4 °C and 13,000 rpm) and the supernatants were discarded. The pellet was gently dried under a hood at room temperature and resuspended with 200 μL of lysis buffer (8 mol/L urea; 5 mmol/L DTT; 30 mmol/L Tris). The protein concentration was quantified by the Bradford quantification method and samples were lyophilized and sent to the Proteomics Lab of the Faculty of Bioscience Engineering and SyBioMa of the Biomedical Sciences Group, KU Leuven, for further proteomics workflow steps. For protein digestion, the samples were re-suspended in 200 μL of lysis buffer and the following process was performed successively, adding the following compounds for each 20 μg of proteins: 20 mmol/L of DTT and 15 min incubation; 50 mmol/L of iodoacetamide, incubation in the dark for 30 min; three times dilution with 150 mmol/L of ammonium bicarbonate; the addition of 0.2 μg of trypsin and overnight incubation at 37 °C; and trifluoracetic acid to a 0.1% final acidification concentration. After that, Pierce™ C18 Spin Columns (Thermo Fisher Scientific) were used for the desalting of the samples, according to the manufacturer's instructions. Peptides separation of the digested samples (1 μg/5 μL^{-1}) was performed via a UPLC-MS/MS system, with the Ultimate 3000 UPLC system (Dionex, ThermoScientific, Waltham, MA, USA) and Q Exactive Orbitrap mass spectrometer (Thermo Scientific) as previously described [27]. Data were obtained with the Xcalibur 3.0.63 software (ThermoScientific).

2.7. Proteomic Data Analysis

Protein identification was performed with raw data conversion by Proteome Discover version 1.4 (Thermo Fisher Scientific) into mgf files and processing with MASCOT version 2.2.06 (Matrix Science, Columbus, OH, USA) against the Uniprot Mus musculus database (55,398 proteins). The false discovery rate (FDR) was calculated to enhance the identification confidence with Scaffold (version 3.6.3; Proteome Software Inc., Portland, OR, USA). Protein quantification was processed using Progenesis LC–MS version 4.1 (Nonlinear Dynamics, Milford, MA, USA), with automatic alignment from a selected reference run based on ion intensity.

Normalized fluorescence data obtained in the proteomic analysis of BALB/c and C57BL/6 mice PECs in the control condition without rSAG2A and not infected with *T. gondii*; cells only infected with the parasite; and cells only treated with rSAG2A were used to calculate the fold change (treated/control). Proteins with a fold change (FC) higher than 1 were considered to be up-modulated (augmented) and proteins with FC < 1 were considered to be down-modulated (suppressed). These proteins were used to select those exclusively augmented or suppressed in the PECs of C57Bl/6 mice under the conditions of rSAG2A use or *T. gondii* infection.

2.8. Identification of Proteins Exclusively Up-Modulated in PECs of C57BL/6 Mice Treated with rSAG2A

To identify exclusively augmented proteins in the PECs from C57BL/6 mice, a qualitative approach was used. Briefly, the FC was calculated based on comparisons between the treated (rSAG2A) and infected (*T. gondii*) groups with their respective controls: (i) BALB/c_rSAG2A/Control; (ii) BALB/c_*T. gondii*/Control; (iii) C57BL/6_rSAG2A/Control; and (iv) C57BL/6_*T. gondii*/Control. When a protein displayed an FC > 1 exclusively in a determined condition, it was considered as exclusive for the specific condition. The proteins with an FC > 1 for two or more conditions were considered to be shared proteins. Based on this principle, a Venn diagram was plotted using the bash/awk routine in the Linux system. Exclusive or up-modulated proteins present in the PECs from C57BL/6 mice treated with rSAG2A (FC > 2) were compared to the PECs of BALB/c mice treated with rSAG2A. Exclusive or up-modulated proteins present in the PECs from C57BL/6 mice infected with *T. gondii* (FC >2) compared to the PECs of BALB/c mice infected with *T. gondii* were highlighted in blue in the Venn diagram.

We also selected down-modulated proteins (FC < 1.0) exclusively found in the PECs from C57BL/6 mice treated with rSAG2A and down-modulated proteins (FC < 1.0) exclusively found in the PECs from C57BL/6 mice infected with *T. gondii*. The fold changes of the selected proteins identified in Supplementary Tables S1–S4 are demonstrated in the heatmap, plotted using the Heatmap function of the Complex Heatmap package in R version 4.3.3 [24,25].

2.9. Prediction of Protein-Protein Interaction Network

To understand what biological processes were regulated in the PECs of C57BL/6 and BALB/c mice treated with the rSAG2A protein, we conducted a study of protein–protein interactions. We focused on groups of exclusive or up-modulated proteins and on down-modulated proteins due to rSAG2A or infection by *T. gondii*. In this sense, two networks were generated for the PECs of C57BL/6 mice treated with rSAG2A: (i) a network of exclusive proteins or those with augmented accumulation when the PECs were treated with rSAG2A; and (ii) a network for the group of down-modulated proteins. Two other networks were generated for the PECs of C57BL/6 mice infected with *T. gondii*: (i) a network of exclusive or up-modulated proteins when the PECs were infected with *T. gondii*; and (ii) a network of proteins down-modulated when the PECs were infected with *T. gondii*. These networks were generated starting from a high-reliability network (0.700) of the Mus musculus interactome in the STRING database [28]. Module and gene ontology analyses were carried out as described by those authors. For these analyses, the plugins from the Cytoscape MCODE [29] and BiNGO (Biological Network Gene Ontology) [30] were used, respectively. To conduct the gene ontology, a notation file was obtained using Ensemble.

To understand which proteins modulated by rSAG2A were more associated with the inflammatory response, we constructed protein-protein interaction networks starting from the cytokines whose gene expression profile was evaluated in the present study. The proteins modulated by rSAG2A were highlighted and the functional enrichment of the biological processes of the main groups were again analyzed with BiNGO.

2.10. Statistical Analysis

Data were analyzed using GraphPad Prism version 5.0 (GraphPad Software, San Diego, CA, USA). Statistical differences between groups were obtained through one-way ANOVA test, followed by Tukey or *t*-test. The results were considered significant with $p < 0.5$.

3. Results

3.1. rSAG2A Reduced T. gondii Parasitism in PECs from Susceptible C57BL/6 Mice, but Not in PECs from Resistant BALB/c Mice

To determine whether recombinant SAG2A modulates the immune response and impacts *T. gondii* infection, we employed an in vitro model using peritoneal exudate cells (PECs) from BALB/c and C57BL/6 mice. The rationale for using this model was the possibility of investigating the response of susceptible and resistant mice to *T. gondii* tachyzoite (2F1 clone) infection. We observed that, in early infection (24 h), PECs from both the resistant and susceptible mice showed similar levels of parasitism ($p > 0.05$). Interestingly, when treated with rSAG2A, the susceptible mice (C57BL/6) exhibited a significant reduction in parasitism ($p < 0.001$), with a decrease of 2000-fold compared to the untreated C57BL/6 mice (Figure 1). These data suggest that the rSAG2A protein can modulate the immune response in PECs to control *T. gondii* infection.

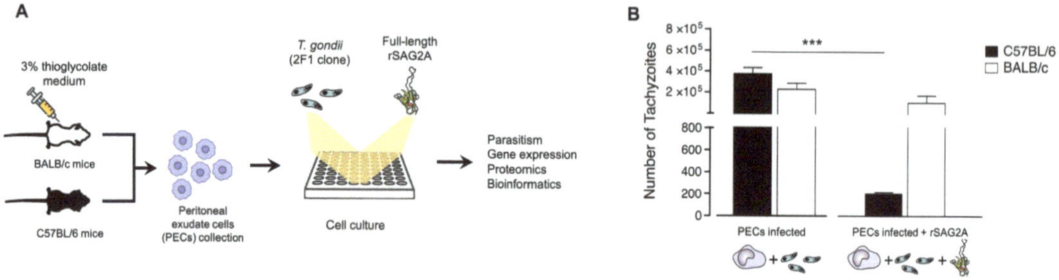

Figure 1. rSAG2A decreases the parasitism in PECs infected with tachyzoites of *T. gondii*. (**A**) PECs from BALB/c and C57BL/6 mice were infected with tachyzoites of *T. gondii* (RH-2FI strain; MOI = 1) for 3 h, followed or not by addition of rSAG2A. (**B**) The number of intracellular tachyzoites was measured by the β -galactosidase colorimetric assay after incubation for 24 h with rSAG2A. Data represent mean + SD of two independent experiments. *** $p < 0.001$ by one-way ANOVA followed by the Tukey post-test.

3.2. rSAG2A Reduced T. gondii Parasitism in PECs from Susceptible C57BL/6 Mice but Not in PECs from Resistant BALB/c Mice

To understand the effect of rSAG2A in modulating the immune response, we treated the PECs with rSAG2A in the presence of absence of *T. gondii* in the resistant and susceptible mice. First, we observed that both the resistant BALB/c and susceptible C57BL/6 mice exhibited the same basal immunological profile in control groups (Figure 2). Interestingly, the PECs infected with *T. gondii* early (24 h) also did not exhibit significative changes in the immunological markers analyzed: *iNOS* (Figure 2A), *Arginase* (Figure 2B), *TNF-α* (Figure 2C), *IL-10* (Figure 2D), *IL-1β* (Figure 2E), *TGF-β* (Figure 2F), and *IL-6* (Figure 2G), when comparing the PECs from the resistant BALB/c and susceptible C57BL/6 mice.

We noticed that rSAG2A alone was not sufficient to modulate *iNOS* (Figure 2A), *TNF-α* (Figure 2C), *IL-1β* (Figure 2E), and *IL-6* (Figure 2G), all genes of pro-inflammatory cytokines. However, rSAG2A alone increased the expression of anti-inflammatory cytokines, *Arginase* (Figure 2B; $p < 0.01$) and *TGF-β* (Figure 2F; $p < 0.01$), while not altering *IL-10* (Figure 2D) expression in the BALB/c compared to C57BL/6 mice. These data indicate an anti-inflammatory property of rSAG2A in resistant BALB/c mice. We next sought to understand if this modulation could affect *T. gondii*-infected PECs. We observed that rSAG2A changed the immune response of the PECs infected with *T. gondii*, but the effects differed between the resistant and susceptible mice.

First, we note that rSAG2A enhanced the levels of both pro- and anti-inflammatory cytokines genes, *iNOS* (Figure 2A; $p < 0.001$), *TNF-α* (Figure 2C; $p < 0.001$), *IL-10* (Figure 2D; $p < 0.001$), *TGF-β* (Figure 2F; $p < 0.001$), and *IL-6* (Figure 2G; $p < 0.001$) in the C57BL/6 mice. However, the BALB/c mice only presented an increase in anti-inflammatory cytokine genes, *TGF-β* (Figure 2F; $p < 0.001$) and *Arginase* (Figure 2B; $p < 0.001$).

Next, we compared the resistant BALB/c and susceptible C57BL/6 mice treated with rSAG2A and infected with *T. gondii*. Our data showed that the C57BL/6 mice exhibited high expression levels of *iNOS* (Figure 2A; $p < 0.01$), *TNF-α* (Figure 2C; $p < 0.05$), *IL-10* (Figure 2D; $p < 0.05$), and *IL-6* (Figure 2G; $p < 0.01$) compared to the BALB/c mice, indicating a more inflammatory profile. Conversely, the BALB/c mice exhibited high expression levels of *Arginase* (Figure 2B; $p < 0.001$), a marker of alternatively activated macrophages (M2-like). These data indicate that rSAG2A stimulates different profiles in resistant and susceptive mice (Figure 2H) and suggest that the increase in pro-inflammatory cytokine levels helps PECs from C57BL/6 mice to control parasitism (Figure 1). We hypothesized that this effect could be a result of protein accumulation and alterations in the immunological pathways in C57BL/6 mice. To test our hypothesis, we performed a proteomic analysis and a protein–protein interaction network with bioinformatics tools.

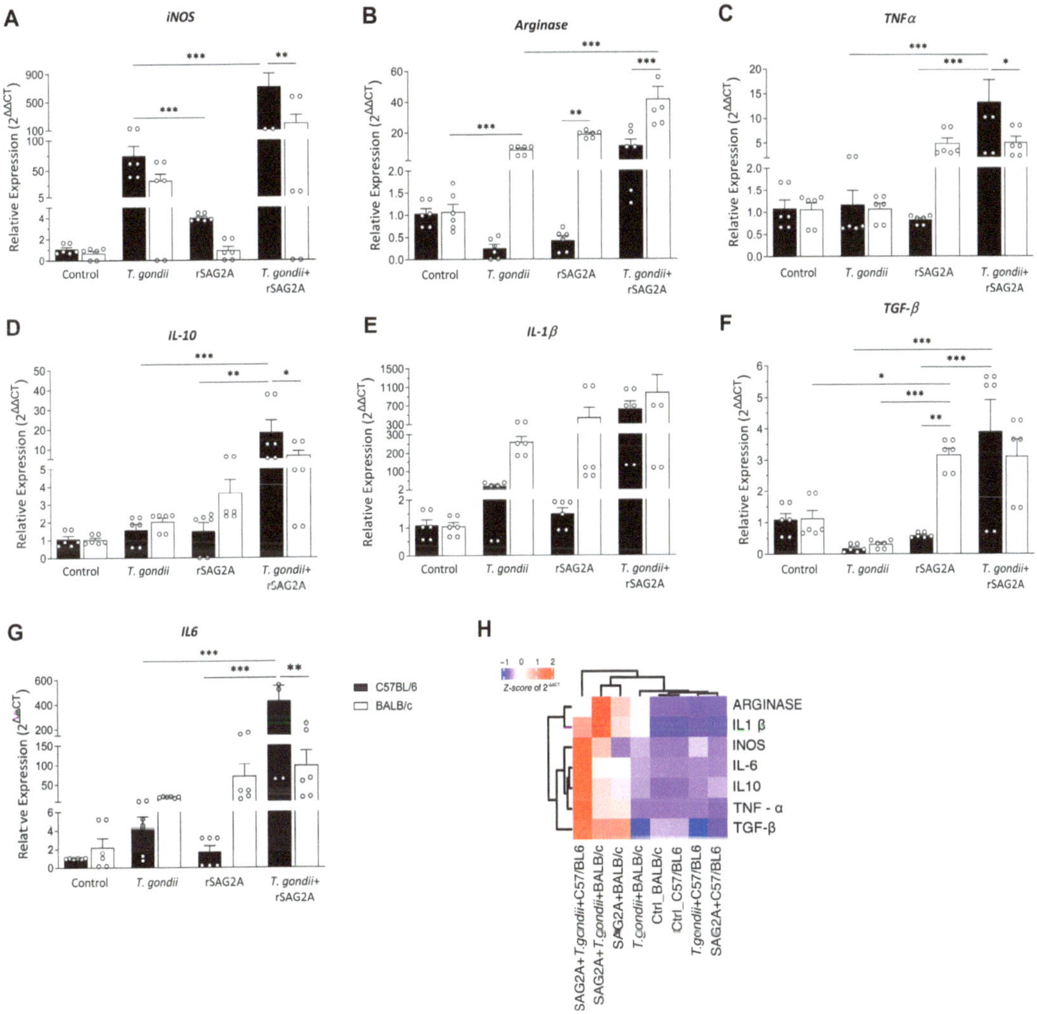

Figure 2. rSAG2A alters the pro-inflammatory and anti-inflammatory cytokines expression in PECs infected with *T. gondii*. Cytokine transcript accumulation was evaluated during the presence of parasites for 3 h in PECs from both BALB/c and C57BL/6 mice or treated with rSAG2A during 24 h, through qPCR. (**A**) INOS, (**B**) Arginase, (**C**) TNF-α, (**D**) IL-10, (**E**) IL-1β, (**F**) TGF-β, and (**G**) IL-6. (**H**) Hierarchical clustering for the same mean relative quantification (RQ) values for the same genes, represented in z score scale, in the different conditions studied. Data represent mean + SD of two independent experiments using a pool of PECs n = 4 animals/experiment. * $p < 0.05$, ** $p < 0.01$, and *** $p < 0.001$ between control and rSAG2A of 6 replicates/group by ANOVA followed by the Tukey post-test.

3.3. rSAG2A Induced Accumulation of Proteins Associated with Oxidative Stress and Immune Pathways in PECs of C57BL/6 Mice

Our results indicated that rSAG2A may act on PECs' microbicidal mechanisms at the transcription level (Figure 2). To understand the effects of rSAG2A in both the susceptible and resistant mice, we next analyzed the differentially expressed proteins using proteomics. In our analysis, proteins were considered as exclusive to a group based on the obtained FC values: FC > 1 indicates being exclusive for a group, and FC > 2 indicates protein augmentation (see Section 2).

We identified six exclusive proteins from the C57BL/6 mice infected with *T. gondii*, 16 from the C57BL/6 mice treated with rSAG2A, 12 from the BALB/c mice infected with *T. gondii*, and 42 for the BALB/c mice treated with rSAG2A (Figure 3). Additionally, we observed 50 proteins augmented in the C57BL/6 mice treated with rSAG2A: (i) 15 were shared among BALB/c_rSAG2A, BALB/c_*T. gondii*, and C57BL/6_rSAG2A mice; (ii) 26 were common to all rSAG2A-incubated PECs from both mice lineages; (iii) 5 were shared among the BALB/c mice treated with rSAG2A, BALB/c mice only infected with *T. gondii*, and C57BL/6 mice infected with *T. gondii*; and (iv) 4 proteins were identified as b3int shared between the BALB/c and C57BL/6 mice treated with rSAG2A (Figure 3, indicated with an asterisk). Interestingly, BALB/c mice treated with rSAG2A or infected with *T. gondii* shared 319 common proteins, two-fold higher than C57BL/6 when comparing the treated and infected animals, which shared only 155 proteins. A total of 211 proteins were shared between the BALB/c and C57BL/6 infected mice.

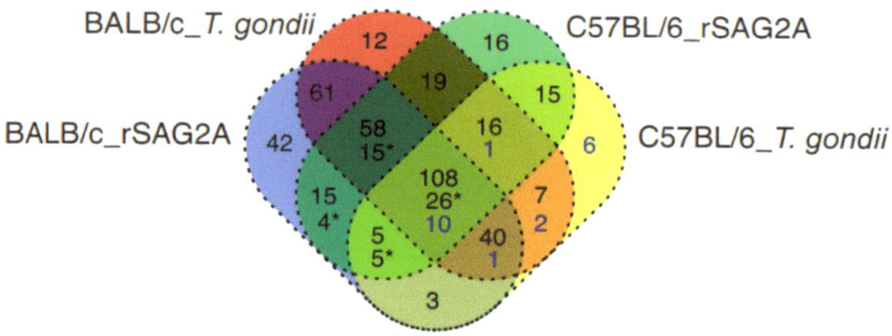

Figure 3. Venn diagram of proteins shared by PECs of mice infected with *T. gondii* or treated with rSAG2A. Venn diagram showing the number of proteins in common and not in common between the different groups. PECs of the BALB/c mice treated with rSAG2A (BALB/c_rSAG2A), BALB/c mice infected with *T. gondii* (BALB/c_*T. gondii*), C57BL/6 mice treated with rSAG2A (C57BL/6_rSAG2A), and C57BL/6 mice infected with *T. gondii* (C57BL/6_*T. gondii*). The asterisk values represent proteins augmented with fold change > 2 exclusively in PECs of C57BL/6 mice treated with rSAG2A.

We used hierarchical clustering and FC to identify the proteins exclusively up- or down-modulated in the treated and infected C57BL/6 PECs compared to BALB/c (Figure 4). We identified 66 proteins exclusively up- and 24 exclusively down-modulated in the C57LB/6 PECs treated with rSAG2A compared to the BALB/c treated PECs (Figure 4A). This profile was significantly different from that observed in the C57BL/6 PECs infected with *T. gondii*. Only 37 proteins were exclusively up- and 19 exclusively down-modulated in the C57LB/6 PECs infected with *T. gondii* compared to the BALB/c infected PECs (Figure 4B). These proteins have been related to immune responses or cellular metabolism, such as annexin A7, prostaglandin reductase 1, proteasome subunit alpha type-1, the 26S proteasome non-ATPase regulatory subunit 2, and the hematopoietic lineage cell-specific protein. Among the down-modulated proteins were vinculin, tropomyosin alpha-4 chain, trans-aldolase, and the 40S ribosomal protein S7 (Supplementary Tables S1–S4).

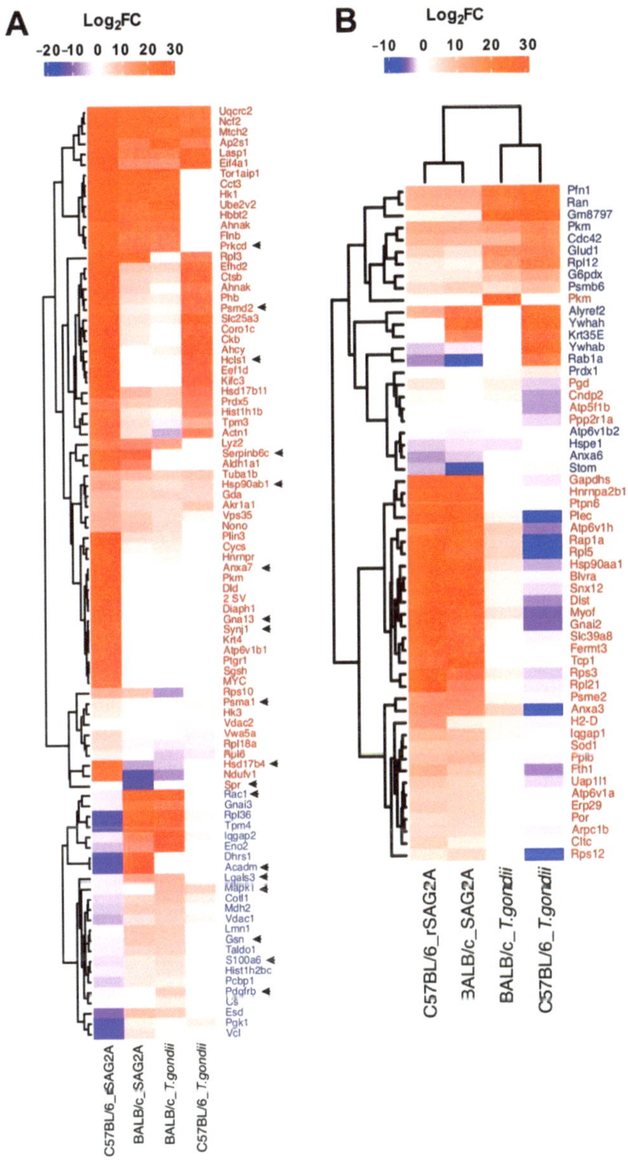

Figure 4. Accumulation profile and hierarchical clustering for exclusive/up-modulated or down-modulated proteins in C57BL/6 mice PECs treated with rSAG2A (**A**) or infected with *T. gondii* (**B**), compared to BALB/c PECs in the same conditions. Proteins in red are those exclusive or that had a Log2FC (compared to the other conditions) at least twice as high in C57BL/6 PECs treated with rSAG2A (red names in **A**) or in C57BL/6 PECs infected with *T. gondii* (red names in **B**). Proteins in blue are those that were repressed only in C57BL/6 PECs treated with rSAG2A (blue names in **A**) or in C57BL/6 PECs infected with *T. gondii* (blue names in **B**). Heatmap colors are represented by Log2FC scale, which indicates the accumulation of proteins in Log2 fold change scale; blue shades indicate the repressed proteins and red shades indicate proteins that increased in relation to their respective controls. Black arrows represent proteins present in the cytokine network.

To understand how rSAG2A affected parasitism in the C57BL/6 PECs, we applied a protein–protein network analysis to identify the pathways and biological processes impacted by rSAG2A and *T. gondii* infection (Figure 5). The integrated network of proteins with a reduced expression from C57BL/6 mice only treated with rSAG2A displayed 760 nodes and 15,850 connectors with a high fidelity. When *T. gondii* was present, 1384 nodes and 41,467 connectors were observed. Cluster analysis of these nodes allowed for the identification of 6 connected protein clusters in the rSAG2A group and 14 clusters when *T. gondii* was present. The main biological processes enriched in these clusters were related to protein metabolism, signaling, oxidative processes like fatty acid beta-oxidation, cell development, chromatin modification, and cytoskeleton organization for rSAG2A (Figure 5A), and glucose metabolism, proton transport, response to stress, cholesterol biosynthesis, and RNA splicing for PECs infected with *T. gondii* (Figure 5B).

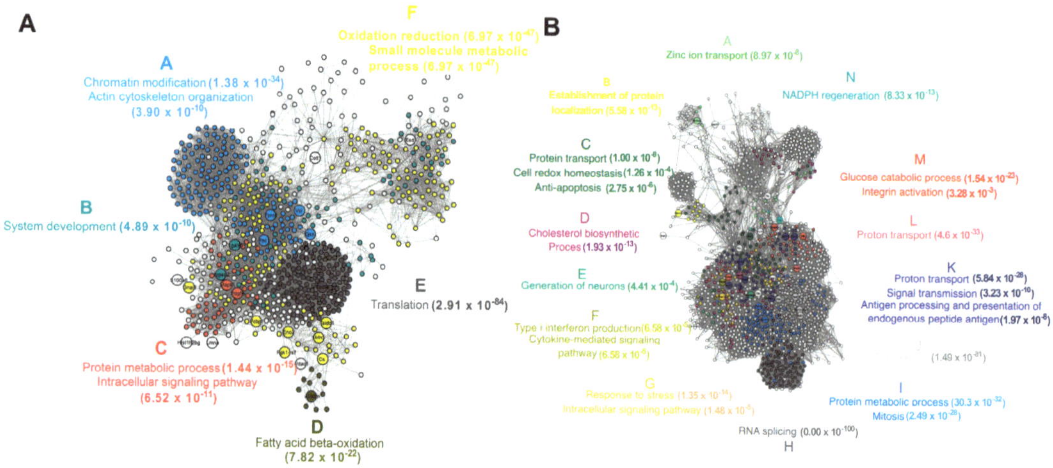

Figure 5. Protein–protein interaction network for exclusive and augmented proteins from PECs of C57BL/6 mice treated with rSAG2A (**A**) or infected with *T. gondii* (**B**). The bigger nodes represent the proteins identified through proteomic analysis. Smaller nodes were aggregated with networks using the STRING database. The different colors represent different clusters identified in the module analysis. The biological processes enriched in each cluster are described with the respective colors around the networks.

The network of exclusive and augmented proteins from the C57BL/6 PECs treated with rSAG2A presented 1391 nodes and 38,537 connectors, as well as 11 clusters (Figure 6A), and when the parasite was present, 1019 nodes and 25,681 connectors were observed (Figure 6B). The biological processes enriched in these clusters were related to (i) actin filament bundle assembly; (ii) the CDP-choline pathway and generation of neurons; (iii) membrane invagination; (iv) heparan sulfate proteoglycan metabolism; (v). the innate immune response; (vi) cell redox homeostasis, oxygen and reactive oxygen species, and the electron transport chain; and (vii) the catabolism of fatty acids, among others, when cells were treated with rSAG2A.

Furthermore, we observed a specific regulation in the intracellular signaling pathways of networks in the C57BL/6 mice treated with rSAG2A, indicating a heightened protein expression of pathways related to the immune response, including the production of intermediate oxygen species. This metabolic profile is typical of an inflammatory immune response and may explain the increased host resistance against *T. gondii* replication in the intracellular environment. Moreover, the presence of other inflammatory proteins in the C57BL/6 mice pretreated with rSAG2A could work together with reactive oxygen species, culminating in pathogen elimination and explaining the reduced intracellular parasitism.

In summary, these data indicate that the activation of the studied inflammatory mediators induced by rSAG2A could help to control the infection by *T. gondii* in PECs from susceptible C57BL/6 mice.

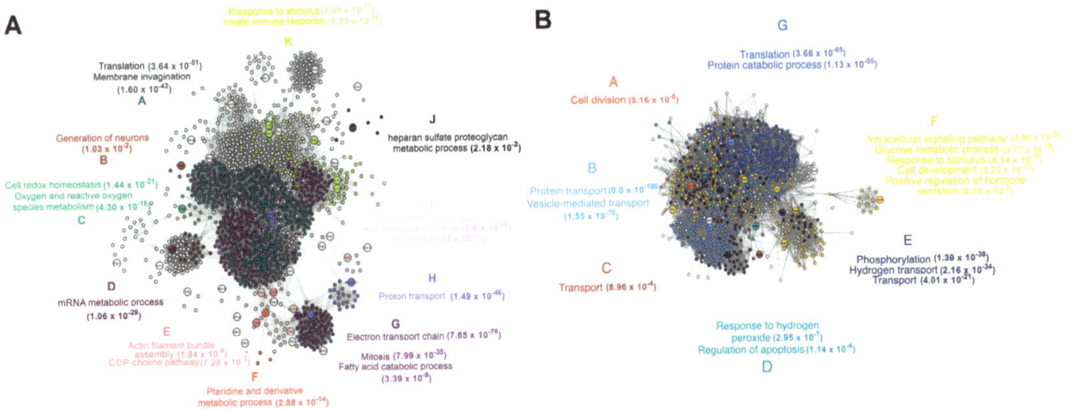

Figure 6. Network of protein–protein interactions for down-modulated proteins in PECs of C57BL/6 mice treated with rSAG2A (**A**) or infected with *T. gondii* (**B**). The bigger nodes represent the proteins identified through proteomic analysis. Smaller nodes were aggregated with networks using the STRING database. The different colors represent different clusters identified in the module analysis. The biological processes enriched in each cluster are described with the respective colors around the networks.

3.4. Major Genes and Responses Associated with Pro-Inflammatory Cytokines

The protein–protein network interaction associated with cytokines (TNF, IL-6, IL-10, and TGF-β1) and iNOS was analyzed and found to be composed of 173 proteins categorized into five different groups (1 to 5) related to important biological processes (Figure 7). Incubation with rSAG2A showed that the proteins Lgals3, Hsp90ab1, and Mapk1 had direct interactions with TNF, iNOS2, and TGF-β1, respectively. Additionally, the Pdgfrb protein interacted with both IL-6 and IL-10, while Serpinb6c interacted with TNF, IL-6, and IL-10.

The proteins Lgals3, Pdgfrb, and Mapk1 were less abundant when PECs from the C57BL/6 mice were treated with rSAG2A. On the other hand, the levels of the proteins Serpinb6c and Hsp90ab1 were at least two times higher in the same conditions. We also observed a reduction in S100 calcium-binding protein (S100A6). The S100A6 of the host cell is important for the invasion of *T. gondii* by reducing or blocking its functional epitopes induced by parasite invasion [31]. Thus, the reduction in S100A6 could be involved in the decrease in parasite proliferation by the inhibition of parasite invasion.

In group 5, the proteins that deserve emphasis were those associated with processes such as the production of metabolic precursors and energy, responses to nutrient levels, lipid modifications, the regulation of responses related with receptors, and endocytosis. Among the proteins associated with this group, Hk3, Dld, Phb, Gm10053, Slc25a3, Uqcrc2, Mtch2, Synj1, Ndufv1, Ap2s1, Hsd17b4, Vdac2, and Hk1 were abundant during incubation with rSAG2A. Moreover, the proteins Taldo1, Vdac1, Eno2, Mdh2, Acadm, Cs, and Pgk1-rs7 were present in group 5.

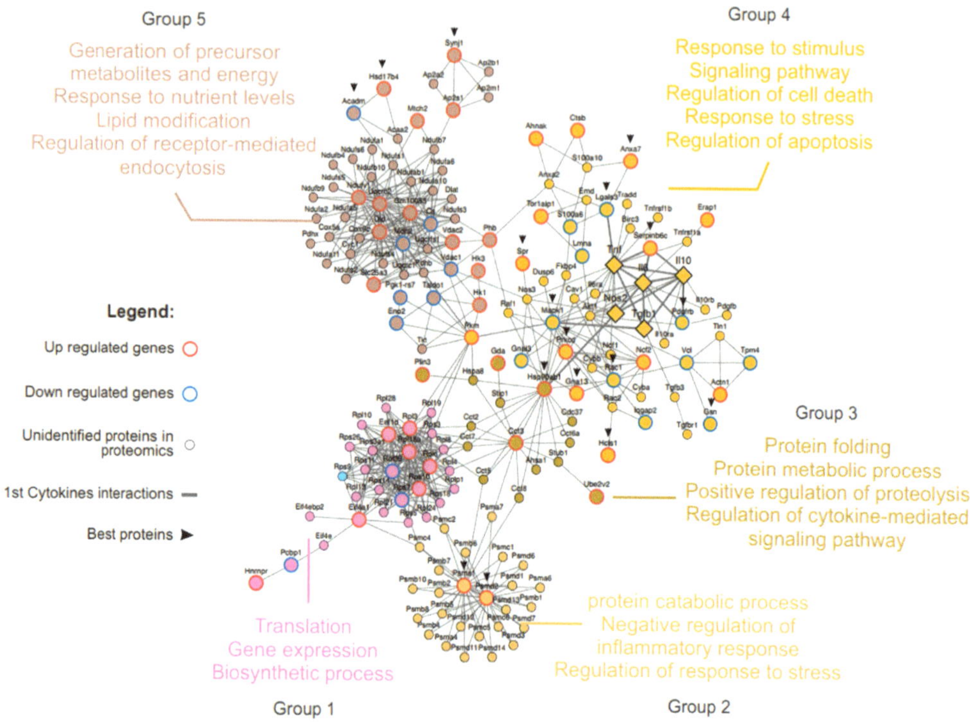

Figure 7. Protein interaction network related with inflammatory and regulatory cytokines. The large circles represent proteins whose abundance was altered in PECs treated with rSAG2A. Diamonds represent the inflammatory and regulatory cytokines, besides INOs. Small circles represent proteins not identified during the proteomic analysis. The fat connectors highlight first-degree interaction with the cytokines and the slender connectors the interactions in distinct degrees. The arrows indicate the most relevant proteins in the study. The red and blue borders represent up-modulated and down-modulated proteins, respectively.

4. Discussion

Our knowledge about *T. gondii* surface antigens can shed light on evasion mechanisms, vaccine development, the parasite life cycle, and effective immune response. The literature has shown that surface antigen glycoproteins (SAGs) activate the immune system, and different SAG antigens can exhibit distinct modulatory mechanisms during *T. gondii* infection. For example, recombinant surface antigen SAG1 (rSAG1) reduced fetal infection in resistant mice, but not in susceptible animals, although similar levels of cytokines such as IL-4, IL-10, and IFN-γ were observed in maternal sera during gestation [32]. SAG2A has been previously suggested as a promising candidate to modulate *T. gondii* infection. The activity of macrophages and dendritic cells from C57BL/6 bone marrow stem cells could be modulated by SAG2A, and the immunomodulatory effect of SAG2A has been associated with the C-terminal portion [14]. Our group previously suggested that the rSAG2A, a recombinant protein with the presence of a disordered structure in the C-terminal region, is a promising candidate to modulate *T. gondii* infection [22].

Here, to underscore the effect of rSAG2A in the immune response of mammals, we used resistant (BALB/c) and susceptible (C57BL/6) mouse lineages to collect peritoneal exudate cells (PECs) and treated them with rSAG2A in the presence or absence of *T. gondii* tachyzoites. Our data showed that rSAG2A can induce an anti-inflammatory profile in resistant mice in the absence (increase *Arginase* and *TGF-β*) and presence (increase

Arginase) of *T. gondii* tachyzoites. Conversely, although rSAG2A alone did not induce a pro-inflammatory profile in the susceptible mice in the absence of *T. gondii* tachyzoites, rSAG2A enhanced the pro-inflammatory profile in the infected PECs, mediated by *iNOS*, *TNF-α*, and *IL-6*. Resistance and susceptibility to *T. gondii* in mouse lineages have been attributed to various factors, including the immune response, T-cell mediated response, and dendritic cell polarization [8,33]. Our data suggest that rSAG2A can enhance the natural response observed in both the C57BL/6 (pro-inflammatory) and BALB/c (anti-inflammatory) lineages. Nonetheless, these modulations in genes involved in the microbicidal mechanisms of macrophages promote the control of *T. gondii* parasitism in C57BL/6 PECs with a low dosage (10 μg/mL^{-1}). We hypothesized that the ability of rSAG2A to induce *IL-10* expression in addition to pro-inflammatory cytokines in the C57Bl/6 PECs could help to balance the immune response and avoid extensive tissue damage by controlling excessive inflammation [34].

Macêdo-Junior and colleagues [14] previously demonstrated that a truncated form of the rSAG2AΔ135 protein, lacking its C-terminal end, was also able to induce high levels of iNOS and IL-12 production in a dose-dependent manner in bone-marrow-derived monocytes from C57BL/6 susceptible mice. Furthermore, the immune response is complex, and the microbicidal mechanisms and regulatory mechanisms of macrophages require the activation of different receptors and signaling pathways. To understand if and how rSAG2A controls biological processes and signaling pathways, we used proteomic analysis and bioinformatics tools.

We identified that *T. gondii* infection modulated different exclusive proteins in the resistant and susceptible mice: 6 in C57BL/6 and 12 in BALB/c. However, incubation with rSAG2A increased these exclusive modulations in each model, from 6 to 16 in the C57BL/6 and from 12 to 42 in the BALB/c PECs. As observed by us, this modulation was not restricted to pro- or anti-inflammatory proteins, as suggested in our mRNA expression analysis. In fact, Lee and colleagues [33] previously showed that *T. gondii* infection modulates not only the cytokines in dendritic cells, but also the surface receptors and molecules involved in antigen presentation and T-cell activation, such as major histocompatibility complex (MHC) II, CD40, CD80, and CD86. Among the proteins exclusively modulated, we noticed that the C57BL/6 PECs were more susceptible to rSAG2A and presented more up- and down-modulated proteins than the BALB/c PECs. These up-modulated proteins included Hsp90, aldehyde dehydrogenase (ALDH) 1A, prostaglandins reductase 1, lysozyme type C, and 26S proteasome non-ATPase regulatory subunit 2, among others.

During *T. gondii* infection, Hsp90, a chaperon protein, is down-modulated and helps the parasite to invade cells and enhance tachyzoite transition to the bradyzoite stage, allowing bradyzoites to evade the immune response [35]. We suggested that the enhancement of Hsp90 after treatment with rSAG2A may control parasitism in PECs by controlling bradyzoite formation. Additionally, Sugi and colleagues [36] showed, in a single-cell transcriptome analysis, that the bradyzoites derived from the ME-49 strain had anti-inflammatory activity in human foreskin fibroblasts, down modulating the NF-κB and IFN-γ response, which helped to prevent the parasite from being eliminated from the body.

The activation of cytotoxic mechanisms through the ALDH pathway is common during *T. gondii* infection in both human and mouse cells. We observed that rSAG2A up-modulated not only ALDH1A, but also prostaglandin reductase 1 production in C57BL/6 PECs. Prostaglandin reductase 1 is a key enzyme for the degradation of prostaglandins, which are messenger lipids that regulate processes such as cell survival and inflammation [37]. The increase in lysozyme type C, a family of proteins known to have microbicidal activities, and the up-regulation of the 26S proteasome non-ATPase regulatory subunit 2 and ALDH1A can positively regulate cell-mediated cytotoxicity and the production of reactive oxygen species (ROS) [38,39], thus controlling parasitism. Additionally, we noticed that biological processes crucial to parasite survival and the immune response were modulated by rSAG2A in the C56BL/6 PECs, including fatty acid beta-oxidation, oxygen and reactive oxygen species metabolism, interferon production, and cytokine-mediated signaling pathways.

5. Conclusions

Together, our study not only suggests that recombinant SAG2A controls *T. gondii* parasitism in susceptible C57BL/6 PECs but supports that this control is closely related to modulation in pro-inflammatory cytokines and the enhanced expression of proteins involved in the cytotoxic response and control of the parasitic life cycle. Ou work may also serve as a valuable source for future therapeutic strategies for toxoplasmosis using surface antigen glycoproteins.

Supplementary Materials: The following supporting information can be downloaded at: https://www.mdpi.com/article/10.3390/microorganisms12112366/s1, Table S1: Exclusive or up-modulated proteins present in PECs from C57BL/6 mice treated with rSAG2A (FC >2 compared to PECs of BALB/c treated with rSAG2A); Table S2: Exclusive or up-modulated proteins present in PECs from C57BL/6 mice infected with *T. gondii* (FC > 2 compared to PECs of BALB/c infected with *T. gondii*); Table S3: Proteins down-modulated (FC < 1.0) exclusively found in PECs from C57BL/6 mice treated with rSAG2A; Table S4: Proteins down-modulated (FC < 1.0) exclusively found in PECs from C57BL/6 mice infected with *T. gondii*.

Author Contributions: Conceptualization, T.A.R.d.S., J.L.-S., N.M.S. and C.P.P.; methodology, T.A.R.d.S., M.C.d.O., E.M.d.A.S., J.P.d.C.-J. and A.L.C.S.; validation, T.A.R.d.S., M.C.d.O. and J.A.L.-S.; formal analysis T.A.R.d.S., E.M.d.A.S. and T.A.d.O.M.; investigation, T.A.R.d.S., M.C.d.O., E.M.d.A.S. and J.L.-S.; resources, T.W.P.M., J.R.M., N.M.S. and C.P.P., writing—original draft preparation, T.A.R.d.S., E.M.d.A.S., U.R.d.S., É.A.M., J.L.-S.; writing—review and editing T.A.R.d.S., E.M.d.A.S., U.R.d.S., É.A.M., M.M.F. and J.L.-S.; supervision, J.L.-S. and C.P.P.; project administration, C.P.P.; funding acquisition, T.W.P.M., J.R.M., N.M.S. and C.P.P. All authors have read and agreed to the published version of the manuscript.

Funding: T.A.R.d.S., A.L.C.S., and M.C.d.O. were funded by Coordenação de Aperfeiçoamento Pessoal de Nível Superior (CAPES, process No 0001). This research was supported by Banco do Nordeste (BNB/Fundeci). This work was also supported by Conselho Nacional de Pesquisa Científica e Tecnológica (CNPq) and CPP received a Productivity Grant from CNPq (PQ) (Process 303765/2019-4).

Data Availability Statement: The raw data supporting the conclusions of this article will be made available by the authors on request.

Acknowledgments: We thank Horlei Vitória for her help during some laboratorial analysis. We also would like to thank the REBIR-UFU for maintenance and animal welfare.

Conflicts of Interest: The authors declare no conflicts of interest.

References

1. Foroutan, M.; Ghaffarifar, F.; Sharifi, Z.; Dalimi, A.; Jorjani, O. Rhoptry Antigens as *Toxoplasma gondii* Vaccine Target. *Clin. Exp. Vaccine Res.* **2019**, *8*, 4–26. [CrossRef]
2. Fernandes, F.D.; Samoel, G.V.A.; Guerra, R.R.; Bräunig, P.; Machado, D.W.N.; Cargnelutti, J.F.; Sangioni, L.A.; Vogel, F.S.F. Five Years of the Biggest Outbreak of Human Toxoplasmosis in Santa Maria, Brazil: A Review. *Parasitol. Res.* **2024**, *123*, 76. [CrossRef]
3. Finamor, L.P.S.; Madalosso, G.; Levi, J.E.; Zamora, Y.F.; Kamioka, G.A.; Marinho, P.; Nascimento, H.; Muccioli, C.; Belfort, R. Toxoplasmosis Outbreak in São Paulo, Brazil: Epidemiology and Visual Outcome. *Arq. Bras. Oftalmol.* **2024**, *87*, e20220374. [CrossRef]
4. Balbino, L.S.; Bernardes, J.C.; Ladeia, W.A.; Martins, F.D.C.; Nino, B.d.S.L.; Mitsuka-Breganó, R.; Navarro, I.T.; Pinto-Ferreira, F. Epidemiological Study of Toxoplasmosis Outbreaks in Brazil. *Transbound. Emerg. Dis.* **2022**, *69*, 2021–2028. [CrossRef]
5. Wang, J.-L.; Zhang, N.-Z.; Li, T.-T.; He, J.-J.; Elsheikha, H.M.; Zhu, X.-Q. Advances in the Development of Anti-*Toxoplasma gondii* Vaccines: Challenges, Opportunities, and Perspectives. *Trends Parasitol.* **2019**, *35*, 239–253. [CrossRef]
6. Tenter, A.M.; Heckeroth, A.R.; Weiss, L.M. Toxoplasma Gondii: From Animals to Humans. *Int. J. Parasitol.* **2000**, *30*, 1217–1258. [CrossRef]
7. Dunay, I.R.; Gajurel, K.; Dhakal, R.; Liesenfeld, O.; Montoya, J.G. Treatment of Toxoplasmosis: Historical Perspective, Animal Models, and Current Clinical Practice. *Clin. Microbiol. Infect.* **2018**, *31*, e00057-17. [CrossRef]
8. Rezaei, F.; Sarvi, S.; Sharif, M.; Hejazi, S.H.; Pagheh, A.s.; Aghayan, S.A.; Daryani, A. A Systematic Review of *Toxoplasma gondii* Antigens to Find the Best Vaccine Candidates for Immunization. *Microb. Pathog.* **2019**, *126*, 172–184. [CrossRef]
9. Sana, M.; Rashid, M.; Rashid, I.; Akbar, H.; Gomez-Marin, J.E.; Dimier-Poisson, I. Immune Response against Toxoplasmosis—Some Recent Updates RH: *Toxoplasma gondii* Immune Response. *Int. J. Immunopathol. Pharmacol.* **2022**, *36*, 3946320221078436. [CrossRef] [PubMed]

10. Alizadeh, P.; Ahmadpour, E.; Daryani, A.; Kazemi, T.; Spotin, A.; Mahami-Oskouei, M.; Flynn, R.J.; Azadi, Y.; Rajabi, S.; Sandoghchian, S. IL-17 and IL-22 Elicited by a DNA Vaccine Encoding ROP13 Associated with Protection against *Toxoplasma gondii* in BALB/c Mice. *J. Cell. Physiol.* **2019**, *234*, 10782–10788. [CrossRef]
11. Zhang, D.; Jiang, N.; Chen, Q. ROP9, MIC3, and SAG2 Are Heparin-Binding Proteins in *Toxoplasma gondii* and Involved in Host Cell Attachment and Invasion. *Acta Trop.* **2019**, *192*, 22–29. [CrossRef]
12. Zheng, B.; Lou, D.; Ding, J.; Zhuo, X.; Ding, H.; Kong, Q.; Lu, S. GRA24-Based DNA Vaccine Prolongs Survival in Mice Challenged With a Virulent *Toxoplasma gondii* Strain. *Front. Immunol.* **2019**, *10*, 418. [CrossRef]
13. Lu, G.; Zhou, J.; Zhou, A.; Han, Y.; Guo, J.; Song, P.; Zhou, H.; Cong, H.; Hou, M.; Wang, L.; et al. SAG5B and SAG5C Combined Vaccine Protects Mice against *Toxoplasma gondii* Infection. *Parasitol. Int.* **2017**, *66*, 596–602. [CrossRef]
14. Macêdo, A.G.; Cunha, J.P.; Cardoso, T.H.; Silva, M.V.; Santiago, F.M.; Silva, J.S.; Pirovani, C.P.; Silva, D.A.; Mineo, J.R.; Mineo, T.W. SAG2A Protein from *Toxoplasma gondii* Interacts with Both Innate and Adaptive Immune Compartments of Infected Hosts. *Parasites Vectors* **2013**, *6*, 163. [CrossRef]
15. Béla, S.R.; Oliveira Silva, D.A.; Cunha-Júnior, J.P.; Pirovani, C.P.; Chaves-Borges, F.A.; Reis de Carvalho, F.; Carrijo de Oliveira, T.; Mineo, J.R. Use of SAG2A Recombinant *Toxoplasma gondii* Surface Antigen as a Diagnostic Marker for Human Acute Toxoplasmosis: Analysis of Titers and Avidity of IgG and IgG1 Antibodies. *Diagn. Microbiol. Infect. Dis.* **2008**, *62*, 245–254. [CrossRef]
16. Santana, S.S.; Silva, D.A.O.; Vaz, L.D.; Pirovani, C.P.; Barros, G.B.; Lemos, E.M.; Dietze, R.; Mineo, J.R.; Cunha-Junior, J.P. Analysis of IgG Subclasses (IgG1 and IgG3) to Recombinant SAG2A Protein from *Toxoplasma gondii* in Sequential Serum Samples from Patients with Toxoplasmosis. *Immunol. Lett.* **2012**, *143*, 193–201. [CrossRef]
17. Theisen, T.C.; Boothroyd, J.C. Transcriptional Signatures of Clonally Derived Toxoplasma Tachyzoites Reveal Novel Insights into the Expression of a Family of Surface Proteins. *PLoS ONE* **2022**, *17*, e0262374. [CrossRef]
18. Cunha-Júnior, J.P.; Silva, D.A.O.; Silva, N.M.; Souza, M.A.; Souza, G.R.L.; Prudencio, C.R.; Pirovani, C.P.; Cezar M Cascardo, J.; Barbosa, B.F.; Goulart, L.R.; et al. A4D12 Monoclonal Antibody Recognizes a New Linear Epitope from SAG2A *Toxoplasma gondii* Tachyzoites, Identified by Phage Display Bioselection. *Immunobiology* **2010**, *215*, 26–37. [CrossRef]
19. Seeber, F.; Boothroyd, J.C. Escherichia Coli Beta-Galactosidase as an in Vitro and in Vivo Reporter Enzyme and Stable Transfection Marker in the Intracellular Protozoan Parasite Toxoplasma Gondii. *Gene* **1996**, *169*, 39–45. [CrossRef]
20. Hermida, M.D.R.; Malta, R.; de S. Santos, M.D.P.C.; Dos-Santos, W.L.C. Selecting the Right Gate to Identify Relevant Cells for Your Assay: A Study of Thioglycollate-Elicited Peritoneal Exudate Cells in Mice. *BMC Res. Notes* **2017**, *10*, 695. [CrossRef]
21. Teo, C.F.; Zhou, X.W.; Bogyo, M.; Carruthers, V.B. Cysteine Protease Inhibitors Block *Toxoplasma gondii* Microneme Secretion and Cell Invasion. *Antimicrob. Agents Chemother.* **2007**, *51*, 679–688. [CrossRef]
22. Leal-Sena, J.A.; Dos Santos, J.L.; Dos Santos, T.A.R.; de Andrade, E.M.; de Oliveira Mendes, T.A.; Santana, J.O.; Mineo, T.W.P.; Mineo, J.R.; da Cunha-Júnior, J.P.; Pirovani, C.P. *Toxoplasma gondii* Antigen SAG2A Differentially Modulates IL-1β Expression in Resistant and Susceptible Murine Peritoneal Cells. *Appl. Microbiol. Biotechnol.* **2018**, *102*, 2235–2249. [CrossRef]
23. Livak, K.J.; Schmittgen, T.D. Analysis of Relative Gene Expression Data Using Real-Time Quantitative PCR and the 2-ΔΔCT Method. *Methods* **2001**, *25*, 402–408. [CrossRef] [PubMed]
24. Gu, Z. Complex Heatmap Visualization. *iMeta* **2022**, *1*, e43. [CrossRef] [PubMed]
25. R Core Team. A: A Language and Environment for Statistical Computing. Available online: https://www.r-project.org/ (accessed on 2 June 2023).
26. Carpentier, S.C.; Witters, E.; Laukens, K.; Deckers, P.; Swennen, R.; Panis, B. Preparation of Protein Extracts from Recalcitrant Plant Tissues: An Evaluation of Different Methods for Two-Dimensional Gel Electrophoresis Analysis. *Proteomics* **2005**, *5*, 2497–2507. [CrossRef] [PubMed]
27. Vanhove, A.-C.; Vermaelen, W.; Swennen, R.; Carpentier, S.C. A Look behind the Screens: Characterization of the HSP70 Family during Osmotic Stress in a Non-Model Crop. *J. Proteom.* **2015**, *119*, 10–20. [CrossRef] [PubMed]
28. Szklarczyk, D.; Gable, A.L.; Lyon, D.; Junge, A.; Wyder, S.; Huerta-Cepas, J.; Simonovic, M.; Doncheva, N.T.; Morris, J.H.; Bork, P.; et al. STRING V11: Protein-Protein Association Networks with Increased Coverage, Supporting Functional Discovery in Genome-Wide Experimental Datasets. *Nucleic Acids Res.* **2019**, *47*, D607–D613. [CrossRef] [PubMed]
29. Bader, G.D.; Hogue, C.W. V An Automated Method for Finding Molecular Complexes in Large Protein Interaction Networks. *BMC Bioinform.* **2003**, *4*, 2. [CrossRef]
30. Maere, S.; Heymans, K.; Kuiper, M. BiNGO: A Cytoscape Plugin to Assess Overrepresentation of Gene Ontology Categories in Biological Networks. *Bioinformatics* **2005**, *21*, 3448–3449. [CrossRef]
31. Zhou, D.H.; Yuan, Z.G.; Zhao, F.R.; Li, H.L.; Zhou, Y.; Lin, R.Q.; Zou, F.C.; Song, H.Q.; Xu, M.J.; Zhu, X.Q. Modulation of Mouse Macrophage Proteome Induced by *Toxoplasma gondii* Tachyzoites in Vivo. *Parasitol. Res.* **2011**, *109*, 1637–1646. [CrossRef]
32. Letscher-Bru, V.; Pfaff, A.W.; Abou-Bacar, A.; Filisetti, D.; Antoni, E.; Villard, O.; Klein, J.-P.; Candolfi, E. Vaccination with *Toxoplasma gondii* SAG-1 Protein Is Protective against Congenital Toxoplasmosis in BALB/c Mice but Not in CBA/J Mice. *Infect. Immun.* **2003**, *71*, 6615–6619. [CrossRef]
33. Lee, J.H.; Yuk, J.M.; Cha, G.H.; Lee, Y.H. Expression of Cytokines and Co-Stimulatory Molecules in the Toxoplasma Gondii-Infected Dendritic Cells of C57BL/6 and BALB/c Mice. *Parasites Hosts Dis.* **2023**, *61*, 138–146. [CrossRef]
34. Delgado Betancourt, E.; Hamid, B.; Fabian, B.T.; Klotz, C.; Hartmann, S.; Seeber, F. From Entry to Early Dissemination-Toxoplasma Gondii's Initial Encounter With Its Host. *Front. Cell. Infect. Microbiol.* **2019**, *9*, 46. [CrossRef]

35. Sun, H.; Zhuo, X.; Zhao, X.; Yang, Y.; Chen, X.; Yao, C.; Du, A. The Heat Shock Protein 90 of *Toxoplasma gondii* Is Essential for Invasion of Host Cells and Tachyzoite Growth. *Parasite* **2017**, *24*, 22. [CrossRef]
36. Sugi, T.; Tomita, T.; Kidaka, T.; Kawai, N.; Hayashida, K.; Weiss, L.M.; Yamagishi, J. Single Cell Transcriptomes of In Vitro Bradyzoite Infected Cells Reveals *Toxoplasma gondii* Stage Dependent Host Cell Alterations. *Front. Cell. Infect. Microbiol.* **2022**, *12*, 848693. [CrossRef] [PubMed]
37. Mesa, J.; Alsina, C.; Oppermann, U.; Parés, X.; Farrés, J.; Porté, S. Human Prostaglandin Reductase 1 (PGR1): Substrate Specificity, Inhibitor Analysis and Site-Directed Mutagenesis. *Chem. Biol. Interact.* **2015**, *234*, 105–113. [CrossRef]
38. Epaud, R.; Delestrain, C.; Weaver, T.E.; Akinbi, H.T. Bacterial Killing Is Enhanced by Exogenous Administration of Lysozyme in the Lungs. *Respir. Med. Res.* **2019**, *76*, 22–27. [CrossRef]
39. Li, Z.; Srivastava, S.; Yang, X.; Mittal, S.; Norton, P.; Resau, J.; Haab, B.; Chan, C. A Hierarchical Approach Employing Metabolic and Gene Expression Profiles to Identify the Pathways That Confer Cytotoxicity in HepG2 Cells. *BMC Syst. Biol.* **2007**, *1*, 21. [CrossRef] [PubMed]

microorganisms

Brief Report

Oxidative Stress in the Murine Model of Extraparenchymal Neurocysticercosis

Diego Generoso [1], Tatiane de Camargo Martins [1], Camila Renata Corrêa Camacho [1], Manuella Pacífico de Freitas Segredo [1], Sabrina Setembre Batah [2], Alexandre Todorovic Fabro [2], Edda Sciutto [3], Agnès Fleury [3,4], Pedro Tadao Hamamoto Filho [1,*] and Marco Antônio Zanini [1]

1 Botucatu Medical School, UNESP—Universidade Estadual Paulista, Botucatu 18618-686, Brazil
2 Ribeirão Preto Medical School, USP—Universidade de São Paulo, Ribeirão Preto 14049-900, Brazil
3 Instituto de Investigación Biomédicas, UNAM—Universidade Nacional Autónoma de México, Ciudad de México 04510, Mexico
4 INNN—Instituto Nacional de Neurología y Neurocirurgía, Ciudad de México 14269, Mexico
* Correspondence: pedro.hamamoto@unesp.br

Abstract: Oxidative stress is associated with several infectious diseases, as well as the severity of inflammatory reactions. The control of inflammation during parasite destruction is a target of neurocysticercosis treatment, as inflammation is strongly related to symptom severity. In this study, we investigated the presence of malondialdehyde and protein carbonyl, two by-products of reactive oxygen species (ROS), in an experimental model of extraparenchymal neurocysticercosis. Twenty male and twenty female rats were inoculated with 50 cysts of *Taenia crassiceps* in the subarachnoid space of the cisterna magna. Ten animals (five males and five females) were used as controls. Three months after inoculation, their brains were harvested for oxidative stress and histological assessments. Infected animals had higher scores for inflammatory cell infiltrates, malondialdehyde, and protein carbonyl. These results encourage future efforts to monitor oxidative stress status in neurocysticercosis, particularly in the context of controlling inflammation.

Keywords: neurocysticercosis; oxidative stress; malondialdehyde; protein carbonyl; inflammation

1. Introduction

Neurocysticercosis (NCC) is a common parasitic disease of the central nervous system (CNS) in low- and middle-income countries. Despite a recent decrease in its endemicity (particularly in Latin America), NCC remains a relevant public health problem with significant social and economic burdens in endemic countries [1–3]. NCC is also a health challenge in countries receiving immigrants from endemic regions, although its extent and impact remain unclear [4,5].

NCC may present as one of two different forms depending on the site at which the larvae of *Taenia solium* are found within the CNS: parenchymal (P-NCC) and extraparenchymal (EP-NCC). While P-NCC is highly responsive to clinical treatment, with a relatively benign course (except for massive infection), EP-NCC is less responsive to anthelminthic drugs and is more prone to complications such as vasculitis, hydrocephalus, and increased intracranial pressure [6–8]. Inflammation is a key pathophysiological mechanism underlying such complications [9], and oxidative stress has been shown to increase the severity of NCC, primarily in patients with neurocysticercosis-induced hydrocephalus [10]. Oxidative stress is initiated by neutrophil and monocyte activation during inflammation. When activated, these cells release reactive oxygen species (ROS), such as superoxide radicals, hydrogen peroxides, and hydroxyl radicals. Excess ROS, known as oxidative stress, damages cells and tissues by inducing oxidative damage to DNA, proteins, and lipids, resulting in organ dysfunction [11]. Lipids are the primary macromolecules affected by oxidative stress, resulting in toxic products such as malondialdehyde, which can destroy proteins via a mechanism termed protein carbonylation [12]. Oxidative stress and inflammation are involved in the pathophysiology of neurocysticercosis.

Since this mechanism still requires better understanding, the use of experimental models of NCC has allowed for a better understanding of NCC pathophysiology and opened pathways for testing new therapeutic regimens [13–17]. The experimental model of EP-NCC with *Taenia crassiceps* has further been shown to closely resemble the features of human diseases, including brain parenchymal displacement, hydrocephalus, and subarachnoid inflammation [18,19]. Previously, *T. crassiceps* was not considered as a suitable model for studying neurocysticercosis, because the parasite does not invade the brain parenchyma—rather, it displaces the nervous tissue. This concern would make sense for studies focusing on parenchymal neurocysticercosis [13]. Nevertheless, for EP-NCC, the growth of the parasite with the displacement of the brain is precisely what is desired. Therefore, the use of *T. crassiceps* reproduces the human disease and allows for the investigation of several pathophysiologic mechanisms, such as oxidative stress.

In the present study, we aimed to investigate the presence of in situ oxidative stress markers in EP-NCC rats, as well as to compare differences between male and female infected rats.

2. Materials and Methods

In this study, we used female and male Wistar rats (*Rattus norvegicus*) handled according to the ethical guidelines and current legislation for animal research. This project was approved by the local Institutional Review Board for ethics of animal use (CEUA 1330/2019).

Rats aged 6 weeks were housed under 12 h light–dark cycles at room temperature (approx. 21 °C), with free availability of water and conventional commercial food. In total, 50 animals (25 males and 25 females), including 10 rats as controls (5 of each sex), were used. The sample size was estimated considering a mortality rate of 30% following the inoculations and a rate of 66% of successful infections, based on reported data on oxidative stress in neurocysticercosis, with a type II error of 20%, and a level of significance of 5%.

2.1. Study Design

A total of 40 rats (20 males and 20 females) received a subarachnoid injection of 50 cysts of *T. crassiceps* (Figure 1A), and were kept alive for three months post-inoculation. After the observational period, the animals were euthanized with a thiopental overdose, and their brains were harvested following wide craniectomy. The presence of cysts in the subarachnoid space upon brain harvesting confirmed the inclusion of the animal (Figure 1B), while rats that did not present with any cysts (not infected) were discarded. The right hemisphere was embedded in 4% buffered paraformaldehyde, and the left hemisphere was frozen and stored at −80 °C for posterior quantification of oxidative stress markers (malondialdehyde and protein carbonyl). Ten animals (five male and five female) did not receive any intervention and underwent the same protocol for euthanasia and tissue processing at the same age as the inoculated animals (Figure 1C).

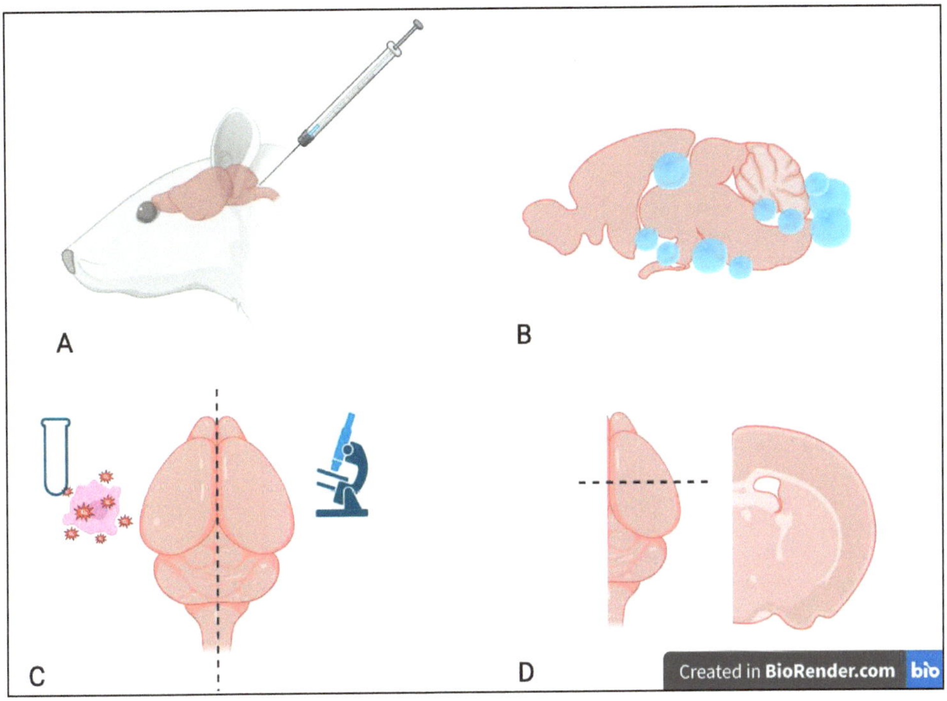

Figure 1. Schematic representation of the experimental procedures. (**A**): The cyst injection was performed at the cisterna magna, i.e., the subarachnoid space between the posterior-interior portion of the cerebellum and the dorsal portion of the medulla. (**B**): Three months after the inoculation, the cysts could be observed throughout the subarachnoid space, located predominantly in the basal convexity of the brain and brainstem. (**C**): The right hemisphere was used for histologic analysis, and the left hemisphere for oxidative stress assays. (**D**): The dorsal and coronal views at the level of tissue histologic assessment.

2.2. Parasites and Inoculations

Cysts of *T. crassiceps* (ORF strain) were maintained in the peritoneal cavity of mice (*Mus musculus*), as described elsewhere [20,21], and aseptically transferred onto Petri dishes with saline 0.9%. Cysts were selected based on their size (0.5 mm) and membrane integrity. Each rat received 50 cysts in 0.2 mL of saline 0.9%, as previously described [16]. Briefly, after intraperitoneal anesthesia with xylazine and ketamine, the skin was incised at the posterior cervical region along the craniocervical junction, and the subarachnoid space of the cisterna magna was punctured with a 24-gauge needle for injection. Finally, the skin was sutured with a 4.0 mononylon thread.

2.3. Histologic Assessment

After 24 h of fixation in 4% buffered paraformaldehyde, the right hemispheres were dehydrated with increasing concentrations of alcohol, clarified in xylene, and paraffinized. Slices with 5 μm of thickness were subsequently taken at the level of the optic chiasm (Figure 1D) and stained with hematoxylin and eosin.

The lymphocyte infiltrates in the basal arachnoid–pia–mater–brain interface were analyzed semi-quantitatively using the following score: 1, absence of lymphocyte infiltration; 2, discrete infiltration (scarce and sparce lymphocytes); 3, moderate infiltration (focally present lymphocytes); and 4, intense infiltration (several lymphocytes diffusely).

2.4. Oxidative Stress Assessment

Because free radicals (reactive oxygen species) are rapidly degraded, they cannot be measured directly; instead, their activity must be measured indirectly by assessing their byproducts. Herein, we assessed the malondialdehyde and protein carbonyl levels. Malondialdehyde is a derivative of the peroxidation of polyunsaturated fatty acids [22], and carbonyl proteins refer to the process by which the oxidation of the side chains of amino acids leads to the formation of reactive ketones and aldehydes [23].

After defrosting the left hemisphere, the samples were homogenized. For malondialdehyde quantification, 250 µL of brain tissue supernatant was mixed with 750 µL of 10% trichloroacetic acid for protein precipitation. The samples were then centrifuged at 3000 rpm for 5 min (Eppendorf ® Centrifuge 5804-R, Hamburg, Germany), and the supernatant was removed. Thiobarbituric acid (TBA) was added in a proportion of 0.67% (1:1) and the samples were heated for 15 min at 100 °C. After cooling, the absorbance was measured at 535 nm using a Spectra Max 190 microplate reader (Molecular Devices, Sunnyvale, CA, USA). The MDA concentration was calculated using the molar extinction coefficient (1.56×10^5 M^{-1} cm^{-1}) and the absorption of the samples, with the final result reported in nmol/g of protein.

To assess protein carbonyl, 100 µg of brain tissue was homogenized in ice-cold PBS solution (1 mL, pH 7.4) using an ULTRA-TURRAX® T25 basic (IKA® Werke, Staufen, Germany) and then centrifuged at 800× *g* for 10 min at 4 °C. The supernatant was incubated with 2,4-dinitrophenylhydrazine (DNPH) for 1 h to derivatize carbonyl proteins. The samples were subsequently deproteinized with trichloroacetic acid and centrifuged. The supernatant was discarded, and the sediment was washed three times with an ethanol/ethyl acetate solution to remove excess DNPH, preventing the quantification of chromophores not bound to carbonyl protein. Finally, 6M guanidine hydrochloride was added to the washed sediment, and the absorbance was measured at a wavelength of 370 nm. The total protein concentration in the samples was quantified using commercial assays to adjust the carbonyl protein values. The concentrations of carbonyl proteins were expressed as µg/mg of protein.

2.5. Statistical Analysis

The data were analyzed in three steps. First, all infected animals (regardless of sex) were compared to the control group. For the parametric data, we used an independent *t*-test. Next, the same tests were used to compare the infected males and females with their respective controls. Finally, we compared infected male and female rats. The analyses were then performed using the Statistical Package for the Social Sciences (SPSS) version 24.0 (IBM Corp., Armonk, NY, USA) and GraphPad Prism v. 8.2.0 (GraphPad Software, La Jolla, CA, USA). Differences were considered statistically significant at $p < 0.05$.

3. Results

Among the twenty male rats, seven died after inoculation and two did not have any cysts, while among the twenty female rats, eight died and two did not develop infection. Therefore, 11 male and 10 female infected rats were included in the final analysis and compared to 10 controls (5 males and 5 females).

Histologic assessment revealed a normal arachnoid–pia–mater–brain interface in the control animals, with a thin and delicate arachnoid layer juxtaposed to the pia–mater, with few detectable lymphocytes. Among the infected animals, variable degrees of lymphocyte infiltration, with many cases of thickened arachnoids and lymphocytes in the foci of infiltration or diffuse distribution, were observed (Figure 2). The median lymphocyte infiltration score was higher in the infected animals than in the control animals (3.0 vs. 1.0, $p < 0.001$). This difference remained significant between males ($p = 0.038$) and females ($p = 0.003$). We found no significant differences between the infected male and female rats ($p = 0.721$).

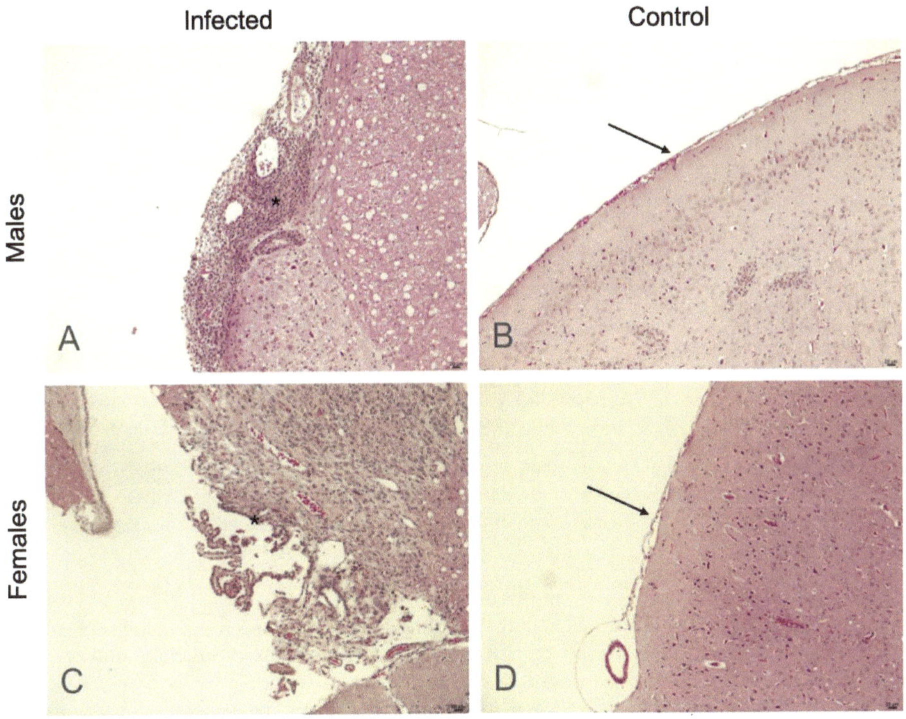

Figure 2. Microscopic view of the basal arachnoid–pia–mater–brain interfaces for male (**A**,**B**) and female (**C**,**D**) rats. In control animals, a thin arachnoid layer (arrows) juxtaposed to the pia–mater with preserved histoarchitecture (**B**,**D**) can be observed. In infected animals (**A**,**C**), the arachnoid is thickened and lymphocytes (*) can be observed in intense (**A**) or discrete (**C**) infiltrations.

Figure 3A shows a significantly higher mean of malondialdehyde levels (15.6 ± 5.8 nmol/mg) in the infected than in the control rats (11.3 ± 3.2 nmol/g), (p = 0.048). Malondialdehyde levels showed no significant difference between males (Figure 3B, p = 0.059) and females (Figure 3C, p = 0.305). Infected males had higher MDA concentrations than females (p = 0.024, Figure 3D).

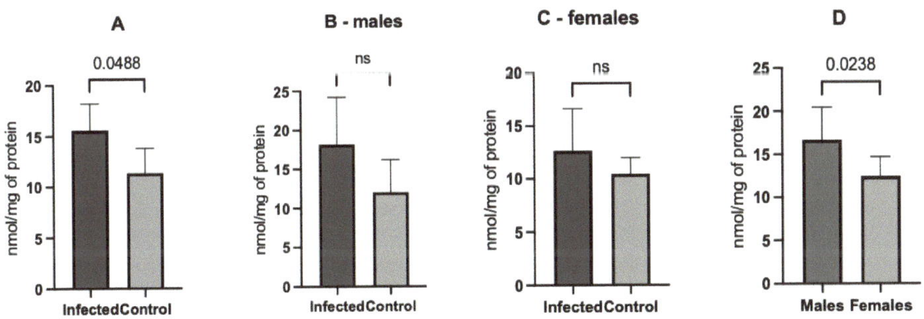

Figure 3. Comparisons of the malondialdehyde concentration between the experimental groups. When considering all infected animals, the mean concentration was higher than the control animals (**A**). However, the differences were no longer significant (ns) when analyzing male (**B**) and female (**C**) animals with their controls. Further, considering the infected animals, males had a higher concentration of malondialdehyde than females (**D**).

Infected animals had a higher concentration of protein carbonyl than controls (789.2 ± 128.7 vs. 582.5 ± 30.3 µg/g, $p < 0.001$); this effect was also seen when comparing only male ($p = 0.007$) and female rats ($p = 0.005$) (Figure 4). No significant differences were observed between the infected males and females ($p = 0.513$).

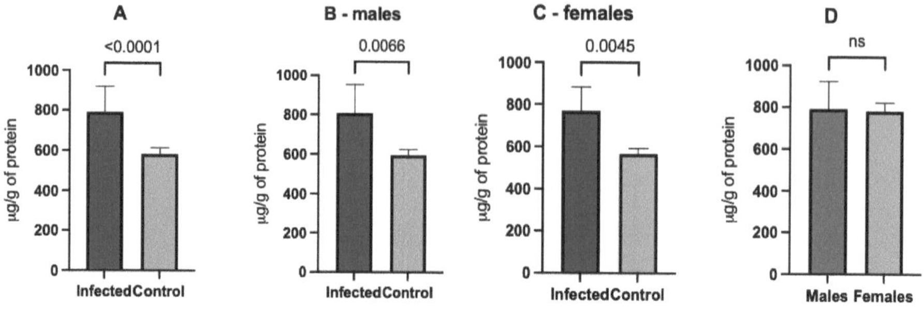

Figure 4. Comparisons of protein carbonyl between the experimental groups. The infected animals presented higher concentrations when comparing all animals (**A**), the males (**B**), and females (**C**) with their respective controls. There was no difference between infected males and females (**D**). ns: non-significant.

4. Discussion

Inflammation occurs throughout the natural history of NCC, even in cases of calcified cysts [9]. Inflammation plays a pivotal role in effective parasite destruction; however, this effect is a double-edged sword, as it is also related to symptom severity. Since the use of anthelminthics can trigger inflammation by increasing the exposure of parasites to immunological surveillance, the concomitant use of corticosteroids is necessary to control the possible increase in severity [24]. As such, the control of inflammation without impairing the effectiveness of cysticides is a significant research target.

Reactive oxygen and nitrogen species produced during inflammation may damage DNA, proteins, and lipids, resulting in consequent lesions in the surrounding cells and tissues [25]. While physiological amounts of these reactive species can be tolerated under healthy conditions, oxidative stress has been clearly demonstrated to impair neurological function under several conditions, such as neurodegenerative and infectious diseases [26–28].

Few studies have addressed the oxidative stress status in patients with NCC. Rodríguez et al. found that patients with NCC had higher levels of lipid peroxidase (LP) in the cerebrospinal fluid than controls, while the LP levels were also significantly correlated with the inflammatory response and severity of the symptoms [12]. Interestingly, even patients receiving steroid therapy exhibited increased LP levels. Prasad et al. studied children with NCC and found higher levels of malondialdehyde, protein carbonyl, and nitrite, accompanied by low levels of antioxidants in the CSF [29]. It has also been reported for different infections that such acute respiratory conditions caused by different viruses where oxidative stress is implicated with lung tissue injury and epithelial barrier dysfunction could also increase the susceptibility to new infections. Increased oxidative stress may also promote tissue fibrosis and metabolic dysfunction, as has been reported in viral hepatitis [30].

Oxidative stress in the brain has been evaluated in animal models of meningitis. Barichello et al. found increased levels of protein carbonyl in the hippocampus and the cortex from 6 to 96 h after the subarachnoid injection of group B Streptococcus [31]. Using a similar approach with *Escherichia coli*, Giridharan et al. also observed increased levels of MDA, protein carbonyl, and superoxide dismutase in the hippocampus [32]. The oxidative stress in the hippocampus may be related to epileptogenesis, and the use of antioxidant drugs may improve long-term outcomes of patients exposed to potential epileptogenic insults, such as traumatic brain injury [33,34]. Additionally, for chronic infections such as NCC, the scenario may be less promising. For instance, in experimental models of toxoplasmosis, the use of antioxidants is associated with increase in the parasitemia [35].

Another possible effect associated with oxidative stress in our experimental model is hydrocephalus, which refers to an active distension of the ventricular system of the brain from an imbalance on CSF production, circulation, and absorption [36]. Hydrocephalus is a common feature of EP-NCC due to a combination of inflammation (triggered by the host response against the parasite) and mechanical obstruction of the CSF pathways (caused by the presence of the cysts in the narrow points of the ventricle and basal cisterns) [7]. We have previously demonstrated that this model leads to hydrocephalus [18], and several experimental models of hydrocephalus have demonstrated that oxidative stress occurs and is related to the severity of neurological lesions [37–39].

In the present study, we demonstrated the occurrence of oxidative stress in a murine model of extraparenchymal neurocysticercosis. The differences between infected and control animals were even more pronounced in protein carbonyl levels, while significant differences were maintained even between male and female rats. Further studies should attempt cyst preservation upon brain harvesting to correlate the intensity of oxidative stress with the parasite load.

In previous studies, female rats showed greater inflammatory features than male rats [40,41]. In contrast, in the present study, we found higher concentrations of malondialdehyde in males. A possible explanation for this relationship between hormones and the modulation of the antioxidant defense system; estrogen and progesterone display antioxidant properties [42]. However, females seem to present more intense inflammatory features in NCC than males [43,44]. The crosstalk between cysticercosis, sex hormones, and oxidative stress should be further explored.

The results obtained in our study should be considered to optimize the treatment of ExP-NCC and reduce associated collateral damage.

In summary, in this study, we found significant evidence to indicate an increase in oxidative stress markers in NCC, one more feature consolidating the similarities between this model of extraparenchymal neurocysticercosis and human infection. Monitoring the oxidative stress status throughout the course of infection could provide relevant information regarding the optimal time to administer anti-helminthics and corticoids, as well as for the withdrawal of corticoids. However, further experimental studies are required to evaluate this possibility from a translational perspective.

Author Contributions: Conceptualization, D.G., P.T.H.F. and M.A.Z.; methodology, D.G., T.d.C.M., C.R.C.C., S.S.B. and A.T.F.; validation, M.P.d.F.S., E.S. and A.F.; formal analysis, C.R.C.C. and P.T.H.F.; investigation, D.G., T.d.C.M. and P.T.H.F.; resources, C.R.C.C. and P.T.H.F.; data curation, D.G. and T.d.C.M.; writing—original draft preparation, P.T.H.F.; writing—review and editing, D.G., C.R.C.C., M.P.d.F.S., S.S.B., A.T.F., E.S., A.F. and M.A.Z.; visualization, all; supervision, M.A.Z.; project administration, P.T.H.F.; funding acquisition, P.T.H.F. All authors have read and agreed to the published version of the manuscript.

Funding: This research was partially funded by CNPQ—Conselho Nacional de Desenvolvimento Científico e Tecnológico (grant number 001); by CAPES—Coordenação de Aperfeiçoamento de Pessoal de Nível Superior (grant number 313047/2023-5), and by FAPESP—Fundação de Amparo à Pesquisa do Estado de São Paulo (grant number 22/03042-5).

Data Availability Statement: The data presented in this study are available on request from the corresponding author. The data are not publicly available due to privacy recommendations.

Conflicts of Interest: The authors declare no conflicts of interest.

References

1. Gripper, L.B.; Welburn, S.C. Neurocysticercosis infection and disease—A review. *Acta Trop.* **2017**, *166*, 218–224. [CrossRef]
2. Rodríguez-Rivas, R.; Flisser, A.; Norcia, L.F.; Hamamoto Filho, P.T.; Bonilla-Aldana, D.K.; Rodriguez-Morales, A.J.; Carpio, A.; Romo, M.L.; Fleury, A. Neurocysticercosis in Latin America: Current epidemiological situation based on official statistics from four countries. *PLOS Negl. Trop. Dis.* **2022**, *16*, e0010652. [CrossRef]

3. Singh, B.B.; Khatkar, M.S.; Gill, J.P.; Dhand, N.K. Estimation of the health and economic burden of neurocysticercosis in India. *Acta Trop.* **2017**, *165*, 161–169. [CrossRef] [PubMed]
4. Serpa, J.A.; White, A.C., Jr. Neurocysticercosis in the United States. *Pathog. Glob. Health* **2012**, *106*, 256–260. [CrossRef] [PubMed]
5. Abraham, A.; Schmidt, V.; Kaminski, M.; Stelzle, D.; De Meijere, R.; Bustos, J.; Sahu, P.S.; Garcia, H.H.; Bobić, B.; Cretu, C.; et al. Epidemiology and surveillance of human (neuro)cysticercosis in Europe: Is enhanced surveillance required? *Trop. Med. Int. Health* **2020**, *25*, 566–578. [CrossRef] [PubMed]
6. Marcin Sierra, M.; Arroyo, M.; Cadena Torres, M.; Ramírez Cruz, N.; García Hernández, F.; Taboada, D.; Galicia Martínez, Á.; Govezensky, T.; Sciutto, E.; Toledo, A.; et al. Extraparenchymal neurocysticercosis: Demographic, clinicoradiological, and inflammatory features. *PLoS Negl. Trop. Dis.* **2017**, *11*, e0005646. [CrossRef]
7. Hamamoto Filho, P.T.; Zanini, M.A.; Fleury, A. Hydrocephalus in neurocysticercosis: Challenges for Clinical Practice and Basic Research Perspectives. *World Neurosurg.* **2019**, *126*, 264–271. [CrossRef]
8. Abanto, J.; Blanco, D.; Saavedra, H.; Gonzales, I.; Siu, D.; Pretell, E.J.; Bustos, J.A.; Garcia, H.H.; Cysticercosis Working Group in Peru. Mortality in parenchymal and subarachnoid neurocysticercosis. *Am. J. Trop. Med. Hyg.* **2021**, *105*, 176–180. [CrossRef]
9. Hamamoto Filho, P.T.; Fragoso, G.; Sciutto, E.; Fleury, A. Inflammation in neurocysticercosis: Clinical relevance and impact on treatment decisions. *Expert. Rev. Anti-Infect. Ther.* **2021**, *19*, 1503–1518. [CrossRef]
10. Bartsch, H.; Nair, J. Accumulation of lipid peroxidation-derived DNA lesions: Potential lead markers for chemoprevention of inflammation-driven malignancies. *Mutat. Res.* **2005**, *591*, 34–44. [CrossRef]
11. Halliwell, B. Antioxidants in human health and disease. *Annu. Rev. Nutr.* **1996**, *16*, 33–50. [CrossRef]
12. Rodríguez, U.; Ríos, C.; Corona, T.; Talayero, B.; Ostrosky-Wegman, P.; Herrera, L.A. Lipid peroxidation in the cerebrospinal fluid of patients with neurocysticercosis. *Trans. R. Soc. Trop. Med. Hyg.* **2008**, *102*, 1025–1031. [CrossRef]
13. Alvarez, J.I.; Mishra, B.B.; Gundra, U.M.; Mishra, P.K.; Teale, J.M. Mesocestoides corti intracranial infection as a murine model for neurocysticercosis. *Parasitology* **2010**, *137*, 359–372. [CrossRef] [PubMed]
14. Matos-Silva, H.; Reciputti, B.P.; Paula, E.C.; Oliveira, A.L.; Moura, V.B.L.; Vinaud, M.C.; Oliveira, M.A.P.; Lino-Júnior, L.J. Experimental encephalitis caused by Taenia crassiceps cysticerci in mice. *Arq. Neuropsiquiatr.* **2012**, *70*, 287–292. [CrossRef] [PubMed]
15. Fleury, A.; Trejo, A.; Cisneros, H.; García-Navarrete, R.; Villalobos, N.; Hernández, M.; Villeda Hernández, J.; Hernández, B.; Rosas, G.; Bobes, R.J.; et al. Taenia solium: Development of an Experimental Model of Porcine neurocysticercosis. *PLoS Negl. Trop. Dis.* **2015**, *9*, e0003980. [CrossRef] [PubMed]
16. Verastegui, M.R.; Mejia, A.; Clark, T.; Gavidia, C.M.; Mamani, J.; Ccopa, F.; Angulo, N.; Chile, N.; Carmen, R.; Medina, R.; et al. Novel rat model for neurocysticercosis using Taenia solium. *Am. J. Pathol.* **2015**, *185*, 2259–2268. [CrossRef]
17. de Lange, A.; Mahanty, S.; Raimondo, J.V. Model systems for investigating disease processes in neurocysticercosis. *Parasitology* **2019**, *146*, 553–562. [CrossRef]
18. Hamamoto Filho, P.T.; Fogaroli, M.O.; Oliveira, M.A.C.; Oliveira, C.C.; Batah, S.S.; Fabro, A.T.; Vulcano, L.C.; Bazan, R.; Zanini, M.A. A rat model of neurocysticercosis-induced hydrocephalus: Chronic progressive hydrocephalus with mild clinical impairment. *World Neurosurg.* **2019**, *132*, e535–e544. [CrossRef]
19. Espinosa-Cerón, A.; Méndez, A.; Hernández-Aceves, J.; Juárez-González, J.C.; Villalobos, N.; Hernández, M.; Díaz, G.; Soto, P.; Concha, L.; Pérez-Osorio, I.N.; et al. Standardizing an experimental murine model of extraparenchymal neurocysticercosis that immunologically resembles human infection. *Brain Sci.* **2023**, *13*, 1021. [CrossRef]
20. Chernin, J. Maintenance of the metacestodes of *Taenia crassiceps* inn rats. *J. Helminthol.* **1975**, *49*, 91–92. [CrossRef]
21. Willms, K.; Zurabian, R. *Taenia crassiceps*: In vivo and in vitro models. *Parasitology* **2010**, *137*, 335–346. [CrossRef] [PubMed]
22. Del Rio, D.; Stewart, A.J.; Pellegrini, N. A review of recent studies on malondialdehyde as toxic molecule and biological marker of oxidative stress. *Nutr. Metab. Cardiovasc. Dis.* **2005**, *15*, 316–328. [CrossRef] [PubMed]
23. Suzuki, Y.J.; Carini, M.; Butterfield, D.A. Protein carbonylation. *Antioxid. Redox Signal* **2010**, *12*, 323–325. [CrossRef]
24. Toledo, A.; Osorio, R.; Matus, C.; Martinez Lopez, Y.; Ramirez Cruz, N.; Sciutto, E.; Fragoso, G.; Arauz, A.; Carrillo-Mezo, R.; Fleury, A. Human extraparenchymal neurocysticercosis: The control of inflammation favors the Host...but Also the Parasite. *Front. Immunol.* **2018**, *9*, 2652. [CrossRef]
25. Munk, M.; Poulsen, F.R.; Larsen, L.; Nordström, C.H.; Nielsen, T.H. Cerebral metabolic changes related to oxidative metabolism in a model of bacterial meningitis induced by lipopolysaccharide. *Neurocrit Care* **2018**, *29*, 496–503. [CrossRef]
26. Trist, B.G.; Hare, D.J.; Double, K.L. Oxidative stress in the aging substantia nigra and the etiology of Parkinson's disease. *Aging Cell* **2019**, *18*, e13031. [CrossRef]
27. Principi, N.; Esposito, S. Bacterial meningitis: New treatment options to reduce the risk of brain damage. *Expert. Opin. Pharmacother.* **2020**, *21*, 97–105. [CrossRef]
28. Bai, R.; Guo, J.; Ye, X.Y.; Xie, Y.; Xie, T. Oxidative stress: The core pathogenesis and mechanism of Alzheimer's disease. *Ageing Res. Rev.* **2022**, *77*, 101619. [CrossRef] [PubMed]
29. Prasad, R.; Anil, M.O.P.; Mishra, O.P.; Mishra, S.P.; Upadhyay, R.S.; Singh, T.B. Oxidative stress in children with neurocysticercosis. *Pediatr. Infect. Dis. J.* **2012**, *31*, 1012–1015. [CrossRef]
30. Ivanov, A.V.; Bartosch, B.; Isaguliants, M.G. Oxidative stress in infection and consequent disease. *Oxid. Med. Cell Longev.* **2017**, *2017*, 3496043. [CrossRef]

31. Barichello, T.; Lemos, J.L.; Generoso, J.S.; Cipriano, A.L.; Milioli, G.L.; Marcelino, D.M.; Vuolo, F.; Petronilho, F.; Dal-Pizzol, F.; Vilela, M.C.; et al. Oxidative Stress, Cytokine/Chemokine and Disruption of Blood–Brain Barrier in Neonate Rats After Meningitis by *Streptococcus agalactiae*. *Neurochem. Res.* **2011**, *36*, 1922–1930. [CrossRef]
32. Giridharan, V.V.; Simões, L.R.; Dagostin, V.S.; Generoso, J.S.; Rezin, G.T.; Florentino, D.; Muniz, J.P.; Collodel, A.; Petronilho, F.; Quevedo, J.; et al. Temporal changes of oxidative stress markers in *Escherichia coli* K1-induced experimental meningitis in a neonatal rat model. *Neurosci. Lett.* **2017**, *653*, 288–295. [CrossRef] [PubMed]
33. Geronzi, U.; Lotti, F.; Grosso, S. Oxidative stress in epilepsy. *Expert. Rev. Neurother.* **2018**, *18*, 427–434. [CrossRef] [PubMed]
34. Eastman, C.L.; D'Ambrosio, R.; Ganesh, T. Modulating neuroinflammation and oxidative stress to prevent epilepsy and improve outcomes after traumatic brain injury. *Neuropharmacology* **2020**, *172*, 107907. [CrossRef] [PubMed]
35. Szewczyk-Golec, K.; Pawłowska, M.; Wesołowski, R.; Wróblewski, M.; Mila-Kierzenkowska, C. Oxidative Stress as a Possible Target in the Treatment of Toxoplasmosis: Perspectives and Ambiguities. *Int. J. Mol. Sci.* **2021**, *22*, 5705. [CrossRef]
36. Rekate, H. The definition and classification of hydrocephalus: A personal recommendation to stimulate debate. *Cerebrospinal Fluid. Res.* **2008**, *5*, 2. [CrossRef]
37. Di Curzio, D.L.; Turner-Brannen, E.; Del Bigio, M.R. Oral antioxidant therapy for juvenile rats with kaolin-induced hydrocephalus. *Fluid. Barriers CNS* **2014**, *11*, 23. [CrossRef]
38. Wang, C.; Wang, X.; Tan, C.; Wang, Y.; Tang, Z.; Zhang, Z.; Liu, J.; Xiao, G. Novel therapeutics for hydrocephalus: Insights from animal models. *CNS Neurosci. Ther.* **2021**, *27*, 1012–1022. [CrossRef]
39. Silva, S.C.; Beggiora, P.S.; Catalão, C.H.R.; Dutra, M.; Matias Júnior, I.; Santos, M.V.; Machado, H.R.; Lopes, L.S. Hyperbaric Oxygen Therapy Associated with Ventricular-Subcutaneous Shunt Promotes Neuroprotection in Young Hydrocephalic Rats. *Neuroscience* **2022**, *488*, 77–95. [CrossRef]
40. Moreira, C.A.A.; Murayama, L.H.V.M.; Martins, T.C.; Oliveira, V.T.; Generoso, D.; Machado, V.M.V.; Batah, S.S.; Fabro, A.T.; Bazan, R.; Zanini, M.A.; et al. Sexual dimorphism in the murine model of extraparenchymal neurocysticercosis. *Parasitol. Res.* **2023**, *122*, 2147–2154. [CrossRef]
41. Ambrosio, J.R.; Valverde-Islas, L.; Nava-Castro, K.E.; Palacios-Arreola, M.I.; Ostoa-Saloma, P.; Reynoso-Ducoing, O.; Escobedo, G.; Ruíz-Rosado, A.; Dominguez-Ramírez, L.; Morales-Montor, J. Androgens exert a cysticidal effect upon Taenia crassiceps by disrupting flame cell morphology and function. *PLoS ONE* **2015**, *10*, e0127928. [CrossRef]
42. Chainy, G.B.N.; Sahoo, D.K. Hormones and oxidative stress: An overview. *Free Radic. Res.* **2020**, *54*, 1–26. [CrossRef] [PubMed]
43. Del Brutto, O.H.; García, E.; Talámas, O.; Sotelo, J. Sex-related severity of inflammation in parenchymal brain cysticercosis. *Arch. Intern. Med.* **1988**, *148*, 544–546. [CrossRef] [PubMed]
44. Fleury, A.; Dessein, A.; Preux, P.M.; Dumas, M.; Tapia, G.; Larralde, C.; Sciutto, E. Symptomatic human neurocysticercosis—Age, sex and exposure factors relating with disease heterogeneity. *J. Neurol.* **2004**, *251*, 830–837. [CrossRef] [PubMed]

microorganisms

MDPI

Review

Plasmodium cynomolgi: What Should We Know?

Fauzi Muh [1], Ariesta Erwina [1], Fadhila Fitriana [1], Jadidan Hada Syahada [1], Angga Dwi Cahya [2], Seongjun Choe [3], Hojong Jun [4], Triwibowo Ambar Garjito [5], Josephine Elizabeth Siregar [6] and Jin-Hee Han [4,*]

[1] Department of Epidemiology and Tropical Diseases, Faculty of Public Health, Universitas Diponegoro, Semarang 50275, Indonesia; fauzimuh010@lecturer.undip.ac.id (F.M.); ariestaerwina@alumni.undip.ac.id (A.E.); fadhilafitriana@alumni.undip.ac.id (F.F.); jadidanhada@alumni.undip.ac.id (J.H.S.)
[2] Department of Environmental Health, Faculty of Public Health, Universitas Diponegoro, Semarang 50275, Indonesia; 13angga@alumni.undip.ac.id
[3] Department of Parasitology, School of Medicine, Chungbuk National University, Cheongju 28644, Republic of Korea; vetazmo@gmail.com
[4] Department of Environmental Biology and Tropical Medicine, School of Medicine, Kangwon National University, Chuncheon 24341, Republic of Korea; goodseed87@gmail.com
[5] Vector-Borne and Zoonotic Research Group, Research Center for Public Health and Nutrition, National Research and Innovation Agency Indonesia, Salatiga 50721, Indonesia; triw018@brin.go.id
[6] Eijkman Research Center for Molecular Biology, National Research and Innovation Agency, Jalan Raya Bogor Km. 46, Cibinong, Bogor 16911, Indonesia; jose001@brin.go.id
* Correspondence: han.han@kangwon.ac.kr

Abstract: Even though malaria has markedly reduced its global burden, it remains a serious threat to people living in or visiting malaria-endemic areas. The six *Plasmodium* species (*Plasmodium falciparum*, *Plasmodium vivax*, *Plasmodium malariae*, *Plasmodium ovale curtisi*, *Plasmodium ovale wallikeri* and *Plasmodium knowlesi*) are known to associate with human malaria by the *Anopheles* mosquito. Highlighting the dynamic nature of malaria transmission, the simian malaria parasite *Plasmodium cynomolgi* has recently been transferred to humans. The first human natural infection case of *P. cynomolgi* was confirmed in 2011, and the number of cases is gradually increasing. It is assumed that it was probably misdiagnosed as *P. vivax* in the past due to its similar morphological features and genome sequences. Comprehensive perspectives that encompass the relationships within the natural environment, including parasites, vectors, humans, and reservoir hosts (macaques), are required to understand this zoonotic malaria and prevent potential unknown risks to human health.

Keywords: *P. cynomolgi*; zoonosis; malaria

1. Introduction

In the late 19th century, over 30 different *Plasmodium* species had been discovered, affecting human and non-human primates [1,2]. While there was a previous dominant belief that the natural transmission of simian malaria parasites to humans was rare, it has been firmly established for a significant period that specific non-human primates can acquire malaria, and this infection can also impact humans [3]. Various simian malaria species, *P. cynomolgi*, *P. knowlesi*, *P. inui*, *P. simium*, and *P. brasilianum*, have been intentionally transmitted in experiments and have been found to cause infections in humans through mosquito vectors [4,5]. Among these species, *P. cynomolgi*, the prevailing malaria parasite detected in Old-World monkeys within Southeast Asia, is recognized for its natural ability to cause zoonotic infections in humans [1,6–9]. As of now, there is no recorded evidence of *P. cynomolgi* being naturally transmitted from one human to another. Therefore, it is crucial to understand that the natural reservoir hosts for *P. cynomolgi*, long-tailed (*Macaca fascicularis*) and pig-tailed (*Macaca nemestrina*) macaques, are now living in close proximity to the human population, meaning humans are at a higher risk of contracting a *P. cynomolgi* infection [1,10]. Furthermore, numerous mosquito species that play a role in transmitting other types of human malaria can also serve as vectors for *P. cynomolgi*. This situation could potentially give rise to concerns regarding the transmission of the parasite between different species in regions where both the natural hosts and humans share the same habitat [4]. The potential for non-human malaria to spread to humans is increasing, primarily due to deforestation, significant shifts in ecological conditions, the presence of suitable hosts and vectors, and adaptive alterations in the biology of parasites due to climate changes [11,12]. In numerous Southeast Asian nations like Thailand, cases of *P. cynomolgi* infections are commonly found in pig-tailed and long-tailed macaques [9].

Furthermore, there is a well-documented instance of *P. cynomolgi* malaria naturally transmitting to a human in eastern Malaysia [3]. Additional monitoring in Northern Sabah, Malaysia, and Western Cambodia found asymptomatic human infections of *P. cynomolgi*, but they were not very common [7]. A symptomatic *P. cynomolgi* infection was confirmed in an individual who had recently returned to Denmark from Southeast Asia [13]. In Thailand, nine malaria-infected patients were discovered to be co-infected with both hidden *P. cynomolgi* and other *Plasmodium* species during the diagnosis of malaria patients presenting symptoms [9]. Typical clinical signs of *P. cynomolgi* infections usually involve symptoms like other clinical features of malaria such as headaches, reduced appetite, muscle discomfort, and nausea [3]. These symptoms typically manifest only during episodes of fever, are of moderate intensity, and can be effectively managed with basic medications used for the antimalarial regimen for *P. vivax*. [14]. The most notable physical observations consist of an enlarged spleen and liver [8]. *P. cynomolgi*, a simian malaria parasite, has emerged as a recent contributor to human malaria cases and continues to pose public health concerns in specific regions. In this review, we assess the updated literature on various aspects of *P. cynomolgi*, covering biology and genomes, epidemiology, its natural hosts and vectors, pathogenesis, and diagnostic approaches.

2. *Plasmodium cynomolgi* Biology

P. cynomolgi was initially studied and documented in Germany by Mayer in 1907. This study involved an imported *Macaca cynomolgus* from Java [15]. In 1937, H.W. Mulligan re-studied and re-described it as the Mulligan strain or M strain [16]. In the late 1950s, the *P. cynomolgi* Bastianielli strain or B strain was found by P.C.C. Garnham [17]. Other new *P. cynomolgi* strains (e.g., Berok, Cambodian, Gombak, Ceylonensis, and Smithsonian) were isolated from monkeys and mosquitoes [18–23]. Among studied strains in *Macaca mulatta*, the B strain typically exhibited higher mean parasitemia compared to the Mulligan (M) strain [24,25]. A study conducted earlier, between June 2013 and December 2017, at Kapit Hospital in Malaysia, involving 1047 blood samples from malaria patients, revealed that *P. cynomolgi* parasites were detected in the blood samples of two patients, making up approximately 1.5% to 4.7% of the total malaria parasite count [1]. The parasites *P. cynomolgi* had differentiated based on the morphological features of infected erythrocytes, including Schüffner's stippling. Erythrocytes infected by the *P. cynomolgi* parasite undergo enlargement and can sometimes become distorted, with trophozoites of *P. cynomolgi* displaying single, double, or triple chromatin dots [1]. Research on *P. cynomolgi*'s M and B strains revealed liver incubation periods of 15 to 20 days and 16 to 37 days, respectively. The erythrocytic cycle takes 48 h, with a human prepatent period of 19 days. In *Macaca speciosa* and *Macaca mulatta*, commonly known hosts, the prepatent period spans from 7 to 16 days. [15]. Additionally, the asexual erythrocytic cycle of *P. cynomolgi* lasts for 48 h. [15,26]. In the early asexual stage of *P. cynomolgi*, infected red blood cells noticeably enlarge as the young parasite grows to almost half the size of the original host cell [25]. As the parasite undergoes its developmental stages, there is an augmentation in the prominence of Schüffner's stippling and pigmentation. An observable resemblance emerges with *P. vivax* during the later trophozoite stage, where both trophozoites and schizonts exhibit the presence of Schüffner's dots [15]. At maturity, *P. cynomolgi* produces an average of 16 merozoites, typically ranging from 14 to 20. The biology of *P. cynomolgi* closely resembles that of *non-Laverania* species, with shorter incubation and pre-patent periods compared to *P. falciparum* [8]. Obtained from different berok monkeys (K2, K3, and K4), the Berok K4 line culture was better than the others because it had a multiplication rate of two-fold to four-fold over more than five cycles (Figure 1) [25].

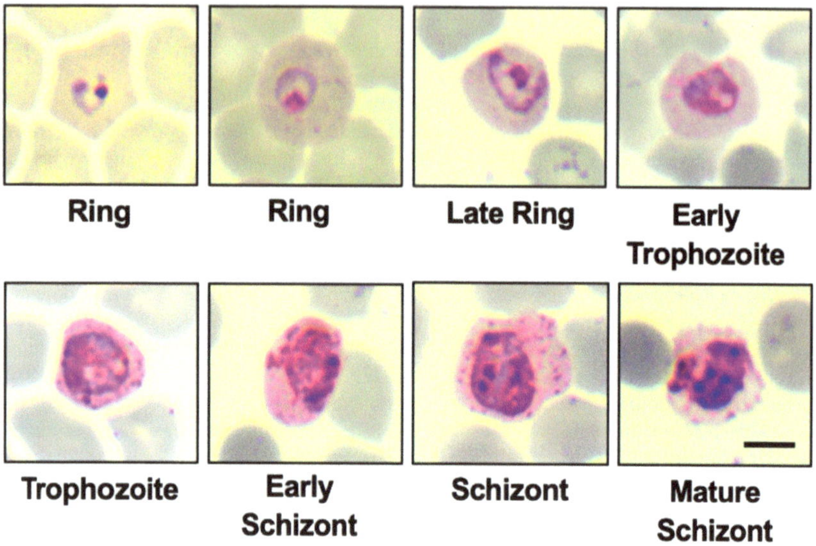

Figure 1. The morphology of *P. cynomolgi* K4 line is shown in each development stage cultured in 100% of rhesus macaque RBCs (Fauzi Muh et al., original unpublished data). Scale bar indicates 5 μm.

Multigene families are found to be very common in *P. cynomolgi* vs. *P. vivax*, and *P. cynomolgi* vs. *P. knowlesi* [27,28]. An analysis of 192 conserved ribosomal, translational, and transcriptional genes indicates that *P. cynomolgi* and *P. vivax* are closely related [28]. *P. cynomolgi* B strain (PcyB) genomes, isolated from monkeys in Malaysia, revealed 5722 genes with 90% of those genes (4613) being orthologs in *P. vivax* and *P. knowlesi* [28]. Meanwhile, the macaque-infected *P. cynomolgi* M strain (PcyM) has 966 new genes compared to PcyB [27]. In the development of a model for the study of other *Plasmodium* species, genes there were 214 genes found to be identical to both *P. cynomolgi* and *P. vivax*, and between *P. cynomolgi* and *P. knowlesi*, there were 100 genes found to be identical [27]. Consequently, the *P. cynomolgi* and *P. vivax* lineage typically contains a greater quantity of genes within multigene families compared to *P. knowlesi*, which indicates that repeated gene duplication occurred in the ancestor lines of *P. vivax* and *P. cynomolgi* or that some deletions may happen in *P. knowlesi*, such as the *vir*, *kir*, *SICAvar*, *Duffy binding protein* (*dbp*) and *reticulocyte binding protein* (*rbp*) genes [28]. Understanding the invasion biology of *Plasmodium* species is important. The gene families of *erythrocyte binding-like* (*ebl*) and *reticulocyte binding-like* (*rbl*) gene families are found to encode the parasite ligands required for successful invasion into red blood cells [29,30]. DBPs, molecules that interact with the Duffy antigen receptor for chemokines (DARC) found on the surface of both human and monkey erythrocytes, make up one of the *ebl* genes that encode the EBL ligands. *P. cynomolgi* has three *ebl* genes (*dbp1*, *dbp2*, *ebp*), in contrast to *P. vivax* and *P. knowlesi*, which have two (*dbp* and *ebp*) and three (*dbp-α*, *dbp-β*, *and dbp-γ*) genes, respectively. The *dbp* genes are thought to be the important ligands that invade the host's red blood cells. The presence of more than one *dbp* in *P. cynomolgi* and *P. knowlesi* can also be thought to be responsible for infecting both human and monkey erythrocytes [31]. *P. cynomolgi* from different strains (Berok, Gombak, Cambodian, Rossan, Cylonesis, Smithsonian, B and M strain) showed

92% DNA identity or two very similar *dbp* genes [31]. The divergence of *rbp* genes has been thought to be involved in the species-specific erythrocyte invasion mechanism in different *Plasmodium* species [28]. The *rbl* genes encode large ligand proteins that exist on the apical membrane of invasive merozoites, such as *rbp* genes [32]. The variation in *rbp* genes has been shown between interspecies of *P. cynomolgi* strains, such as *rbp1b*, which exists in Berok and Gombak strains but is absent in the M/B, Rossan, Smithsonian, Ceylon, Langur and Cambodian strains [28,31]. Furthermore, *rbp2a* was absent in the Gombak and Berok strains. Thus, it is thought that the presence or absence of *rbp1b* or *rbp2a* compensate for each other when infecting the host's red blood cells [31]. In the *P. cynomolgi* B strain, approximately 256 *pir* (plasmodium-interspersed repeat) superfamily or *cyir* (cynomolgi-interspersed repeat genes) are found [28]. Whilst in the *P. cynomolgi* M strain, a total of 1373 *cyir* genes are found [27]. *Cyir* genes are thought to have function related to immune evasion or antigenic variation [28,33–35]. The high variability of the genome in *P. cynomolgi* could be affected by natural host adaptation.

3. Natural Hosts and Vectors of *P. cynomolgi*

One study that inspected blood samples taken from *Macaca* monkeys in SEA countries mainly from Malaysia, found that, in its natural condition, the infections of *P. cynomolgi* in *Macaca* monkeys have been documented both as single infections and in combination with other simian malaria parasites (e.g., *P. vivax*, *P. inui*, *P. coatneyi*, and *P. fieldi*) [8]. Regarding the existence of *P. cynomolgi* in great apes, such as the orangutan (*Pongo pygmaeus*), one study found that the transmission of *P. cynomolgi* could possibly happen between the *Macaca* genera and the orangutan in Kalimantan Indonesia, whilst another study found that the orangutan is not a host of *P. cynomolgi* [36]. A systematic review of the prevalence of simian malaria in Malaysia, collected from seven studies conducted between 2000 and 2021, examined blood samples from *Macaca* genus and described that, from four studies, *P. cynomolgi* was commonly found in the *Macaca* species in Malaysia with an average prevalence of 33.05%. Studies also describe the type of infection among macaques in Malaysia. *P. cynomolgi* is commonly found in mono-infection and mix-infection with *P. inui* (dual infection), with *P. knowlesi* and *P. coatneyi* (triple infection), and with *P. knowlesi*, *P. coatneyi*, and *P. inui* (quadruple infection) [37]. Extensive research on wild long-tailed macaques (*M. fascicularis*) and pig-tailed macaques (*M. nemestrina*) in the region revealed that these species are hosts of a total of six simian malaria parasites, including *P. cynomolgi*, *P. coatneyi*, *P. fieldi*, *P. inui*, *P. knowlesi*, and *P. simiovale* (found in *Macaca sinica*) [15]. This discovery came after a large focus on humans [12,38]. *P. inui*, *P. knowlesi*, and *P. cynomolgi* were the three most common parasites among the 108 macaques studied (82%, 78% and 56%, respectively). In addition to *P. knowlesi*, *P. inui* and *P. cynomolgi* are two other simian parasites with zoonotic potential that have been demonstrated by unintentional and deliberate infections [1,15]. A study in the Philippines examined blood samples from 40 wild *Macaca fascicularis* (long-tailed macaque), as the natural host of *P. cynomolgi*, and described the prevalence of *Macaca fascicularis* monkeys infected by *P. cynomolgi* as 23.2% [39]. Another study describing the distribution of infected *M. fascicularis* with *P. cynomolgi* in countries of Southeast Asia (the Philippines, Indonesia, Cambodia, Singapore, and Laos) stated that *P. cynomolgi* was the most widespread parasite among all the sample populations with a prevalence of 53.3% [10].

The successful transmission of zoonotic malaria largely hinges on the ecological behavior and geographical prevalence of capable vectors. The changes in the habitat of vectors may have altered the bio-ecology of *Anopheles* mosquitoes [40,41]. The increase in zoonotic malaria is due to large deforestation required for changes to agriculture and human settlements, causing the mosquito vectors to live in close proximity to the host, both humans and macaque monkeys [42]. A change in land use, occupation, and settlements is often associated with proximity to infected vectors. Furthermore, mosquitoes that readily feed on humans and macaque monkeys harboring the parasites should live in a shared habitat with the reservoir hosts and the humans in order to cause the infection in the human population [43]. *P. cynomolgi* is mostly transmitted by *Anopheles leucosphyrus* subgroup mosquitoes [44]. A previous study conducted in seven states in Peninsular Malaysia revealed the infection of *P. cynomolgi* in *Anopheles introlatus*, and *An. Latens* [45]. The mosquito vectors can be infected by mono-*P. cynomolgi* infection or together with *P. inui* and *P. fieldi* [45]. In Sabah and Sarawak Malaysia, *An. balabacensis* were found to be infected with both single *P. cynomolgi* or together with *P. inui*, *P. knowlesi*, or *P. fieldi*. The *P. cynomolgi* that infects mosquitos, monkeys and humans showed the identity of nucleotide as 99.7–100%. This means that *P. cynomolgi* has a close relationship to those three isolates, monkey–human–mosquito, concluding that habitat sharing is the main factor of successful of *P. cynomolgi* transmission to the human population [46,47].

An investigation of the first natural *P. cynomolgi* infection by the Malaysian Vector Borne Disease Control Program found that the predominant mosquito within the patient's housing area was *Anopheles cracens* [3]. In Thailand, *An. introlatus* has been shown to possess the *P. cynomolgi* in the salivary gland. *An. introlatus* is thought to be a responsible vector of *P. cynomolgi* in the Southern part of Thailand [9,48]. A study from Vietnam found that *P. cynomolgi* can be detected together with other simian malaria parasites in the primary mosquito vectors, *An. dirus* and *An. Minimus* [49]. This evidence can be a potential threat that the vector might be able to transmit to the human population, similar to *P. knowlesi* [50,51]. Furthermore, there was a study comparing the susceptibility of the *P. cynomolgi* B strain infection in *An. dirus*, *An. takasagoensis*, *An. maculatus*, and *An. philipinensis*. *An. dirus* showed the highest infectivity of *P. cynomolgi*, whereas *An. philipinensis* was shown to be the least susceptible [52].

4. Epidemiology

Transmission of *P. cynomolgi* through mosquito bites has been reported to be geographically spreading around Southeast Asia, with one case imported to Denmark (Figure 2 and Table 1). It is estimated that around 0–1.4% of people are infected with *P. cynomolgi* globally. Several factors that are associated with the spread of *P. cynomolgi* in natural infection are the presence of suitable vectors and non-primate hosts in a shared habitat, globalization, climate change, and deforestation [8]. *P. cynomolgi* infections usually appeared along with *P. falciparum* or *P. vivax* in the blood samples. *P. cynomolgi* may have the same Anopheline vectors or may have other vectors with a comparable zoophilic and anthropophilic tendency [1,15]. The existence of vectors plays an important role in the successful zoonotic malaria transmission of *P. cynomolgi*. It significantly relies on the bionomics and geographic distribution of the vectors, as well as the natural hosts of the parasite. Living in close proximity to hosts, particularly people working as farmers or in agriculture near the forests, or tourists traveling to a macaque endemic area, have a high risk of being exposed to *P. cynomolgi*. In a meta-analysis and systematic review of human *P. cynomolgi* infection cases in Malaysia ($n = 8$), a high proportion of natural hosts (macaque) was found to be infected with *P. cynomolgi*, accounting for 37.42% [53]. The existence of macaque (commonly pig-tailed and long-tailed species) infected with *P. cynomolgi* and other simian malaria parasites in Thailand, mainly in the southern region of the country, has been proven in a previous study [54]. The occurrence of transmission between humans and hosts is suspected to have happened because most of the patients infected with *P. cynomolgi* share the same environment with the host and mosquito. Patients co-infected with *P. cynomolgi*, *P. knowlesi*, and *P. vivax*, from a study in the Yala Province, had a history of living in a neighborhood surrounded by a group of domesticated pig-tailed and long-tailed macaques [9]. A study in Malaysia investigated the infection of simian malaria parasites among indigenous communities living in the forest fringe and found a mixed infection of *P. cynomolgi* with *P. inui*. These indigenous people were at high risk of simian malaria because of the high chance of exposure to monkeys, as the natural host, and mosquito bites [11].

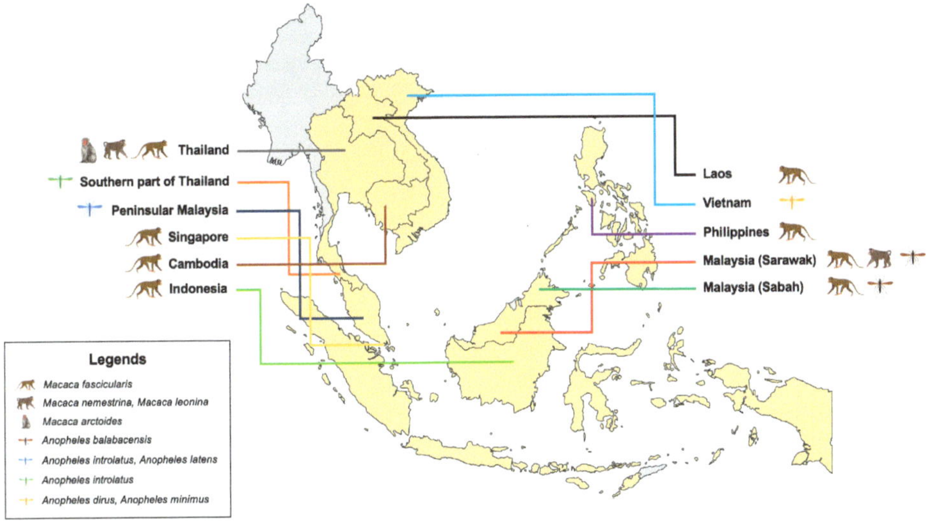

Figure 2. The geographical distribution of natural hosts and vectors of *P. cynomolgi*.

Table 1. Natural human infection of *P. cynomolgi*.

Study Locations	Samples (*n*)	Human Malaria Positive (*n*)	Type of Infection	
			Mono-Infection (*n*)	Mixed Infection (*n*)
Thailand (2007–2016) (Putaporntip, C. et al., 2021 [9])	1359	1180	0	*P. cynomolgi* + *P. vivax* (7); *P. cynomolgi* + *P. falciparum* (1); *P. cynomolgi* + *P. falciparum* + *P. vivax* + *P. knowlesi* (1)
Malaysia (first case 2011) (Ta, T. H. et al., 2014 [3])	-	1	1	0
Malaysia (2011–2014) (Yap, N. J. et al., 2021 [11])	288	90	9	0
Cambodia (2013–2016) (Imwong, M. et al., 2019 [7])	14,732	1361	11	*P. cynomolgi* + *P. vivax* (2)
Malaysia (2013–2017) (Raja, T. N. et al., 2020 [1])	-	1047	0	*P. cynomolgi* + *P. knowlesi* (6)
Malaysia (2015) (Grignard, L. et al., 2019 [55])	876	54	2	0
Denmark (2018) (Hartmeyer, G. N. et al., 2019 [13])	-	1	1	0
Thailand (2008–2016) (Putaporntip, C. et al., 2010 [54])	5271	4195	2	*P. cynomolgi* + *P. vivax* (2); *P. cynomolgi* + *P. vivax* (1); *P. cynomolgi* + *P. vivax* (1); *P. cynomolgi* + *P. vivax* (3); *P. cynomolgi* + *P. vivax* (8); *P. cynomolgi* + *P. falciparum* (2); *P. cynomolgi* + *P. knowlesi* (2)
Thailand (2015) (Sai-ngam, P. et al., 2022 [14])	3	3	2	*P. cynomolgi* + *P. vivax* (1)

5. Clinical Presentation

Malaria infections resulting from *P. cynomolgi* manifest numerous symptoms that overlap with those caused by other malaria species. Nevertheless, different strains of this parasite show some slight differences in their clinical signs. The striking feature of deliberately induced *P. cynomolgi* malaria infections in humans was the clear presence of substantial clinical symptoms, even when the level of parasites was low. These symptoms followed a sequence of cephalgia, anorexia, myalgia, and nausea [56]. Importantly, they typically occurred only during fever or febrile, were of moderate severity, and were easily treatable with antimalarial regimens for *P. vivax* [14]. Additionally, common physical findings included a splenomegaly and a hepatomegaly [8].

P. cynomolgi infections are frequently found in macaque monkeys, including *Macaca fascicularis* (long-tailed macaque), *Macaca nemestrina*, and *Macaca leonina* (pig-tailed macaque) (Figure 2). In Java, *P. cynomolgi* was first discovered by Halberstadter and von Prowazek in 1907 from long-tailed macaques (*M. fascicularis*). Similar to *P. vivax*, *P. cynomolgi* was reported to have recurrent relapses in rhesus monkeys. This exoerythrocytic source was later identified as dormant "hypnozoites" in the liver. As a result, it was used as the animal model for recurrent malaria. *P. cynomolgi* first infected humans accidentally with the *P. cynomolgi* B strain in 1960. This infection was later acquired naturally by humans in Southeast Asia from various macaque monkeys [57–61].

The first known case of naturally acquired *P. cynomolgi* malaria in humans was reported in 2011. The patient was a 39-year-old woman from the east coast of Peninsular Malaysia with no previous history of malaria who did not travel to any other malaria-endemic areas. The clinical symptoms of the patient were non-specific and mimicked a flu-like syndrome with a febrile condition. The patient took oral chloroquine for medication and recovered within a week.

Additional symptoms, including muscle pain, general malaise, headache, fever, and abdominal pain were reported by a traveler who had returned from a Southeast Asian country. A blood test revealed elevated C reactive protein, S-alanine aminotransferase (S-ALAT), thrombocytopenia, and low platelet levels [13]. The prepatent period, defined as the duration between infection and the onset of symptoms, ranges from 7 to 16 days. The incubation period, which is the time between infection and the development of the disease, ranges from 15 to 20 days, with some differences seen in different *P. cynomolgi* strains. On the other hand, patients infected with *P. cynomolgi* typically experience, at worst, mild and non-life-threatening symptoms [3,7,13,55]. An experimental study described that both the *P. cynomolgi* M and B strains had similar significant symptoms, which included high fever, headaches, loss of appetite, muscle pain, and nausea. However, there were differences in how long these symptoms lasted, how often fever episodes occurred, and the degree of spleen enlargement. Individuals infected with the M strain exhibited prolonged symptoms, a higher incidence of tertian fever, and an increased likelihood of developing splenomegaly when compared to those infected with the B strain. Additionally, only those intentionally infected with the M strain reported experiencing chills and vomiting [56].

6. *P. cynomolgi* Confirmation and Diagnosis

P. cynomolgi exhibits phenotypic and phylogenetic resemblances to *P. vivax*, posing difficulties in differentiation between the two when examining blood smears using routine microscopy. Often, routine microscopy can lead to the misdiagnosis of *P. knowlesi as P. malariae* and *P. cynomolgi* as *P. vivax* [62]. In such situations, it becomes essential to precisely assess and comprehend the prevalence and transmission patterns of non-human *Plasmodium* species, particularly *P. cynomolgi*, within human populations using a more sensitive diagnostic tool [3,6,63]. Molecular methods are the most accurate diagnostic test for morphologically identical species [64]. The first case of natural infection of *P. cynomolgi* in humans was confirmed by molecular detection, using nested PCR. The sequencing result of 785 nucleotides showed that 99.9% of the genes were similar to the *P. cynomolgi* M-strain from Malaysia [3,15]. A recently developed test to confirm the presence of *P. cynomolgi* used a combined lateral flow with a recombinase polymerase amplification (RPA-LFD) [6]. This assay uses the designed *18S rRNA* primers. From a total of 30 *Plasmodium*-positive blood samples from wild macaques, nested PCR detected positivity in 11 out of these samples, and 9 positives were detected by RPA-LFD assay. Of the 19 negative samples by nested PCR, RPA-LFD assay showed 18 true negatives. The limit of detection (LoD) of this assay also showed 22.14 copies/μL, which was said to be 10 times greater than qPCR or RPA-AGE (Agarose Gel Electrophoresis) assay. It is concluded that the latest developed test has given 81.82% sensitivity and 94.74% specificity in detecting the target DNA of *P. cynomolgi* [6]. However, this assay may require specialized techniques and require a longer time for sample preparation. This technique may not be suitable for use in a resource-limited setting. Another test, qPCR, is also performed to confirm the *P. cynomolgi* targeting the *18S rRNA* gene. The LoD observed was 0.075 ng/μL. However, using the suspected clinical isolates ($n = 250$ human blood samples), the qPCR was not able to confirm the presence of *P. cynomolgi* [63]. The inability to detect *P. cynomolgi* in the clinical samples may be due to the low DNA content in the samples, or this may be the true negative result of *P. cynomolgi*. Thus, a large number of clinical samples may be required to validate the use of qPCR for the detection of the presence of *P. cynomolgi* DNA. Furthermore, loop-mediated isothermal amplification (LAMP) is an alternative assay that reduces the downsides of a PCR diagnosis-based assay. The LAMP assay is employed with a focus on species-specific targeting of the mitochondrial genes [65–72]. Studies have indicated that LAMP exhibits greater sensitivity and specificity compared to ELISA and microscopy, with a reported accuracy of 95.6% and 100%, respectively [73].

7. *P. cynomolgi* for *P. vivax* Malaria Research

P. cynomolgi is a species genetically related to *P. vivax*, which is one of the malaria parasite species in humans. *P. cynomolgi* naturally infects monkeys, especially long-tailed macaques (Figure 2). Although *P. cynomolgi* usually does not cause serious illness in humans, this parasite has many similarities with *P. vivax*, especially relating to phenotype, biology, and genetic characteristics [27,74,75]. The lack of a long-term in vitro culture method for *P. vivax* has severely limited our understanding of the *P. vivax* invasion biology and drug and vaccine development. However, the putative *P. vivax* drug-resistance marker *mdr1* Y976F was investigated by using *P. cynomolgi* Berok in an in vitro culture model. By conducting genetic manipulations in *P. cynomolgi* and observing the ensuing alterations in its characteristics, this model is able to identify the specific impact of the Y976F mutation within the *Pcymdr1* gene on drug sensitivity [75].

P. vivax merozoites predominantly invade human reticulocytes, thereby limiting the development of an in vitro culture. The inability to establish a continuous culture system for *P. vivax* affects the development of a growth inhibition assay (GIA). To address this limitation, a previous study utilized a surrogate species like *P. cynomolgi* as a model for *P. vivax* [76]. *P. cynomolgi* is commonly used in monkey studies for drug discovery and understanding the biological characteristics. The capacity to infect human reticulocytes using *P. cynomolgi* creates valuable opportunities for studying invasion mechanisms and changes in red blood cells. A previous study showed success in growing the blood stages of two *P. cynomolgi* strains, particularly the Berok strain. Initially, from a wild *M. nemestrina* in Malaysia in the 1960s, the Berok strain was maintained by blood or sporozoite inoculation in monkeys. This strain was not cloned, suggesting it might contain various parasite types that could vary during infection or across hosts [21]. It is unclear if Berok parasites from different monkeys differ genetically or as variations within a group. Addressing initial setbacks persistently could likely lead to cultivating other *P. cynomolgi* lines or macaque parasite species. The study highlighted a valuable surrogate for *P. vivax* to expand research possibilities with lab-cultured parasites. Research findings confirm that lab-grown Berok K4 parasites resemble those from monkeys and maintain infectivity to produce infective sporozoites in mosquitoes. This study has shown the possible use of a *P. cynomolgi* model for *P. vivax* research to understand the morphology, characteristics, and behavior [25]. The previous study has shown that *P. cynomolgi* in in vitro culture was totally restricted to human reticulocytes. Meanwhile, the in vitro cultures of *P. cynomolgi* using monkey RBCs were not restricted to only reticulocytes [77]. The *P. cynomolgi* has two invasion pathways (DARC-dependent and independent) to invade the rhesus macaques [78]. However, the invasion into human RBCs by *P. cynomolgi* depends on the DARC-dependent pathway [77].

It serves as a great model to assess vaccines before clinical trials, contributing significantly to developing new strategies to control the widespread and challenging *P. vivax* [77].

8. Treatment

The importance of malaria management lies in preventing the transmission of the disease and decreasing the immediate risk to the host. One aspect of this management is treating malaria patients with specific medications. Currently, there is no single therapy that can eradicate *P. cynomolgi* at each respective lifecycle stage. Some studies reported that *P. cynomolgi*-infected patients were treated with a combination of existing antimalarial drugs, such as atovaquone plus proguanil or chloroquine plus primaquine. Moreover, another study found that *P. cynomolgi* co-infected with *P. falciparum* patients were treated using artesunate plus mefloquine [9,13,14].

The significant morphological and biological characteristics of both *P. vivax* and *P. cynomolgi* are the dormant liver stages (hypnozoites), which are responsible for relapses due to their reactivation within several weeks to years after the initial infection. It is essential to prevent relapses of *P. cynomolgi* through safer radical curative compounds that efficiently kill hypnozoites. Similar to *P. vivax*, previous research has identified primaquine as a potential treatment for *P. cynomolgi* infections against hypnozoites. However, primaquine can cause acute hemolytic anemia in malaria patients with glucose-6-phospate dehydrogenase (G6PD) deficiency [4]. Additionally, the CDC's recommendation of a primaquine treatment schedule of 30 mg/day for 14 days for non-G6PD-deficient patients can lead to primaquine resistance due to limited patient compliance. A previous study was conducted to identify a new potent non-8-aminoquinoline compound that efficiently kills the early developmental forms of hypnozoites in vitro using a drug assay. The result shows that the activity of KAF156 was limited to schizont; meanwhile, KAI 407 showed activity against both liver stages schizonts and hypnozoites forms, like primaquine [79]. Another study found that MMV019721, which is an acetyl-CoA synthetase inhibitor that affects histone acetylation, selectively kills *P. vivax* and *P. cynomolgi* hypnozoites [80].

Author Contributions: Conceptualization and funding acquisition, F.M. and J.-H.H.; writing—original draft, A.E., F.F., J.H.S., F.M. and J.-H.H.; visualization, A.D.C.; writing—review and editing, S.C., H.J., T.A.G., J.E.S., F.M. and J.-H.H. All authors have read and agreed to the published version of the manuscript.

Funding: This work was supported by the Basic Science Research Program through the National Research Foundation of Korea (NRF) funded by the Ministry of Education (RS-2023-00240627) (J-H.H.); Program Riset dan Inovasi untuk Indonesia Maju (RIIM), and The National Research and Innovation Agency, Indonesia (BRIN) (No. 37/II.7/HK/2023) (F.M).

Data Availability Statement: All data that support this study are available upon request to the corresponding author.

Conflicts of Interest: The authors declare no conflicts of interest.

References

1. Raja, T.N.; Hu, T.H.; Kadir, K.A.; Mohamad, D.S.A.; Rosli, N.; Wong, L.L.; Hii, K.C.; Simon Divis, P.C.; Singh, B. Naturally Acquired Human Plasmodium cynomolgi and P. knowlesi Infections, Malaysian Borneo. *Emerg. Infect. Dis.* **2020**, *26*, 1801–1809. [CrossRef] [PubMed]
2. Ramasamy, R. Zoonotic malaria—Global overview and research and policy needs. *Front. Public Health* **2014**, *2*, 123. [CrossRef] [PubMed]
3. Ta, T.H.; Hisam, S.; Lanza, M.; Jiram, A.I.; Ismail, N.; Rubio, J.M. First case of a naturally acquired human infection with Plasmodium cynomolgi. *Malar. J.* **2014**, *13*, 68. [CrossRef] [PubMed]
4. Bykersma, A. The New Zoonotic Malaria: *Plasmodium cynomolgi*. *Trop. Med. Infect. Dis.* **2021**, *6*, 46. [CrossRef] [PubMed]
5. Eyles, D.E.; Coatney, G.R.; Getz, M.E. *Vivax*-Type Malaria Parasite of Macaques Transmissible to Man. *Science* **1960**, *131*, 1812–1813. [CrossRef] [PubMed]
6. Rattaprasert, P.; Chavananikul, C.; Fungfuang, W.; Chavalitshewinkoon-Petmitr, P.; Limudomporn, P. Combining isothermal recombinase polymerase amplification with lateral flow assay for diagnosis of *P. cynomolgi* malaria infection. *PLoS Neglected Trop. Dis.* **2023**, *17*, e0011470. [CrossRef] [PubMed]

7. Imwong, M.; Madmanee, W.; Suwannasin, K.; Kunasol, C.; Peto, T.J.; Tripura, R.; von Seidlein, L.; Nguon, C.; Davoeung, C.; Day, N.P.J.; et al. Asymptomatic Natural Human Infections With the Simian Malaria Parasites Plasmodium cynomolgi and Plasmodium knowlesi. *J. Infect. Dis.* **2019**, *219*, 695–702. [CrossRef] [PubMed]
8. Kojom Foko, L.P.; Kumar, A.; Hawadak, J.; Singh, V. Plasmodium cynomolgi in humans: Current knowledge and future directions of an emerging zoonotic malaria parasite. *Infection* **2023**, *51*, 623–640. [CrossRef] [PubMed]
9. Putaporntip, C.; Kuamsab, N.; Pattanawong, U.; Yanmanee, S.; Seethamchai, S.; Jongwutiwes, S. Plasmodium cynomolgi co-infections among symptomatic malaria patients, Thailand. *Emerg. Infect. Dis.* **2021**, *27*, 590. [CrossRef]
10. Zhang, X.; Kadir, K.A.; Quintanilla-Zariñan, L.F.; Villano, J.; Houghton, P.; Du, H.; Singh, B.; Smith, D.G. Distribution and prevalence of malaria parasites among long-tailed macaques (Macaca fascicularis) in regional populations across Southeast Asia. *Malar. J.* **2016**, *15*, 450. [CrossRef]
11. Yap, N.J.; Hossain, H.; Nada-Raja, T.; Ngui, R.; Muslim, A.; Hoh, B.P.; Khaw, L.T.; Kadir, K.A.; Simon Divis, P.C.; Vythilingam, I.; et al. Natural Human Infections with Plasmodium cynomolgi, P. inui, and 4 other Simian Malaria Parasites, Malaysia. *Emerg. Infect. Dis.* **2021**, *27*, 2187–2191. [CrossRef] [PubMed]
12. Lee, K.S.; Cox-Singh, J.; Singh, B. Morphological features and differential counts of Plasmodium knowlesi parasites in naturally acquired human infections. *Malar. J.* **2009**, *8*, 73. [CrossRef] [PubMed]
13. Hartmeyer, G.N.; Stensvold, C.R.; Fabricius, T.; Marmolin, E.S.; Hoegh, S.V.; Nielsen, H.V.; Kemp, M.; Vestergaard, L.S. Plasmodium cynomolgi as Cause of Malaria in Tourist to Southeast Asia, 2018. *Emerg. Infect. Dis.* **2019**, *25*, 1936–1939. [CrossRef] [PubMed]
14. Sai-ngam, P.; Pidtana, K.; Suida, P.; Poramathikul, K.; Lertsethtakarn, P.; Kuntawunginn, W.; Tadsaichol, S.; Arsanok, M.; Sornsakrin, S.; Chaisatit, C.; et al. Case series of three malaria patients from Thailand infected with the simian parasite, Plasmodium cynomolgi. *Malar. J.* **2022**, *21*, 142. [CrossRef] [PubMed]
15. Coatney, G.R.; Collins, W.E.; Warren, M.; Contacos, P.G. *The Primate Malarias*; US National Institute of Allergy and Infectious Diseases: Rockville, MD, USA, 1971.
16. Mulligan, H. Descriptions of Two Species of Monkey Pías-medium isolated from Silenus irus. *Arch. Protistenkd.* **1935**, *84*, 285–314.
17. Garnham, P. A new sub-species of Plasmodium cynomolgi. *Riv. Parassitol.* **1959**, *20*, 273–278.
18. Bennett, G.F.; Warren, M.; Cheong, W. Biology of the simian malarias of Southeast Asia. III. Sporogony of the Cambodian strain of Plasmodium cynomolgi. *J. Parasitol.* **1966**, *52*, 632–638. [CrossRef] [PubMed]
19. Bennett, G.F.; Warren, M.; Cheong, W. Biology of the simian malarias of Southeast Asia. IV. Sporogony of four strains of Plasmodium cynomolgi. *J. Parasitol.* **1966**, *52*, 639–646. [CrossRef] [PubMed]
20. Bennett, G.F.; Warren, M.; Cheong, W. Biology of the simian malarias of southeast Asia. II. The susceptibility of some Malaysian mosquitoes to infection with five strains of Plasmodium cynomolgi. *J. Parasitol.* **1966**, *52*, 625–631. [CrossRef]
21. Collins, W.E.; Warren, M.; Galland, G.G. Studies on infections with the Berok strain of Plasmodium cynomolgi in monkeys and mosquitoes. *J. Parasitol.* **1999**, *85*, 268–272. [CrossRef]
22. Dissanaike, A.; Nelson, P.; Garnham, P. Two new malaria parasites *Plasmodium cynomolgi ceylonensis* subsp. nov. and *Plasmodium fragile* sp. nov. from monkeys in Ceylon. *Ceylon J. Med. Sci.* **1965**, *14*, 1–9.
23. Shortt, H.E.; Garnham, P.C. The exoerythrocytic parasites of Plasmodium cynomolgi. *Trans. R. Soc. Trop. Med. Hyg.* **1948**, *41*, 705–716. [CrossRef]
24. Hang, J.W.; Tukijan, F.; Lee, E.Q.; Abdeen, S.R.; Aniweh, Y.; Malleret, B. Zoonotic Malaria: Non-Laverania Plasmodium Biology and Invasion Mechanisms. *Pathogens* **2021**, *10*, 889. [CrossRef] [PubMed]
25. Chua, A.C.Y.; Ong, J.J.Y.; Malleret, B.; Suwanarusk, R.; Kosaisavee, V.; Zeeman, A.-M.; Cooper, C.A.; Tan, K.S.W.; Zhang, R.; Tan, B.H.; et al. Robust continuous in vitro culture of the Plasmodium cynomolgi erythrocytic stages. *Nat. Commun.* **2019**, *10*, 3635. [CrossRef]
26. Singh, B.; Daneshvar, C. Human infections and detection of Plasmodium knowlesi. *Clin. Microbiol. Rev.* **2013**, *26*, 165–184. [CrossRef] [PubMed]
27. Pasini, E.M.; Böhme, U.; Rutledge, G.G.; Voorberg-Van der Wel, A.; Sanders, M.; Berriman, M.; Kocken, C.H.; Otto, T.D. An improved Plasmodium cynomolgi genome assembly reveals an unexpected methyltransferase gene expansion. *Wellcome Open Res.* **2017**, *2*, 42. [CrossRef]
28. Tachibana, S.-I.; Sullivan, S.A.; Kawai, S.; Nakamura, S.; Kim, H.R.; Goto, N.; Arisue, N.; Palacpac, N.M.Q.; Honma, H.; Yagi, M.; et al. Plasmodium cynomolgi genome sequences provide insight into Plasmodium vivax and the monkey malaria clade. *Nat. Genet.* **2012**, *44*, 1051–1055. [CrossRef] [PubMed]
29. Lopaticki, S.; Maier, A.G.; Thompson, J.; Wilson, D.W.; Tham, W.H.; Triglia, T.; Gout, A.; Speed, T.P.; Beeson, J.G.; Healer, J.; et al. Reticulocyte and erythrocyte binding-like proteins function cooperatively in invasion of human erythrocytes by malaria parasites. *Infect. Immun.* **2011**, *79*, 1107–1117. [CrossRef]
30. Iyer, J.; Grüner, A.C.; Rénia, L.; Snounou, G.; Preiser, P.R. Invasion of host cells by malaria parasites: A tale of two protein families. *Mol. Microbiol.* **2007**, *65*, 231–249. [CrossRef]
31. Sutton, P.L.; Luo, Z.; Divis, P.C.S.; Friedrich, V.K.; Conway, D.J.; Singh, B.; Barnwell, J.W.; Carlton, J.M.; Sullivan, S.A. Characterizing the genetic diversity of the monkey malaria parasite Plasmodium cynomolgi. *Infect. Genet. Evol.* **2016**, *40*, 243–252. [CrossRef]

32. Rayner, J.C.; Huber, C.S.; Galinski, M.R.; Barnwell, J.W. Rapid evolution of an erythrocyte invasion gene family: The Plasmodium reichenowi Reticulocyte Binding Like (RBL) genes. *Mol. Biochem. Parasitol.* **2004**, *133*, 287–296. [CrossRef] [PubMed]
33. Cunningham, D.; Lawton, J.; Jarra, W.; Preiser, P.; Langhorne, J. The pir multigene family of Plasmodium: Antigenic variation and beyond. *Mol. Biochem. Parasitol.* **2010**, *170*, 65–73. [CrossRef] [PubMed]
34. al-Khedery, B.; Barnwell, J.W.; Galinski, M.R. Antigenic variation in malaria: A 3′ genomic alteration associated with the expression of a P. knowlesi variant antigen. *Mol. Cell* **1999**, *3*, 131–141. [CrossRef] [PubMed]
35. Lapp, S.A.; Korir, C.C.; Galinski, M.R. Redefining the expressed prototype SICAvar gene involved in Plasmodium knowlesi antigenic variation. *Malar. J.* **2009**, *8*, 181. [CrossRef] [PubMed]
36. Singh, B.; Simon Divis, P.C. Orangutans not infected with Plasmodium vivax or P. cynomolgi, Indonesia. *Emerg. Infect. Dis.* **2009**, *15*, 1657–1658. [CrossRef] [PubMed]
37. Sam, J.; Shamsusah, N.A.; Ali, A.H.; Hod, R.; Hassan, M.R.; Agustar, H.K. Prevalence of simian malaria among macaques in Malaysia (2000–2021): A systematic review. *PLoS Negl. Trop. Dis.* **2022**, *16*, e0010527. [CrossRef] [PubMed]
38. Nada Raja, T.; Hu, T.H.; Zainudin, R.; Lee, K.S.; Perkins, S.L.; Singh, B. Malaria parasites of long-tailed macaques in Sarawak, Malaysian Borneo: A novel species and demographic and evolutionary histories. *BMC Evol. Biol.* **2018**, *18*, 49. [CrossRef] [PubMed]
39. Gamalo, L.E.; Dimalibot, J.; Kadir, K.A.; Singh, B.; Paller, V.G. Plasmodium knowlesi and other malaria parasites in long-tailed macaques from the Philippines. *Malar. J.* **2019**, *18*, 147. [CrossRef] [PubMed]
40. Fornace, K.M.; Diaz, A.V.; Lines, J.; Drakeley, C.J. Achieving global malaria eradication in changing landscapes. *Malar. J.* **2021**, *20*, 69. [CrossRef]
41. Hawkes, F.M.; Manin, B.O.; Cooper, A.; Daim, S.; R, H.; Jelip, J.; Husin, T.; Chua, T.H. Vector compositions change across forested to deforested ecotones in emerging areas of zoonotic malaria transmission in Malaysia. *Sci. Rep.* **2019**, *9*, 13312. [CrossRef]
42. Phang, W.K.; bin Abdul Hamid, M.H.; Jelip, J.; binti Mudin, R.N.; Chuang, T.-W.; Lau, Y.L.; Fong, M.Y. Spatial-temporal Distribution and Demography of Knowlesi Malaria in Peninsular Malaysia, 2011 to 2018. *Res. Sq.* **2020**, *19*, 9271. [CrossRef]
43. Thompson, R.A. Parasite zoonoses and wildlife: One health, spillover and human activity. *Int. J. Parasitol.* **2013**, *43*, 1079–1088. [CrossRef]
44. van de Straat, B.; Sebayang, B.; Grigg, M.J.; Staunton, K.; Garjito, T.A.; Vythilingam, I.; Russell, T.L.; Burkot, T.R. Zoonotic malaria transmission and land use change in Southeast Asia: What is known about the vectors. *Malar. J.* **2022**, *21*, 109. [CrossRef] [PubMed]
45. Jeyaprakasam, N.K.; Low, V.L.; Pramasivan, S.; Liew, J.W.K.; Wan-Sulaiman, W.-Y.; Vythilingam, I. High transmission efficiency of the simian malaria vectors and population expansion of their parasites Plasmodium cynomolgi and Plasmodium inui. *PLoS Neglected Trop. Dis.* **2023**, *17*, e0011438. [CrossRef]
46. Chua, T.H.; Manin, B.O.; Daim, S.; Vythilingam, I.; Drakeley, C. Phylogenetic analysis of simian Plasmodium spp. infecting Anopheles balabacensis Baisas in Sabah, Malaysia. *PLoS Neglected Trop. Dis.* **2017**, *11*, e0005991. [CrossRef]
47. Ang, J.X.D.; Kadir, K.A.; Mohamad, D.S.A.; Matusop, A.; Divis, P.C.S.; Yaman, K.; Singh, B. New vectors in northern Sarawak, Malaysian Borneo, for the zoonotic malaria parasite, Plasmodium knowlesi. *Parasites Vectors* **2020**, *13*, 472. [CrossRef]
48. Yanmanee, S.; Seethamchai, S.; Kuamsab, N.; Karaphan, S.; Suwonkerd, W.; Jongwutiwes, S.; Putaporntip, C. Natural vectors of Plasmodium knowlesi and other primate, avian and ungulate malaria parasites in Narathiwat Province, Southern Thailand. *Sci. Rep.* **2023**, *13*, 8875. [CrossRef] [PubMed]
49. Chinh, V.D.; Masuda, G.; Hung, V.V.; Takagi, H.; Kawai, S.; Annoura, T.; Maeno, Y. Prevalence of human and non-human primate Plasmodium parasites in anopheline mosquitoes: A cross-sectional epidemiological study in Southern Vietnam. *Trop. Med. Health* **2019**, *47*, 9. [CrossRef] [PubMed]
50. Marchand, R.P.; Culleton, R.; Maeno, Y.; Quang, N.T.; Nakazawa, S. Co-infections of Plasmodium knowlesi, P. falciparum, and P. vivax among Humans and Anopheles dirus Mosquitoes, Southern Vietnam. *Emerg. Infect. Dis.* **2011**, *17*, 1232. [CrossRef] [PubMed]
51. Maeno, Y.; Quang, N.T.; Culleton, R.; Kawai, S.; Masuda, G.; Nakazawa, S.; Marchand, R.P. Humans frequently exposed to a range of non-human primate malaria parasite species through the bites of Anopheles dirus mosquitoes in South-central Vietnam. *Parasites Vectors* **2015**, *8*, 376. [CrossRef]
52. Klein, T.A.; Harrison, B.A.; Dixon, S.V.; Burge, J.R. Comparative susceptibility of Southeast Asian Anopheles mosquitoes to the simian malaria parasite Plasmodium cynomolgi. *J. Am. Mosq. Control Assoc.* **1991**, *7*, 481–487.
53. Kotepui, M.; Masangkay, F.R.; Kotepui, K.U.; Milanez, G.D.J. Preliminary review on the prevalence, proportion, geographical distribution, and characteristics of naturally acquired Plasmodium cynomolgi infection in mosquitoes, macaques, and humans: A systematic review and meta-analysis. *BMC Infect. Dis.* **2021**, *21*, 259. [CrossRef]
54. Putaporntip, C.; Jongwutiwes, S.; Thongaree, S.; Seethamchai, S.; Grynberg, P.; Hughes, A.L. Ecology of malaria parasites infecting Southeast Asian macaques: Evidence from cytochrome b sequences. *Mol. Ecol.* **2010**, *19*, 3466–3476. [CrossRef] [PubMed]
55. Grignard, L.; Shah, S.; Chua, T.H.; William, T.; Drakeley, C.J.; Fornace, K.M. Natural Human Infections With Plasmodium cynomolgi and Other Malaria Species in an Elimination Setting in Sabah, Malaysia. *J. Infect. Dis.* **2019**, *220*, 1946–1949. [CrossRef]
56. Contacos, P.G.; Elder, H.A.; Coatney, G.R.; Genther, C. Man to man transfer of two strains of Plasmodium cynomolgi by mosquito bite. *Am. J. Trop. Med. Hyg.* **1962**, *11*, 186–193. [CrossRef] [PubMed]

57. Chaturvedi, R.; Biswas, S.; Bisht, K.; Sharma, A. The threat of increased transmission of non-knowlesi zoonotic malaria in humans: A systematic review. *Parasitology* **2023**, *150*, 1167–1177. [CrossRef]
58. Most, H. Plasmodium cynomolgi malaria: Accidental human infection. *Am. J. Trop. Med. Hyg.* **1973**, *22*, 157–158. [CrossRef]
59. Kuvin, S.F.; Beye, H.K.; Stohlman, F., Jr.; Contacos, P.G.; Coatney, G.R. Clinical and physiological responses in sporozoite-induced B strain Plasmodium cynomolgi and Plasmodium vivax infections in normal volunteers. *Trans. R. Soc. Trop. Med. Hyg.* **1962**, *56*, 371–378. [CrossRef]
60. Kuvin, S.F.; Beye, H.K.; Stohlman, F., Jr.; Contacos, P.G.; Coatney, G.R. Malaria in Man: Infection by Plasmodium vivax and the B Strain of Plasmodium cynomolgi. *JAMA* **1963**, *184*, 1018–1020. [CrossRef]
61. Druilhe, P.; Trape, J.F.; Leroy, J.P.; Godard, C.; Gentilini, M. Two accidental human infections by Plasmodium cynomolgi bastianellii. A clinical and serological study. *Ann. Soc. Belg. Med. Trop.* **1980**, *60*, 349–354.
62. Cox-Singh, J.; Davis, T.M.; Lee, K.S.; Shamsul, S.S.; Matusop, A.; Ratnam, S.; Rahman, H.A.; Conway, D.J.; Singh, B. Plasmodium knowlesi malaria in humans is widely distributed and potentially life threatening. *Clin. Infect. Dis.* **2008**, *46*, 165–171. [CrossRef] [PubMed]
63. Das, R.; Vashisht, K.; Pandey, K.C. A novel multiplex qPCR assay for clinical diagnosis of non-human malaria parasites-Plasmodium knowlesi and Plasmodium cynomolgi. *Front. Vet. Sci.* **2023**, *10*, 1127273. [CrossRef] [PubMed]
64. Singh, B.; Kim Sung, L.; Matusop, A.; Radhakrishnan, A.; Shamsul, S.S.; Cox-Singh, J.; Thomas, A.; Conway, D.J. A large focus of naturally acquired Plasmodium knowlesi infections in human beings. *Lancet* **2004**, *363*, 1017–1024. [CrossRef] [PubMed]
65. Dinzouna-Boutamba, S.-D.; Yang, H.-W.; Joo, S.-Y.; Jeong, S.; Na, B.-K.; Inoue, N.; Lee, W.-K.; Kong, H.-H.; Chung, D.-I.; Goo, Y.-K.; et al. The development of loop-mediated isothermal amplification targeting alpha-tubulin DNA for the rapid detection of Plasmodium vivax. *Malar. J.* **2014**, *13*, 248. [CrossRef] [PubMed]
66. Britton, S.; Cheng, Q.; Grigg, M.J.; Poole, C.B.; Pasay, C.; William, T.; Fornace, K.; Anstey, N.M.; Sutherland, C.J.; Drakeley, C.; et al. Sensitive Detection of Plasmodium vivax Using a High-Throughput, Colourimetric Loop Mediated Isothermal Amplification (HtLAMP) Platform: A Potential Novel Tool for Malaria Elimination. *PLoS Neglected Trop. Dis.* **2016**, *10*, e0004443. [CrossRef] [PubMed]
67. Mohon, A.N.; Elahi, R.; Khan, W.A.; Haque, R.; Sullivan, D.J.; Alam, M.S. A new visually improved and sensitive loop mediated isothermal amplification (LAMP) for diagnosis of symptomatic falciparum malaria. *Acta Trop.* **2014**, *134*, 52–57. [CrossRef] [PubMed]
68. Yongkiettrakul, S.; Jaroenram, W.; Arunrut, N.; Chareanchim, W.; Pannengpetch, S.; Suebsing, R.; Kiatpathomchai, W.; Pornthanakasem, W.; Yuthavong, Y.; Kongkasuriyachai, D. Application of loop-mediated isothermal amplification assay combined with lateral flow dipstick for detection of Plasmodium falciparum and Plasmodium vivax. *Parasitol. Int.* **2014**, *63*, 777–784. [CrossRef] [PubMed]
69. Poon, L.L.; Wong, B.W.; Ma, E.H.; Chan, K.H.; Chow, L.M.; Abeyewickreme, W.; Tangpukdee, N.; Yuen, K.Y.; Guan, Y.; Looareesuwan, S.; et al. Sensitive and Inexpensive Molecular Test for Falciparum Malaria: Detecting Plasmodium falciparum DNA Directly from Heat-Treated Blood by Loop-Mediated Isothermal Amplification. *Clin. Chem.* **2006**, *52*, 303–306. [CrossRef] [PubMed]
70. Han, E.-T.; Watanabe, R.; Sattabongkot, J.; Khuntirat, B.; Sirichaisinthop, J.; Iriko, H.; Jin, L.; Takeo, S.; Tsuboi, T. Detection of Four *Plasmodium* Species by Genus- and Species-Specific Loop-Mediated Isothermal Amplification for Clinical Diagnosis. *J. Clin. Microbiol.* **2007**, *45*, 2521–2528. [CrossRef]
71. Piera, K.A.; Aziz, A.; William, T.; Bell, D.; González, I.J.; Barber, B.E.; Anstey, N.M.; Grigg, M.J. Detection of Plasmodium knowlesi, Plasmodium falciparum and Plasmodium vivax using loop-mediated isothermal amplification (LAMP) in a co-endemic area in Malaysia. *Malar. J.* **2017**, *16*, 29. [CrossRef]
72. Lau, Y.-L.; Lai, M.-Y.; Fong, M.-Y.; Jelip, J.; Mahmud, R. Loop-Mediated Isothermal Amplification Assay for Identification of Five Human Plasmodium Species in Malaysia. *Am. Soc. Trop. Med. Hyg.* **2016**, *94*, 336–339. [CrossRef] [PubMed]
73. Malpartida-Cardenas, K.; Moser, N.; Ansah, F.; Pennisi, I.; Ahu Prah, D.; Amoah, L.E.; Awandare, G.; Hafalla, J.C.R.; Cunnington, A.; Baum, J.; et al. Sensitive Detection of Asymptomatic and Symptomatic Malaria with Seven Novel Parasite-Specific LAMP Assays and Translation for Use at Point-of-Care. *Microbiol. Spectr.* **2023**, *11*, e0522222. [CrossRef] [PubMed]
74. Joyner, C.; Moreno, A.; Meyer, E.V.; Cabrera-Mora, M.; Kissinger, J.C.; Barnwell, J.W.; Galinski, M.R. Plasmodium cynomolgi infections in rhesus macaques display clinical and parasitological features pertinent to modelling vivax malaria pathology and relapse infections. *Malar. J.* **2016**, *15*, 451. [CrossRef] [PubMed]
75. Shen, F.-h.; Ong, J.J.Y.; Sun, Y.-f.; Lei, Y.; Chu, R.-l.; Kassegne, K.; Fu, H.-t.; Jin, C.; Han, E.-T.; Russell, B.; et al. A Chimeric Plasmodium vivax Merozoite Surface Protein Antibody Recognizes and Blocks Erythrocytic P. cynomolgi Berok Merozoites In Vitro. *Infect. Immun.* **2021**, *89*, e00645-20. [CrossRef] [PubMed]
76. Ong, J.J.Y.; Russell, B.; Han, J.-H. Plasmodium cynomolgi Berok Growth Inhibition Assay by Thiol-reactive Probe Based Flow Cytometric Measurement. *Bio-Protocol* **2021**, *11*, e4147. [CrossRef] [PubMed]
77. Kosaisavee, V.; Suwanarusk, R.; Chua, A.C.Y.; Kyle, D.E.; Malleret, B.; Zhang, R.; Imwong, M.; Imerbsin, R.; Ubalee, R.; Sámano-Sánchez, H.; et al. Strict tropism for CD71(+)/CD234(+) human reticulocytes limits the zoonotic potential of Plasmodium cynomolgi. *Blood* **2017**, *130*, 1357–1363. [CrossRef] [PubMed]

78. Tachibana, S.; Kawai, S.; Katakai, Y.; Takahashi, H.; Nakade, T.; Yasutomi, Y.; Horii, T.; Tanabe, K. Contrasting infection susceptibility of the Japanese macaques and cynomolgus macaques to closely related malaria parasites, Plasmodium vivax and Plasmodium cynomolgi. *Parasitol. Int.* **2015**, *64*, 274–281. [CrossRef] [PubMed]
79. Zeeman, A.M.; van Amsterdam, S.M.; McNamara, C.W.; Voorberg-van der Wel, A.; Klooster, E.J.; van den Berg, A.; Remarque, E.J.; Plouffe, D.M.; van Gemert, G.J.; Luty, A.; et al. KAI407, a potent non-8-aminoquinoline compound that kills Plasmodium cynomolgi early dormant liver stage parasites in vitro. *Antimicrob. Agents Chemother.* **2014**, *58*, 1586–1595. [CrossRef]
80. Maher, S.P.; Bakowski, M.A.; Vantaux, A.; Flannery, E.L.; Andolina, C.; Gupta, M.; Antonova-Koch, Y.; Argomaniz, M.; Cabrera-Mora, M.; Campo, B.; et al. A Drug Repurposing Approach Reveals Targetable Epigenetic Pathways in Plasmodium vivax Hypnozoites. *bioRxiv* **2024**. [CrossRef]

microorganisms

MDPI

Brief Report

A Pilot Study for the Characterization of *Bacillus* spp. and Analysis of Possible *B. thuringiensis*/*Strongyloides stercoralis* Correlation

Elena Pomari [1,*], Pierantonio Orza [1], Milena Bernardi [1], Fabio Fracchetti [2], Ilenia Campedelli [2], Patrick De Marta [2], Alessandra Recchia [3], Paola Paradies [3] and Dora Buonfrate [1]

1 Department of Infectious, Tropical Diseases and Microbiology, IRCCS Sacro Cuore Don Calabria Hospital, Negrar di Valpolicella, 37024 Verona, Italy; pierantonio.orza@sacrocuore.it (P.O.); milena.bernardi@sacrocuore.it (M.B.); dora.buonfrate@sacrocuore.it (D.B.)

2 Microbion srl, San Giovanni Lupatoto, 37057 Verona, Italy; f.fracchetti@microbion.it (F.F.); i.campedelli@microbion.it (I.C.); info@microbion.it (P.D.M.)

3 Department of Precision and Regenerative Medicine and Ionian Area (DiMePRe-J), Veterinary Section, University of Bari "Aldo Moro", Valenzano, 70010 Bari, Italy; alessandra.recchia1@uniba.it (A.R.); paola.paradies@uniba.it (P.P.)

* Correspondence: elena.pomari@sacrocuore.it; Tel.: +39-04-56-013-095

Abstract: Differentiating between *Bacillus* species is relevant in human medicine. *Bacillus thuringiensis* toxins might be effective against *Strongyloides stercoralis*, a nematode causing relevant human morbidity. Our first objective was to evaluate genomic and MALDI-TOF identification methods for *B. thuringiensis*. Our secondary objective was to evaluate a possible negative selection pressure of *B. thuringiensis* against *S. stercoralis*. PCR and Sanger were compared to MALDI-TOF on a collection of 44 *B. cereus* group strains. *B. thuringiensis* toxin genes were searched on 17 stool samples from *S. stercoralis*-infected and uninfected dogs. Metagenomic *16S* rRNA was used for microbiome composition. The inter-rate agreement between PCR, Sanger, and MALDI-TOF was 0.631 k (p-value = 6.4×10^{-10}). *B. thuringiensis* toxins were not found in dogs' stool. *Bacteroidota* and *Bacillota* were the major phyla in the dogs' microbiome (both represented >20% of the total bacterial community). *Prevotella* was underrepresented in all *Strongyloides*-positive dogs. However, the general composition of bacterial communities was not significantly linked with *S. stercoralis* infection. The genomic methods allowed accurate differentiation between *B. thuringiensis* and *B. cereus*. There was no association between *B. thuringiensis* and *S. stercoralis* infection, but further studies are needed to confirm this finding. We provide the first descriptive results about bacterial fecal composition in dogs with *S. stercoralis* infection.

Keywords: *Strongyloides stercoralis*; *Bacillus thuringiensis*; cry5; DNA; microbiota; dog

1. Introduction

Bacillus species are spore-forming, Gram-positive bacteria that include *Bacillus thuringiensis*, *B. cereus*, *B. cytotoxicus*, *B. anthracis*, *B. pseudomycoides*, *B. weihenstephanensis*, *B. toyonensis*, and *B. mycoidesare*. Being highly similar in genotype and phenotype, these bacteria are classified as part of the *B. cereus* group in taxonomy [1,2]. Unambiguous differentiation between *Bacillus* species can be relevant for human and animal health, but this is not usually possible with routine diagnostics [3]. For instance, *B. thuringiensis* is used as a biocontrol agent in agriculture, permitting a reduction in the amount of chemical products [4]. This is possible through the production of various crystal protein toxins (Cry5, Cry6, Cry12, Cry13, Cry14, Cry21, and Cry55) that exhibit a substantial biological activity against various insects, nematodes, and other pathogenic pests [5–8]. Previous studies reported *B. thuringiensis*-derived crystal protein Cry5B to be effective against a broad range of gastrointestinal parasitic hookworms [9] and *Strongyloides stercoralis* [10]. In particular, Cry5B was tested in vitro and in vivo on hamsters and dogs, demonstrating good efficacy against *Ancylostoma ceylanicum*, *Ancylostoma caninum*, and *Necator americanus* [9]. Regarding *S. stercoralis*, multiple stages, including the first larval stage (L1s), infective stage (iL3s), free-living adult stage, and parasitic female stage, were all susceptible to Cry5B, as indicated by the impairment of motility and decreased viability in vitro [10]. *Strongyloides stercoralis* is a soil-transmitted helminth that affects around 600 million people globally, causes a neglected tropical disease [11–13], and occurs in humans, non-human primates, dogs, cats, and wild canids [11]. Its zoonotic potential is under study [14–16]. Indeed, some authors have considered strongyloidiasis a zoonotic disease, while others have argued that the different hosts carry host specialized populations of *S. stercoralis*. In particular, clinical manifestations of *S. stercoralis* in humans and dogs range from asymptomatic to severe infection, with symptoms and signs involving mostly the intestines, respiratory tract, and skin [17]. Moreover, a potentially fatal syndrome (hyperinfection/dissemination) can develop; this is associated with immunosuppression in humans, while it can occur also in immunocompetent dogs [17]. The implementation of public health strategies for the control of *S. stercoralis* in endemic areas has been recommended by the WHO [18]. For instance, the strategies can include regular parasitological examinations of dogs and inspections of park soil, kennels, and dog shelters. Moreover, the strategies primarily entail mass drug administration of ivermectin, but caution should be paid to the possible emergence of drug resistance, considering the already-existing resistance to this drug observed in veterinary medicine [19,20]. Hence, additional strategies that could help to contain environmental contamination with *S. stercoralis* larvae might be useful for integration with mass drug administration. The primary objective of this study was to assess a MALDI-TOF and genomic lab pipeline for the identification of *B. thuringiensis* and its genes involved in the synthesis of the Cry toxins, such as cry5Ab, cry5Ac, and cry5Ba genes [21]. Our secondary objective was to investigate a hypothetical negative selective pressure of these toxins against *S. stercoralis* and a possible different bacterial composition in the gut microbiota of *S. stercoralis*-infected and uninfected dogs.

2. Materials and Methods

2.1. Study Setting and Population

A total of 44 bacterial strains (Table S1) from our collection were used for the genotyping and MALDI-TOF analyses. In particular, among this collection, 3 strains were used as positive controls for *B. thuringiensis* (NRRL B-18400, NRRL B-18765, and DSM 2046T) and 6 for *B. cereus* (LMG 6923T, LMG 12334, LMG 12335, LMG 17615, NCTC 11143, and ATCC 11778). Moreover, in order to explore the potential effect of *B. thuringiensis* on *S. stercoralis*, we analyzed 17 dog stool specimens (Table S2), a cohort enrolled in our previous study investigating the epidemiology of strongyloidiasis in Southern Italy [22]. The dogs' samples were collected from a kennel and three farms located in Apulia Region and all samples were tested for *S. stercoralis* infection using both Baermann method and Real-Time PCR (rtPCR), as previously described [22]. In particular, three dogs resulted to be infected by *S. stercoralis* using both methods and another one tested positive only by rtPCR (see Table S2).

2.2. Mass Spectrometry

The *B. cereus* group strains were cultured on Columbia Agar with Sheep Blood (PB5039A Thermofisher, Monza, Italy) and incubated at 37 °C for 16/24 h. The mass spectrometry identification was performed by MALDI-TOF using the instrument Maldi Biotyper MBT smart (Bruker, Milan, Italy) with the MBT IVD Library DB 8326 March 2019 J and the software MBT Compass IVD v 4.1.100 (Bruker, Milan, Italy) following the manufacturer's instructions.

2.3. DNA Extraction/Purification

For the *B. cereus* group strains, the total genomic DNA was extracted and purified from a 2 mL overnight culture using the Wizard Genomic DNA purification kit (Promega Corporation, Madison, WI, USA), following the manufacturer's instructions. For the stool samples, the DNA was isolated from 200 mg using a Qiamp Fast DNA stool mini kit (Qiagen, Milan, Italy), according to the manufacturer's instructions. The samples were eluted in 30 µL of elution buffer. The quality and quantity of DNA was analyzed using a NanoDrop One/Onec Spectrophotometer (Thermofisher, Monza, Italy) and a Qubit 4 Fluorometer (Thermofisher, Monza, Italy). The isolated DNA was stored at −80 °C until PCR and sequencing.

2.4. Genotyping of B. cereus Group Strains

The *B. cereus* group strains (n = 44) were first characterized by *Bacillus* species-specific PCR for a fragment of gyrase B gene (*gyrB*), followed by Sanger sequencing (Eurofins Genomics, Ebersberg, Germany) as previously described [23]. Moreover, the *B. cereus* group strains (n = 44) and the dog stools (n = 17) were analyzed using specific PCRs for the gene *ces* [24] coding *B. cereus* emetic toxin cereulide and for the genes *cry5Ab*, *cry5Ac*, and *cry5Ba* [21] coding for *B. thuringiensis* Cry toxins. In this analysis, we included the positive controls *B. thuringiensis* NRRL B-18400 and NRRL B-18765 for *cry* genes and *B. cereus* LMG 12,334 for *ces*. For all PCR experiments, the amplification products were loaded on a 1.5% agarose gel and visualized by exposure to ultraviolet (UV) light.

The Sanger sequences of the *gyrB* gene were aligned with Clustal X software v2.0 [25], obtaining a final consensus length of 645 nucleotides for the 44 strains under analysis and the type strain of the species *B. mycoides* DSM 2048T and *B. cytotoxicus* NVH 391-98T, for which the sequence was retrieved from the NCBI database. The consensus-sequence alignment was imported on MEGA software version 11 [26], and the Neighbor-Joining algorithm was used for the reconstruction of the phylogenetic tree. The evolutionary distances were computed using the Tamura–Nei model with complete deletion. The *gyrB* sequences of the strain *B. subtilis* subsp. *subtilis* BCRC 10255T was retrieved from the NCBI database and used as the outgroup.

2.5. Sequencing and Bioinformatic Analysis

To reduce the possible bias of biological/dietary/environmental interferents, we chose to analyze samples collected from dogs belonging to the same kennel, with sampling performed on the same day. For this purpose, among the total dogs cohort (n = 17), the metagenomics analysis was conducted on ten stool samples (three positives and seven negatives for *S. stercoralis*) collected from dogs living in the same kennel (Table S2). Libraries were prepared following the *16S* Metagenomic Sequencing Library Preparation protocol (Illumina, San Diego, CA, USA) in two amplification steps: an initial 35 cycle PCR amplification using 16S rDNA V3–V4-specific PCR primers (16S-341F 5′-CCTACGGGNBGCASCAG-3′ and 16S-805R 5′-GACTACNVGGGTATCTAATCC-3′) and a subsequent amplification that integrates relevant flow-cell binding domains and unique indices (NexteraXT Index Kit, FC-131-1001/FC-131-1002). Libraries were sequenced on a NovaSeq instrument (Illumina, San Diego, CA, USA) using 300 bp paired-end mode. Base calling, demultiplexing, and adapter masking were carried out through the Illumina BCL Convert v3.9.3 (https://emea.support.illumina.com/ (accessed on 18 January 2023)). The FASTQ sequences obtained were analyzed firstly using Kraken 2, which examines the k-mers obtained from a sequencing read sample with those produced from the Silva ribosomal RNA Database (release 138.1) available for Kraken 2 [27]. The taxonomic abundance for each taxon was estimated through Bracken [28]. In addition, the reads generated for the ten stool samples were analyzed through DADA2 version 1.8 [29] by the R 3.5.1 environment. DADA2 was run as described in https://benjjneb.github.io/dada2/tutorial.html (accessed on 20 July 2023), applying the following parameters: trimLeft equal to 30 and the truncLen option set to 270 and 200 for the forward and reverse fastq files, respectively. Taxonomic assignment was performed comparing the amplicon sequence variants (ASVs) predicted from DADA2 against the Silva ribosomal RNA Database (release 138.1) using the function assignTaxonomy and addSpecies.

2.6. Statistics

B. cereus and *B. thuringiensis* cases were summarized for each technique using frequencies and percentages. The agreement between MALDI-TOF, PCR, and Sanger was evaluated using Fleiss' kappa coefficient.

3. Results

3.1. Identification Analysis of *B. cereus* Group Strains

MALDI-TOF and the specific *gyrB* PCR permitted us to identify *B. cereus*, *B. thuringiensis*, *B. mycoides*, *B. cytotoxicus*, and *B. subtilis* in the 44 *B. cereus* strains, as reported in Table S1. The results of Sanger sequencing are shown in Table S1 and Figure 1. Of note, one of the positive controls, *B. thuringiensis* NRRL B-18765, was misidentified as *B. cereus* by MALDI-TOF (Table S1). We then compared the data obtained from MALDI-TOF and the specific PCR followed by Sanger only on the *B. cereus* and *B. thuringiensis* results (n = 32 strains), excluding positive controls and other *Bacillus* species of the collection. We found 0.631 k (p-value = 6.4×10^{-10}) inter-rate agreement between the three methods. As summarized in Table 1, all three methods identified *B. cereus* in twenty-one strains and *B. thuringiensis* in four strains, whilst seven strains resulted in *B. thuringiensis* by genomic approaches and were misidentified as *B. cereus* by MALDI-TOF.

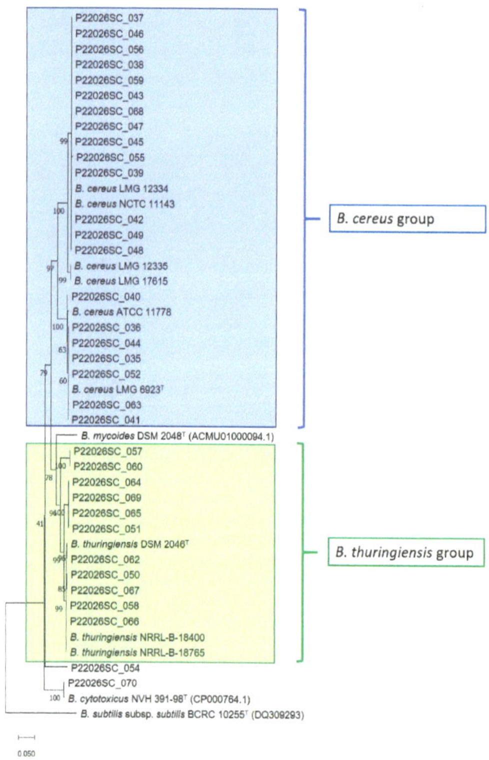

Figure 1. Neighbor-Joining phylogenetic tree based on the *gyrB* gene sequence comparison for the *Bacillus* strains under analysis with the corresponding sequence retrieved for the type strain *B. mycoides* DSM 2048^{T}, *B. cytotoxicus* NVH 391-98^{T}, and *B. subtilis* subsp. *subtilis* BCRC 10255^{T}. The tree was reconstructed through MEGA11 with the Tamura–Nei model and complete deletion treatment for gaps. The accession numbers of the sequence deposited and/or available in NCBI database were reported in brackets.

Table 1. Results of *B. cereus* and *B. thuringiensis* identification with MALDI-TOF, PCR, and Sanger methods. $n = 32$ is the total of analyzed strains.

MALDI-TOF	PCR	Sanger	*n* (%)
B. cereus	*B. cereus*	*B. cereus*	21 (65.62)
B. cereus	*B. thuringiensis*	*B. thuringiensis*	7 (21.88)
B. thuringiensis	*B. thuringiensis*	*B. thuringiensis*	4 (12.50)

3.2. Characterization for cry5Ab, cry5Ac, and cry5Ba by PCR

For a better differentiation, the Bacillus strains were investigated by specific PCRs for the genes *cry5Ab*, *cry5Ac* and *cry5Ba*, which code for the *B. thuringiensis* toxin Cry5B and for the gene *ces* coding for the *B. cereus* emetic toxin cereulide (Table S1). The analysis was performed using control strains *B. thuringiensis* NRRL B18765 and NRRL B-18400 and *B. cereus* LMG 12334, confirming the presence of *cry5Ab*, *cry5Ac*, and *cry5Ba* only in *B. thuringiensis* and the presence of *ces* only in *B. cereus*, as expected (Table S1). Apart from the controls described above, no other strains from the collection showed signals for all the analyzed genes (Table S1). For the study's secondary objective, we tested the dogs' stool for the genes *cry5Ab*, *cry5Ac*, and *cry5Ba*, finding no amplification in all samples (Table S3).

3.3. 16S Illumina

Both Kraken2 | Bracken and DADA2 identified *Bacteroidota* and *Bacillota* as the most abundant phyla (>20% as maximum value, Figure 2, Tables S5 and S6). In particular, four dogs in the present study were characterized by a reduction in *Bacteroidetes*, with a consequently increase in the ratio between *Bacillota* and *Bacteroidota*, presented by KC1A (dogs with frequent episodes of diarrhea), KC3A (dog manifesting some episodes of diarrhea), KC7A (dog positive for *S. stercoralis* infection in 2018), and KC6A (dog positive for *S. stercoralis* infection). The main orders were *Bacteroidales*, *Lactobacillales*, and *Eubacteriales* (>20% as maximum value, Figure 2, Tables S7 and S8). The main genera were *Prevotella*, *Streptococcus*, *Alloprevotella*, *Fusobacterium*, and *Clostridium* (>20% as maximum value, Tables S9 and S10). Regarding the order *Caryophanales* (former *Bacillales*) and the genus *Bacillus*, they were mostly identified in the sample KC10A (*S. stercoralis* negative), with a relative abundance of 2.38% and 2.14%, respectively (Tables S7 and S9). In fact, in all the other samples these taxa were found with a relative abundance lower than 1.5% (Tables S9 and S10), resulting in the least represented bacterial community. This fact could explain why the genus *Bacillus* and, consequently, the order *Caryophanales* were not identified in the samples using the approach based on the DADA2 pipeline (Tables S8 and S10). The general composition of the bacterial communities was not directly linked with *S. stercoralis* infection. However, all *S. stercoralis*-positive samples (KC4A, KC6A, and KC9A) showed very low percentages of the genus *Prevotella*, ranging between 0.17% and 1.06% for the data analyzed with the Kraken2/Bracken approach and between 0.21% and 1.31% according to the DADA2 analysis. In addition, one *S. stercoralis*-negative sample (KC1A) showed similarly low values for *Prevotella*; interestingly, this sample originated from a dog with frequent episodes of diarrhea (Table S2).

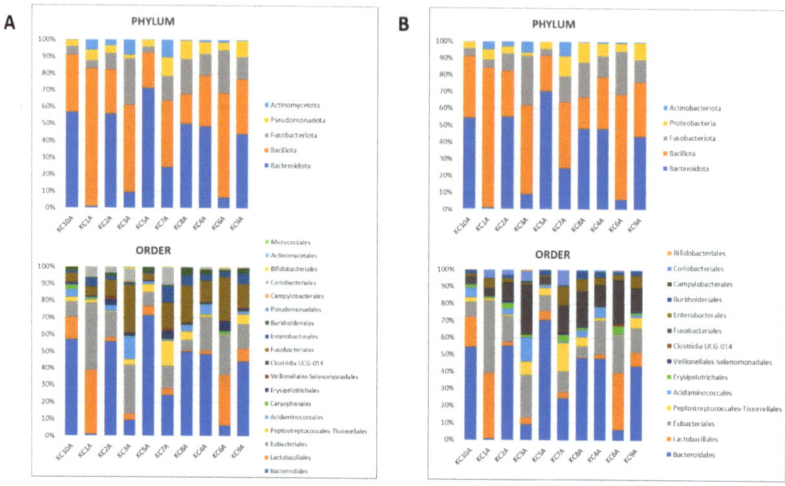

Figure 2. Phylum- and order-level gut microbiota composition in the fecal samples of 10 dogs. (**A**) Data obtained with Kraken2/Bracken. (**B**) Data obtained with DADA2.

4. Discussion

The primary aim of this study was to assess a lab pipeline for the identification of *B. thuringensis* and its characterization for the presence of genes involved in the synthesis of Cry toxins, such as *cry5Ab*, *cry5Ac*, and *cry5Ba* genes [21]. While all six *B. cereus* controls were correctly identified, one out of the three *B. thuringiensis* positive controls (NRRL B-18765) was misidentified as *B. cereus* by MALDI-TOF, highlighting a possible discrepancy with the PCR/Sanger. This observation was deep-rooted extending the analysis to all the collection strains, and the *gyrB* genomic approach allowed differentiation between *B. cereus* and *B. thuringiensis*.

We then characterized all the strains by *ces*- and *cry5*-specific PCRs. As expected, only *B. cereus* had the genetic determinant for the emetic toxin cereulide (ces), whilst two positive *B. thuringiensis* controls (NRRL B-18765 and NRRL B-18400) confirmed all cry5 genes.

Of note, although *B. cereus* and *B. thuringiensis* have different pathogenicity and applications [30], still there are no available guidelines to reliably distinguish species and strains [3]. So far, different approaches such as protein crystallization, pulsed-field gel electrophoresis, and molecular typing were tested [3], but no reliable differentiation has been achieved with good results. A recent work assessed MALDI-TOF mass-spectrometry for differentiating closely related *Bacillus* species [31]. In this work, we chose *gyrB* as the target gene because it is also the target of the putative species-specific PCR developed by Yamada and colleagues (1999) [23] and we combined in-home PCR, Sanger sequencing, and MALDI-TOF approaches with the instruments and library currently used in our laboratory for a comprehensive identification of the strains. Our results highlight that the genomic approach (*gyrB* PCR and Sanger) is more accurate for *B. thuringiensis* identification compared to MALDI-TOF. Based on these results, we proceeded with our secondary objective and we used the approach for genomic characterization in order to investigate the potential presence of *B. thuringiensis* cry5 genes in dog stool. The data obtained from this analysis did not show a possible correlation between the presence of *S. stercoralis* infection and *B. thuringiensis*, as all samples from the dogs were negative for the toxin genes. To the best of our knowledge, to date, there is no evidence of in vivo correlation of *B. thuringiensis* with *S. stercoralis* or other parasites and only evidence of efficacy testing following treatment/administration [9,10]. Moreover, neither the general composition of the bacterial communities nor specific taxa (at the phylum, order, or genus level) were directly associated with *S. stercoralis* infection. In addition to *gyrB*- and *cry5*-specific PCRs, we used a *16S* metagenomics approach in order to detect the fecal bacteria composition. To reduce possible bias of biological/dietary/environmental interferents, we chose to analyze samples collected from dogs belonging to the same kennel, with sampling performed on the same day. However, being a small group, some differences in gender, age, physiological, and health status remain among the dogs leading to possible differences in fecal composition as well. By the way, these characteristics were not taken into account in this work. In order to strengthen the analysis, we conducted the bioinformatics examination using two different tools, in order to investigate how different approaches of read analysis can affect the interpretation of bacteria composition and community structure and try to evaluate the agreement between them. Kraken2 | Brachen is based on alignment-free k-mer searches against a reference genome library [27], while DADA2 infers amplicon sequence variants (ASVs) and the identification at taxonomic level is based on sequence alignment against a reference database [29]. Overall, the two pipelines yielded almost identical compositions and relative frequencies at the phylum level for the samples analyzed, while some differences arose at the order and genus levels specifically for very-low-represented taxa, such as *Caryophanales* (former *Bacillales*) and *Bacillus*. The low sensitivity of DADA2 paired with SILVA for the characterization of samples at the genus level has been recently reported [32]. In addition, the authors confirmed the accuracy of the SILVA database for the investigation of the composition of microbial communities. Another aspect to consider is that the outcomes of DADA2 pipelines are directly dependent on the parameters chosen by the users for read filtering and trimming, which can affect the final identification of the taxa [33].

The two pipelines identified the presence of five different phyla among the samples analyzed, where the most abundant were *Bacteroidota* (former *Bacteroidetes* [34]) and *Bacillota* (former *Firmicutes* [34]). This is consistent with previous reports [35], which characterized the microbiota of 96 healthy dogs. *Bacteroidetes* and *Firmicutes*, together with *Fusobacterium*, represent the most important bacterial phyla in the gastrointestinal tract of dogs [36]. In particular, it is widely accepted that the *Firmicutes/Bacteroidetes* (F/B) ratio has an important role in maintaining intestinal homeostasis [37,38] and the evidence of this study is in accordance with that reported by Chaitman et al. (2020) [39], where the authors observed a significantly lower abundance of *Bacteroidetes* in dogs affected by acute diarrhea.

The overall composition of the fecal microbiome of the *Strongyloides*-infected and uninfected dogs enrolled in our study is in line with what is reported in the literature for dogs either under natural or experimental conditions [40,41]. In spite of the overall similarities in the composition of the fecal microbiota between *Strongyloides*-positive and -negative dogs, *Prevotella* was relatively mostly abundant in the uninfected compared to the infected group. There are no other data in the literature exploring the potential microbiota difference between *Strongyloides*-infected and uninfected dogs. Limited data have been described in chronic *S. stercoralis*-infected humans, showing significantly expanded populations of *Clostridia* and *Leuconostocaceae* [42] and an increasing presence of the *Ruminococcus* torques group [43]. On the other hand, discordant observations have been reported about *Prevotella* abundance in humans, with or without helminth infection [11]. Putatively the dogs enrolled in this study had *S. stercoralis* infection of a long duration due to re-infection from the environment, suggesting a potential correlation with *Prevotella* differential abundance between *Strongyloides*-infected and uninfected dogs.

Some limitations should be highlighted: (i) the small size of the dog cohort, which depended on the availability of samples in the context of our previous study [22] for which they were collected, and (ii) the uninfected dogs were not fully representative of the general canine population.

To conclude, here, we assessed an internal lab pipeline that aimed to characterize a list of *B. cereus* group species for three cry5 toxin genes and for the differential analysis of *B. thuringiensis* and *B. cereus*. Our results suggest that the genomic approach combining specific gyrB PCR and Sanger might be superior for *Bacillus* identification compared to the MALDI-TOF approach.

The specific PCRs and *16S* metagenomics analyses on dog stools showed no significant correlation between *B. thuringiensis* and the fecal microbiome, although a potential *Prevotella* differential abundance between *Strongyloides*-infected and uninfected dogs should be further explored. We provide preliminary descriptive results about fecal bacterial composition in dogs with and without *S. stercoralis* infection. Further investigations are needed in larger cohorts to investigate whether bacterial toxins might have a role in reducing environmental contamination by *S. stercoralis* larvae.

Supplementary Materials: The following supporting information can be downloaded at: https://www.mdpi.com/article/10.3390/microorganisms12081603/s1, Table S1. MALDI-TOF, PCR and Sanger sequencing results for *B. cereus* group strains; Table S2. Dataset of the study dogs; Table S3. PCR results in stool samples; Table S4. Number of reads analyzed for each stool sample by 16S metagenomics approach; Table S5. Kraken2/Bracken—Silva results at phylum level in the fecal samples of 10 dogs; Table S6. DADA2-Silva results at phylum level in the fecal samples of 10 dogs; Table S7. Kraken2/Bracken -Silva results at order level in the fecal samples of 10 dogs; Table S8. DADA2-Silva results at order level in the fecal samples of 10 dogs; Table S9. Kraken2/Bracken -Silva results at genus level in the fecal samples of 10 dogs; Table S10. DADA2-Silva results at genus level in the fecal samples of 10 dogs.

Author Contributions: E.P. conceived the study; E.P., F.F., I.C., M.B., P.O., A.R. and P.P. contributed to data collection; E.P., F.F., I.C., P.D.M., M.B. and P.O. contributed to data analyses; E.P. and D.B. wrote the first draft of the manuscript. All authors read, revised, and approved the final manuscript. All authors have read and agreed to the published version of the manuscript.

Funding: This work received funds from the Italian Ministry of Health "5×1000-2020", project "Bacillus thuringiensis (gruppo *Bacillus cereus*) come possibile antagonista di *Strongyloides stercoralis* e di altri elminti correlati con malattie tropicali neglette (NTD), alla base di una strategia di prevenzione" and "Ricerca Corrente", project L2P11.

Data Availability Statement: All data generated or analyzed during this study are included in this published article (and its Supplementary Materials). The raw read sequences of the dog stool samples analyzed were deposited in the NCBI database under the BioProject ID PRJNA1134405 and the sequence of the gyrB gene obtained for the 44 *Bacillus* strains can be accessed in the NCBI database with the following accession number range: PQ014599–PQ014641.

Conflicts of Interest: Authors Fabio Fracchetti, Ilenia Campedelli, and Patrick De Marta were employed by the company Microbion srl. The remaining authors declare that the research was conducted in the absence of any commercial or financial relationships that could be construed as a potential conflict of interest.

References

1. Bianco, A.; Capozzi, L.; Monno, M.R.; Del Sambro, L.; Manzulli, V.; Pesole, G.; Loconsole, D.; Parisi, A. Characterization of *Bacillus* cereus Group Isolates from Human Bacteremia by Whole-Genome Sequencing. *Front. Microbiol.* **2020**, *11*, 599524. [CrossRef] [PubMed]
2. Baldwin, V.M. You Can't *B. cereus*—A Review of *Bacillus* cereus Strains That Cause Anthrax-Like Disease. *Front. Microbiol.* **2020**, *11*, 1731. [CrossRef] [PubMed]
3. EFSA Panel on Biological Hazards (BIOHAZ). Risks for public health related to the presence of *Bacillus* cereus and other *Bacillus* spp. including *Bacillus thuringiensis* in foodstuffs. *EFSA J.* **2016**, *14*, 4524.
4. Khan, A.R.; Mustafa, A.; Hyder, S.; Valipour, M.; Rizvi, Z.F.; Gondal, A.S.; Yousuf, Z.; Iqbal, R.; Daraz, U. *Bacillus* spp. as Bioagents: Uses and Application for Sustainable Agriculture. *Biology* **2022**, *11*, 1763. [CrossRef] [PubMed]
5. Wei, J.-Z.; Hale, K.; Carta, L.; Platzer, E.; Wong, C.; Fang, S.-C.; Aroian, R.V. *Bacillus thuringiensis* crystal proteins that target nematodes. *Proc. Natl. Acad. Sci. USA* **2003**, *100*, 2760–2765. [CrossRef]
6. Schnepf, E.; Crickmore, N.; Van Rie, J.; Lereclus, D.; Baum, J.; Feitelson, J.; Zeigler, D.R.; Dean, D.H. *Bacillus thuringiensis* and its pesticidal crystal proteins. *Microbiol. Mol. Biol. Rev.* **1998**, *62*, 775–806. [CrossRef]
7. Liang, Z.; Ali, Q.; Wang, Y.; Mu, G.; Kan, X.; Ren, Y.; Manghwar, H.; Gu, Q.; Wu, H.; Gao, X. Toxicity of *Bacillus thuringiensis* Strains Derived from the Novel Crystal Protein Cry31Aa with High Nematicidal Activity against Rice Parasitic Nematode *Aphelenchoides besseyi*. *Int. J. Mol. Sci.* **2022**, *23*, 8189. [CrossRef]
8. Palma, L.; Muñoz, D.; Berry, C.; Murillo, J.; Caballero, P. *Bacillus thuringiensis* toxins: An overview of their biocidal activity. *Toxins* **2014**, *6*, 3296–3325. [CrossRef] [PubMed]

9. Hu, Y.; Nguyen, T.-T.; Lee, A.C.Y.; Urban, J.F.J.; Miller, M.M.; Zhan, B.; Koch, D.J.; Noon, J.B.; Abraham, A.; Fujiwara, R.T.; et al. *Bacillus thuringiensis* Cry5B protein as a new pan-hookworm cure. *Int. J. Parasitol. Drugs Drug Resist.* **2018**, *8*, 287–294. [CrossRef]
10. Charuchaibovorn, S.; Sanprasert, V.; Sutthanont, N.; Hu, Y.; Abraham, A.; Ostroff, G.R.; Aroian, R.V.; Jaleta, T.G.; Lok, J.B.; Nuchprayoon, S. *Bacillus thuringiensis* Cry5B is Active against *Strongyloides stercoralis* in vitro. *Am. J. Trop. Med. Hyg.* **2019**, *101*, 1177–1182. [CrossRef]
11. Buonfrate, D.; Bisanzio, D.; Giorli, G.; Odermatt, P.; Fürst, T.; Greenaway, C.; French, M.; Reithinger, R.; Gobbi, F.; Montresor, A.; et al. The Global Prevalence of *Strongyloides stercoralis* Infection. *Pathogens* **2020**, *9*, 468. [CrossRef]
12. Sprecher, V.P.; Hofmann, D.; Savathdy, V.; Xayavong, P.; Norkhankhame, C.; Huy, R.; Khieu, V.; Sayasone, S.; Hattendorf, J.; Keiser, J. Efficacy and safety of moxidectin compared with ivermectin against *Strongyloides stercoralis* infection in adults in Laos and Cambodia: A randomised, double-blind, non-inferiority, phase 2b/3 trial. *Lancet Infect. Dis.* **2024**, *24*, 196–205. [CrossRef] [PubMed]
13. Bisoffi, Z.; Buonfrate, D.; Montresor, A.; Requena-Méndez, A.; Muñoz, J.; Krolewiecki, A.J.; Gotuzzo, E.; Mena, M.A.; Chiodini, P.L.; Anselmi, M.; et al. *Strongyloides stercoralis*: A Plea for Action. *PLoS Negl. Trop. Dis.* **2013**, *7*, e2214. [CrossRef]
14. Jaleta, T.G.; Zhou, S.; Bemm, F.M.; Schär, F.; Khieu, V.; Muth, S.; Odermatt, P.; Lok, J.B.; Streit, A. Different but overlapping populations of *Strongyloides stercoralis* in dogs and humans-Dogs as a possible source for zoonotic strongyloidiasis. *PLoS Negl. Trop. Dis.* **2017**, *11*, e0005752. [CrossRef] [PubMed]
15. Nagayasu, E.; Aung, M.P.P.T.H.H.; Hortiwakul, T.; Hino, A.; Tanaka, T.; Higashiarakawa, M.; Olia, A.; Taniguchi, T.; Win, S.M.T.; Ohashi, I.; et al. A possible origin population of pathogenic intestinal nematodes, *Strongyloides stercoralis*, unveiled by molecular phylogeny. *Sci. Rep.* **2017**, *7*, 4844. [CrossRef] [PubMed]
16. Beknazarova, M.; Barratt, J.L.N.; Bradbury, R.S.; Lane, M.; Whiley, H.; Ross, K. Detection of classic and cryptic *Strongyloides* genotypes by deep amplicon sequencing: A preliminary survey of dog and human specimens collected from remote Australian communities. *PLoS Negl. Trop. Dis.* **2019**, *13*, e0007241. [CrossRef]
17. Nutman, T.B. Human infection with *Strongyloides stercoralis* and other related *Strongyloides* species. *Parasitology* **2017**, *144*, 263–273. [CrossRef]
18. *2030 Targets for Soil-Transmitted Helminthiases Control Programmes*; World Health Organization: Geneva, Switzerland, 2019.
19. Sengthong, C.; Yingklang, M.; Intuyod, K.; Haonon, O.; Pinlaor, P.; Jantawong, C.; Hongsrichan, N.; Laha, T.; Anutrakulchai, S.; Cha'on, U.; et al. Repeated Ivermectin Treatment Induces Ivermectin Resistance in *Strongyloides ratti* by Upregulating the Expression of ATP-Binding Cassette Transporter Genes. *Am. J. Trop. Med. Hyg.* **2021**, *105*, 1117–1123. [CrossRef]
20. Whittaker, J.H.; Carlson, S.A.; Jones, D.E.; Brewer, M.T. Molecular mechanisms for anthelmintic resistance in strongyle nematode parasites of veterinary importance. *J. Vet. Pharmacol. Ther.* **2017**, *40*, 105–115. [CrossRef]
21. Garcia-Robles, I.; Sánchez, J.; Gruppe, A.; Martinez-Ramirez, A.C.; Rausell, C.; Real, M.D.; Bravo, A. Mode of action of *Bacillus thuringiensis* PS86Q3 strain in hymenopteran forest pests. *Insect Biochem. Mol. Biol.* **2001**, *31*, 849–856. [CrossRef]
22. Paradies, P.; Digiaro, S.; Colella, A.; Greco, B.; Recchia, A.; Prato, M.G.; Mazzi, C.; Losurdo, G.; Di Leo, A.; Formenti, F.; et al. Strongyloidiasis in humans and dogs in Southern Italy: An observational study. *Parasitol. Res.* **2023**, *122*, 2885–2890. [CrossRef]
23. Yamada, S.; Ohashi, E.; Agata, N.; Venkateswaran, K. Cloning and nucleotide sequence analysis of gyrB of *Bacillus cereus, B. thuringiensis, B. mycoides*, and *B. anthracis* and their application to the detection of *B. cereus* in rice. *Appl. Environ. Microbiol.* **1999**, *65*, 1483–1490. [CrossRef]
24. Ehling-Schulz, M.; Vukov, N.; Schulz, A.; Shaheen, R.; Andersson, M.; Märtlbauer, E.; Scherer, S. Identification and partial characterization of the nonribosomal peptide synthetase gene responsible for cereulide production in emetic *Bacillus cereus*. *Appl. Environ. Microbiol.* **2005**, *71*, 105–113. [CrossRef]
25. Larkin, M.A.; Blackshields, G.; Brown, N.P.; Chenna, R.; McGettigan, P.A.; McWilliam, H.; Valentin, F.; Wallace, I.M.; Wilm, A.; Lopez, R.; et al. Clustal W and Clustal X version 2.0. *Bioinformatics* **2007**, *23*, 2947–2948. [CrossRef]
26. Tamura, K.; Stecher, G.; Kumar, S. MEGA11: Molecular Evolutionary Genetics Analysis Version 11. *Mol. Biol. Evol.* **2021**, *38*, 3022–3027. [CrossRef] [PubMed]
27. Wood, D.E.; Lu, J.; Langmead, B. Improved metagenomic analysis with Kraken 2. *Genome Biol.* **2019**, *20*, 257. [CrossRef]
28. Lu, J.; Salzberg, S.L. Ultrafast and accurate 16S rRNA microbial community analysis using Kraken 2. *Microbiome* **2020**, *8*, 124. [CrossRef] [PubMed]
29. Callahan, B.J.; McMurdie, P.J.; Rosen, M.J.; Han, A.W.; Johnson, A.J.A.; Holmes, S.P. DADA2: High-resolution sample inference from Illumina amplicon data. *Nat. Methods* **2016**, *13*, 581–583. [CrossRef] [PubMed]
30. Ehling-Schulz, M.; Lereclus, D.; Koehler, T.M. The *Bacillus* cereus Group: *Bacillus* Species with Pathogenic Potential. *Microbiol. Spectr.* **2019**, *7*, 10–1128. [CrossRef]
31. Chen, M.; Wei, X.; Zhang, J.; Zhou, H.; Chen, N.; Wang, J.; Feng, Y.; Yu, S.; Zhang, J.; Wu, S.; et al. Differentiation of *Bacillus* cereus and *Bacillus thuringiensis* Using Genome-Guided MALDI-TOF MS Based on Variations in Ribosomal Proteins. *Microorganisms* **2022**, *10*, 918. [CrossRef]
32. Odom, A.R.; Faits, T.; Castro-Nallar, E.; Crandall, K.A.; Johnson, W.E. Metagenomic profiling pipelines improve taxonomic classification for 16S amplicon sequencing data. *Sci. Rep.* **2023**, *13*, 13957. [CrossRef] [PubMed]
33. Baltrušis, P.; Halvarsson, P.; Höglund, J. Estimation of the impact of three different bioinformatic pipelines on sheep nemabiome analysis. *Parasit. Vectors* **2022**, *15*, 290. [CrossRef] [PubMed]

34. Oren, A.; Garrity, G.M. Valid publication of the names of forty-two phyla of prokaryotes. *Int. J. Syst. Evol. Microbiol.* **2021**, *71*, 005056. [CrossRef] [PubMed]
35. You, I.; Kim, M.J. Comparison of Gut Microbiota of 96 Healthy Dogs by Individual Traits: Breed, Age, and Body Condition Score. *Animals* **2021**, *11*, 2432. [CrossRef] [PubMed]
36. Pilla, R.; Suchodolski, J.S. The Role of the Canine Gut Microbiome and Metabolome in Health and Gastrointestinal Disease. *Front. Vet. Sci.* **2019**, *6*, 498. [CrossRef] [PubMed]
37. Stojanov, S.; Berlec, A.; Štrukelj, B. The Influence of Probiotics on the Firmicutes/Bacteroidetes Ratio in the Treatment of Obesity and Inflammatory Bowel disease. *Microorganisms* **2020**, *8*, 1715. [CrossRef] [PubMed]
38. Thomson, P.; Santibáñez, R.; Rodríguez-Salas, C.; Flores-Yañez, C.; Garrido, D. Differences in the composition and predicted functions of the intestinal microbiome of obese and normal weight adult dogs. *PeerJ* **2022**, *10*, e12695. [CrossRef]
39. Chaitman, J.; Ziese, A.-L.; Pilla, R.; Minamoto, Y.; Blake, A.B.; Guard, B.C.; Isaiah, A.; Lidbury, J.A.; Steiner, J.M.; Unterer, S.; et al. Fecal Microbial and Metabolic Profiles in Dogs with Acute Diarrhea Receiving Either Fecal Microbiota Transplantation or Oral Metronidazole. *Front. Vet. Sci.* **2020**, *7*, 192. [CrossRef] [PubMed]
40. Kubinyi, E.; Bel Rhali, S.; Sándor, S.; Szabó, A.; Felföldi, T. Gut Microbiome Composition is Associated with Age and Memory Performance in Pet Dogs. *Animals* **2020**, *10*, 1488. [CrossRef]
41. Coelho, L.P.; Kultima, J.R.; Costea, P.I.; Fournier, C.; Pan, Y.; Czarnecki-Maulden, G.; Hayward, M.R.; Forslund, S.K.; Schmidt, T.S.B.; Descombes, P.; et al. Similarity of the dog and human gut microbiomes in gene content and response to diet. *Microbiome* **2018**, *6*, 72. [CrossRef]
42. Jenkins, T.P.; Formenti, F.; Castro, C.; Piubelli, C.; Perandin, F.; Buonfrate, D.; Otranto, D.; Griffin, J.L.; Krause, L.; Bisoffi, Z.; et al. A comprehensive analysis of the faecal microbiome and metabolome of *Strongyloides stercoralis* infected volunteers from a non-endemic area. *Sci. Rep.* **2018**, *8*, 15651. [CrossRef] [PubMed]
43. Tran, N.T.D.; Chaidee, A.; Surapinit, A.; Yingklang, M.; Roytrakul, S.; Charoenlappanit, S.; Pinlaor, P.; Hongsrichan, N.; Anutrakulchai, S.; Cha'on, U.; et al. Chronic *Strongyloides stercoralis* infection increases presence of the *Ruminococcus torques* group in the gut and alters the microbial proteome. *Sci. Rep.* **2023**, *13*, 4216. [CrossRef] [PubMed]
44. Llinás-Caballero, K.; Caraballo, L. Helminths and Bacterial Microbiota: The Interactions of Two of Humans' "Old Friends". *Int. J. Mol. Sci.* **2022**, *23*, 13358. [CrossRef] [PubMed]

microorganisms

MDPI

Article

Cryptosporidium Infections in Neonatal Calves on a Dairy Farm

Michaela Kaduková [1], Andrea Schreiberová [1,*], Pavol Mudroň [2], Csilla Tóthová [2], Pavel Gomulec [2] and Gabriela Štrkolcová [1]

[1] Department of Epizootiology, Parasitology and Protection of One Health, University of Veterinary Medicine and Pharmacy in Košice, Komenského 73, 040 01 Košice, Slovakia; michaela.kadukova@student.uvlf.sk (M.K.); gabriela.strkolcova@uvlf.sk (G.Š.)

[2] Clinic of Ruminants, University of Veterinary Medicine and Pharmacy in Košice, Komenského 73, 040 01 Košice, Slovakia; pavol.mudron@uvlf.sk (P.M.); csilla.tothova@uvlf.sk (C.T.); pavel.gomulec@student.uvlf.sk (P.G.)

* Correspondence: andrea.schreiberova@uvlf.sk; Tel.: +421-905-582-309

Abstract: This study was conducted with the aim of the molecular identification of the protozoan parasite *Cryptosporidium* spp. in calves in the early stage of their development on a dairy farm in Eastern Slovakia. Twenty-five Holstein and Holstein cross calves were included in the study and monitored from their birth to the fifth week of life (1–5 weeks). Fresh fecal samples were collected from the same group of calves each week, except during the fourth week, and with the exception of Sample 8. All samples were analyzed using the Ziehl–Neelsen staining method and coproantigen was tested using the ELISA test as the screening method. Using the ELISA method, the highest incidence of cryptosporidiosis was observed in the second week of life of the calves, while the antigen was detected in 21 (91.6%) calves. Using the Ziehl–Neelsen staining method, the highest incidence was also observed in the second week, with an incidence rate of 62.5%. Positive isolates confirmed by the ELISA test were molecularly characterized. The species and subtypes of *Cryptosporidium* in the positive isolates were identified using PCR and the sequence analysis of the small subunit of the ribosomal 18S RNA (ssu rRNA) and the 60 kDa glycoprotein (gp60) genes of the parasite. The sequence analysis of 29 isolates at the 18S rRNA loci confirmed the presence of two species—*Cryptosporidium parvum* and *Cryptosporidium ryanae*. Out of 29 isolates, 25 were assigned to the species *C. parvum*, with the gp60 locus identified as genotype IIaA17G1R1. Among the individual animal groups, calves are the most common reservoirs of the *C. parvum* zoonotic species. This disease has significant public health implications as contact with livestock and their feces and working with barn manure are major sources of infection, not only for other animals but also for humans.

Keywords: calves; cryptosporidium; ELISA; nested PCR; Ziehl–Neelsen staining method

1. Introduction

Cryptosporidium is a coccidial parasite currently classified in the Alveolata phylum, *Eimeria* suborder, Cryptosporidiidae family. It attacks the respiratory and digestive systems of reptiles, birds, and mammals. Cryptosporidiosis is a zoonotic disease that causes significant economic losses in the cattle population, especially in calves and young animals [1,2]. To date, more than 51 species of *Cryptosporidium* with 120 genotypes have been recognized, with 19 species and 4 genotypes reported in humans, including *Cryptosporidium hominis*, *Cryptosporidium parvum*, *Cryptosporidium meleagridis*, *Cryptosporidium canis*, and *Cryptosporidium felis* as the most common ones [3,4]. The first case of *Cryptosporidium* in calves was reported in 1971 in Oklahoma in an 8-month-old calf with diarrhea, weakness, and a chronic course. The histopathological findings corresponded to villous atrophy with the presence of various developmental stages of *Cryptosporidium* in the epithelium of the small intestine [5]. In cattle, the infection is most commonly caused by species like *C. parvum*, *C. bovis*, *C. ryanae*, and *C. andersoni* [6]. *C. parvum* usually occurs in pre-weaning calves, while *C. bovis* and *C. ryanae* species are mostly diagnosed in post-weaning calves and young cattle. *C. andersoni* is the most common species infecting adult cattle [7–9].

Several studies indicate that diarrhea caused by *Cryptosporidium* species is the main cause of calf mortality in the first three weeks of their life. It is characterized by watery, profuse diarrhea, which may be associated with dehydration, anorexia, poor growth development, and low weight gain, with a high mortality rate in untreated calves, leading to significant economic losses [1,10,11]. Infections in older animals are usually subclinical but can still have a negative impact on production, with lower body condition scores, lower carcass weights, and poor carcass-processing hygiene [12]. Enterotoxigenic *Escherichia coli* K99/F5, rotavirus, and coronavirus, along with *Cryptosporidium*, are the four most important enteropathogens causing neonatal diarrhea in calves worldwide [13–15]. The infection primarily spreads through the fecal–oral route, which is considered the main route of transmission, or indirectly through the consumption of contaminated food or water with infectious oocysts [3,16]. The high shedding intensity of environmentally resistant oocysts leads to strong environmental contamination and increases the risk of infection. Infected calves can excrete up to 6×10^6 oocysts per gram of feces, but even a small number of oocysts is sufficient to cause an infection [17]. It is reported that oocyst shedding begins 4 days after birth and peaks on days 7–18, while its intensity decreases after approximately 3 weeks [18].

The predominant species responsible for up to 70% of human cryptosporidiosis cases is *C. hominis*, while *C. parvum* is considered the most common cause of zoonotic infections in humans. Within the *C. parvum* species, more than 20 subtypes with geographical variations have been identified, with subtypes IIa, IIc, and IId showing host adaptation. Subtype IIc appears to be exclusively anthroponotically transmitted, while subtypes IIa and IIb are zoonotic and are mostly found in cattle, sheep, and goats [3,19].

Cryptosporidiosis is challenging to control in both animals and humans due to the environmentally stable oocysts, low infectious doses, and wide range of susceptible hosts. Among the available methods for the diagnosis of *Cryptosporidium*, staining methods, such as Kinyoun or Ziehl–Neelsen staining, as well as immunological methods in the form of commercially available ELISA tests, are well known. Molecular methods based on PCR and DNA sequencing, PCR-RFLP, qPCR, or multiplex PCR are more sensitive than microscopy and immunological detection methods [6,20].

Currently, there are neither vaccines nor effective chemotherapeutics known for the control of bovine cryptosporidiosis [9,21]. From an economic and health perspective, it is important to increase the awareness among individuals, especially those who work with ruminants, considering that these animals are often reservoirs of zoonotic species.

A case report from Slovakia in 2020 confirmed a case of cryptosporidiosis caused by the species *C. parvum*, subtype IIdA15G1, in a veterinary medicine student who worked on a calf farm in the eastern part of Slovakia and was aware of their inadequate hand hygiene after handling calves [22]. This study presents the 5-week monitoring of *Cryptosporidium* oocyst shedding in calves from birth and demonstrates that the second week of a calf's life is the most critical in terms of the risk of infection. It also focuses on the molecular identification and genotyping of *Cryptosporidium* spp. in calves in the early stage of their development on a dairy farm in Eastern Slovakia. The confirmed genotype IIaA17G1R1 has been identified in Holland and Slovenia in humans, indicating that cattle are a potential source and important reservoir of *C. parvum* infection for both humans and animals in Europe [23,24].

2. Materials and Methods

This research was conducted from September 2022 to October 2022 on a dairy farm breeding dairy cows, located in Eastern Slovakia. There were no other animals on the farm besides cows, which were under veterinary supervision, vaccination schedules against Infectious Bovine Rhinotracheitis (IBR) were followed. The study included 25 neonatal dairy calves (Holstein and Holstein cross). The calves' mothers were impregnated through artificial insemination, and the number of calves included in the study was determined based on the expectation that they would calve in the same week. The calving of the mothers took place in birthing pens, and the calves were separated from their mothers immediately after birth into individual pens, which were regularly cleaned and disinfected. The newborn calves were fed with a commercial milk replacer and then with commercial concentrates, with free access to water. Fresh fecal samples were collected from the same group of calves each week from the ground after defecation, from their birth until the fifth week of life (1st week (0–7 days old), 2nd week (8–14 days old), 3rd week (15–21 days old), 4th week (22–28 days old), 5th week (29–35 days old)), except for the fourth week, when sampling was not possible. During the analyzed period, 24 calves were tested, except for the 1st week, in which 25 calves were examined. Sample 8 was tested in the 1st week only because the calf was then relocated to an isolated area due to another non-infectious disease; therefore, sampling was impossible. Samples were collected into clean plastic containers, labeled with the calf's identification number and date of birth and the date of collection, and transported in a polystyrene thermobox designed for the transportation biological material to the Department of Epizootiology, Parasitology, and Protection of One Health at the University of Veterinary Medicine and Pharmacy in Košice.

Fecal samples were analyzed using a flotation method with a flotation solution of specific gravity 1.3 g/cm^3 to detect the oocysts of protozoan parasites and the eggs of helminths. Additionally, a zinc sulfate solution with specific gravity of 1.18 g/cm^3 was utilized to detect cysts of *Giardia duodenalis* [25,26]. Oocysts of *Cryptosporidium* spp. were diagnosed by performing fecal smears; three smears were performed from each fecal sample, stained using the Ziehl–Neelsen staining method, and examined under a microscope after drying [25,27]. One of the examination methods also included the detection of coproantigen using a commercially available kit, CRYPTOSPORIDIUM (FAECAL), Diagnostic Automation, INC, Calabasas, USA. Specific antigens of *Cryptosporidium* spp. in animal feces were detected according to the manufacturer's instructions. For the ELISA test, positivity was detected above an OD value of 0.15, while all samples below OD 0.149 were considered negative. Specific antigens of *Cryptosporidium* spp. in animal feces were detected according to the manufacturer's instructions. Fecal samples from calves were examined using both methods in each of these weeks; subsequently, positive samples confirmed by the ELISA method were molecularly identified using the nested PCR method.

Molecular Analysis of Cryptosporidium Species

Positive samples detected by the ELISA method were used for DNA extraction using the ZR Fecal DNA MiniPrep Kit (Zymo Research, Tustin, CA, USA) following the manufacturer's instructions. The genomic DNA was stored at −20 °C and later used for PCR analysis.

We performed the molecular species identification of *Cryptosporidium* using the nested PCR amplification of the small ribosomal subunit rRNA gene (18S rRNA). In the first step, we used a set of primers, forward 18-F1 (5′-TTCTAGAGCTAATACATGCG-3′) and reverse 18-R1 (5′-CCCTAATCCTTCGAAACAGGA-3′), resulting in a primary product of 1350 bp. The amplification program included initial denaturation at 94 °C for 3 min, followed by 35 cycles of denaturation at 94 °C for 45 s, annealing at 55 °C for 45 s, extension at 72 °C for 60 s, and final extension at 72 °C for 7 min. In the second reaction, we used the forward primer 18S-F2 (5′-GGAAGGGTTGTATTTATTAGATAAAG-3′) and the reverse primer 18S-R2 (5′-AAGGAGTAAGGAACAACCTCCA-3′). The amplification conditions included initial denaturation at 94 °C for 3 min, followed by 35 cycles of denaturation at 94 °C for 30 s, annealing at 58 °C for 1 min and 30 s, extension at 72 °C for 2 min, and final extension at 72 °C for 7 min, resulting in a secondary product of 840 bp [28].

The subtyping of *Cryptosporidium* was performed using the nested PCR amplification of the gp60 gene. In the first PCR step, primers GP60-F1 (5′-ATAGTCTCCGCTGTATTC-3′) and GP60-R1 (5′-TCCGCTGTATTCTCAGCC-3′) were used. The amplification program included initial denaturation at 95 °C for 3 min, followed by 35 cycles of denaturation at

94 °C for 45 s, annealing at 50 °C for 45 s, extension at 72 °C for 60 s, and final extension at 72 °C for 10 min, resulting in a primary product of 1250 bp. In the second reaction, forward primer GP60-F2 (5′-GGAAGGAACGATGTATCT-3′) and reverse primer GP60-R2 (5′-GCAGAGGAACCAGCATC-3′) were used under the same conditions as in the first step. Amplification yielded a secondary product of 850 bp [29].

The amplified products were visualized on a 1% agarose gel and observed under UV light. All positive PCR products were sent to the commercial laboratory Microsynth Seqlab (Vienna, Austria) or SEQme (Dobříš, Czech Republic) for purification and sequencing in both directions, using identical primers to those that were used in the second steps of both PCR reactions for the 18S rRNA and gp60 genes. The sequencing was performed by the Sanger sequencing method. The resulting sequences were analyzed and edited using the software MEGA X version 10.1.5 build 10191107 [30]. The assembly of the nucleotide sequences was carried out using the Gene Tool Lite 1.0 software (BioTools Inc., Jupiter, FL, USA). The consensus sequences were compared with the sequences deposited in GenBank by applying the nucleotide BLAST algorithm at the National Center for Biotechnology Information (NCBI). The sequences from this study have been deposited in GenBank under unique accession numbers for the 18S rRNA gene (PP897358-PP897365) and the gp60 genes (PP916252-PP91627) for *Cryptosporidium* spp.

The molecular identification of *Cryptosporidium* spp. was further confirmed by phylogenetic analysis using the maximum likelihood method with a minimum of 1000 bootstrap replications. For phylogenetic analyses, the nucleotide sequences obtained in this study and other *Cryptosporidium* spp. sequences from GenBank were used in the software MEGA X [30].

3. Results

3.1. Ziehl–Neelsen Staining Method

In the first week, the microscopic observation of fecal smears did not confirm the presence of *Cryptosporidium* spp. oocysts. In the second week, *Cryptosporidium* spp. was confirmed microscopically in 15 samples (62.5%) out of a total of 24. In the third week, it was detected in two samples; in the fifth week, it was found in one sample. Sampling was not conducted in the fourth week (Figure 1) (Tables 1 and 2).

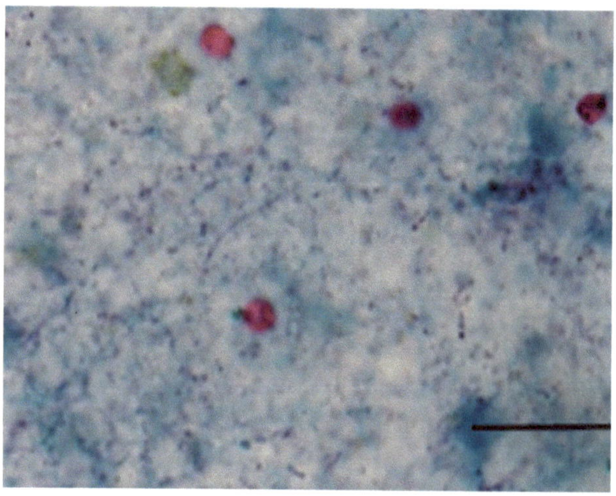

Figure 1. Oocysts of *Cryptosporidium* spp. according to Ziehl–Neelsen staining method; bar—20 μm.

Table 1. Detection of *Cryptosporidium* spp. by Ziehl–Neelsen staining method and ELISA test.

No.	1st Week		2st Week		3st Week		5st Week	
	Ziehl–Neelsen	**Elisa**	**Ziehl–Neelsen**	**Elisa**	**Ziehl–Neelsen**	**Elisa**	**Ziehl–Neelsen**	**Elisa**
1.	-	-	+	+	-	-	-	-
2.	-	-	-	+	-	-	-	-
3.	-	-	-	+	-	-	-	-
4.	-	-	+	+	+	+	-	-
5.	-	-	+	+	-	-	-	-
6.	-	+	+	+	-	-	-	-
7.	-	+	-	-	-	-	-	-
8.	-	0	0	0	0	0	0	0
9.	-	-	+	+	-	-	-	+
10.	-	-	+	-	-	-	-	+
11.	-	-	-	+	-	-	-	-
12.	-	-	-	+	-	-	-	-
13.	-	-	+	+	+	+	+	+
14.	-	-	+	+	-	-	-	-
15.	-	-	+	+	-	-	-	-
16.	-	-	-	+	-	-	-	-
17.	-	-	+	+	-	-	-	-
18.	-	-	+	+	-	-	-	-
19.	-	-	-	+	-	-	-	-
20.	-	-	-	+	-	-	-	-
21.	-	-	+	+	-	-	-	-
22.	-	-	-	+	-	-	-	-
23.	-	-	+	+	-	+	-	-
24.	-	-	+	+	-	-	-	-
25.	-	-	-	+	-	-	-	-

- (negative); + (positive); 0 (no sample).

Table 2. Comparison of Ziehl–Neelsen staining method and ELISA test in each week.

	1st Week (0–7 Days)	2nd Week (7–14 Days)	3rd Week (14–21 Days)	4th Week (21–28 Days)	5th Week (28–35 Days)
Ziehl–Neelsen stain method	0%	62.5%	4.1%	-	4.1%
ELISA test	8%	91.6%	12.5%	-	12.5%

3.2. ELISA Test

In the first week, the presence of the *Cryptosporidium* spp. coproantigen was detected in two calves. Throughout the study period, the highest incidence of cryptosporidiosis was observed in the second week of life, while the antigen was confirmed in 22 (91.6%) calves. In the third and fifth weeks, the coproantigen was detected in only three calves; interestingly, these were different calves (Tables 1 and 2).

Using the flotation method, the fecal samples were repeatedly examined for the presence of *Giardia duodenalis* cysts during all weeks. In the third week of the calves' lives, positive findings of *Giardia duodenalis* cysts were observed. Additionally, in the fifth week, in addition to *Giardia duodenalis*, another protozoan parasite, *Eimeria* spp., was diagnosed (Table 3).

Table 3. Incidence of Coinfections of *Cryptosporidium* spp. detected by ELISA method and other parasites by flotation method.

No.	1st Week	2st Week	3st Week	4st Week	5st Week
1.	-	*Cryptosporidium* spp.	*Giardia duodenalis*	no sampling	-
2.	-	*Cryptosporidium* spp.	-	no sampling	-
3.	-	*Cryptosporidium* spp.	-	no sampling	*Giardia duodenalis*
4.	-	*Cryptosporidium* spp.	*Cryptosporidium* spp.	no sampling	*Eimeria* spp.
5.	-	*Cryptosporidium* spp.	*Giardia duodenalis*	no sampling	*Giardia duodenalis*
6.	*Cryptosporidium* spp.	*Cryptosporidium* spp.	-	no sampling	*Eimeria* spp.
7.	*Cryptosporidium* spp.	-	-	no sampling	*Eimeria* spp.
8.	-	no sampling	no sampling	no sampling	no sampling
9.	-	*Cryptosporidium* spp.	*Giardia duodenalis*	no sampling	*Cryptosporidium* spp.
10.	-	-	*Giardia duodenalis*	no sampling	*Cryptosporidium* spp., *Giardia duodenalis*
11.	-	*Cryptosporidium* spp.	*Giardia duodenalis*	no sampling	*Eimeria* spp.
12.	-	*Cryptosporidium* spp.		no sampling	-
13.	-	*Cryptosporidium* spp.	*Cryptosporidium* spp., *Giardia duodenalis*	no sampling	*Cryptosporidium* spp.
14.	-	*Cryptosporidium* spp.	-	no sampling	-
15.	-	*Cryptosporidium* spp.	-	no sampling	-
16.	-	*Cryptosporidium* spp.	-	no sampling	*Giardia duodenalis*
17.	-	*Cryptosporidium* spp.	-	no sampling	*Eimeria* spp.
18.	-	*Cryptosporidium* spp.	-	no sampling	-
19.	-	*Cryptosporidium* spp.	-	no sampling	-
20.	-	*Cryptosporidium* spp.	-	no sampling	-
21.	-	*Cryptosporidium* spp.	-	no sampling	*Giardia duodenalis*
22.	-	*Cryptosporidium* spp.	*Giardia duodenalis*	no sampling	-
23.	-	*Cryptosporidium* spp.	*Cryptosporidium* spp.	no sampling	*Giardia duodenalis*, *Eimeria* spp.
24.	-	*Cryptosporidium* spp.	*Giardia duodenalis*	no sampling	*Eimeria* spp.
25.	-	*Cryptosporidium* spp.	*Giardia duodenalis*	no sampling	*Eimeria* spp.

3.3. Molecular Identification and Phylogenetic Analysis of Cryptosporidium spp. and Subtypes

The molecular analysis and sequencing of a fragment of the small subunit 18S rRNA gene (ssu rRNA) identified two species of *Cryptosporidium*, namely *C. parvum* (PP897358-PP897363) and *C. ryanae* (PP897364 and PP897365), in isolates taken from pre-weaning calves in various weeks post-birth. In all genomic sequences in the present study, the sequences for the 18S rRNA gene, deposited under numbers PP897358-PP897363, were compared in GenBank to the *C. parvum* MK426796, MK426792, MK241967, MW767058, MF671870, and MF142032 sequences, with the identity ranging from 99.80% to 100%. The sequence numbers PP897364 and PP897365 for 18S rRNA were compared in GenBank to the *C. ryanae* OQ456125, OP132538, MW043439, KT922233, KY711520, and MF074604, with the identity ranging from 99.88% to 100%. For the subgenotyping analysis, the sequence typing of the 60 kDa glycoprotein (gp60) gene was used. In the analysis of the gp60 gene, all sequences in this study (PP916252-PP916276) exhibited 99.64–100% nucleotide identity among sequences KC995129, MH796385, AM988863, and EF073050 of *C. parvum* and the subtype IIaA17G1R1 deposited in GenBank. In two cases, in the second week post-birth, the study detected a co-infection with *C. parvum* and *C. ryanae.* The phylogenetic analysis of the small subunit 18S rRNA gene, constructed using the maximum likelihood method and the Tamura–Nei model, organized the sequences in the present study (PP897358-PP897365) as well as the sequences available in GenBank (MK426796, MK426792, MK241967, MW767058, MF671870, MF142032, OQ456125, OP132538, MW043439, KT922233, KY711520, MF074604) into two distinct clusters for *C. parvum* and *C. ryanae* (Figure 2).

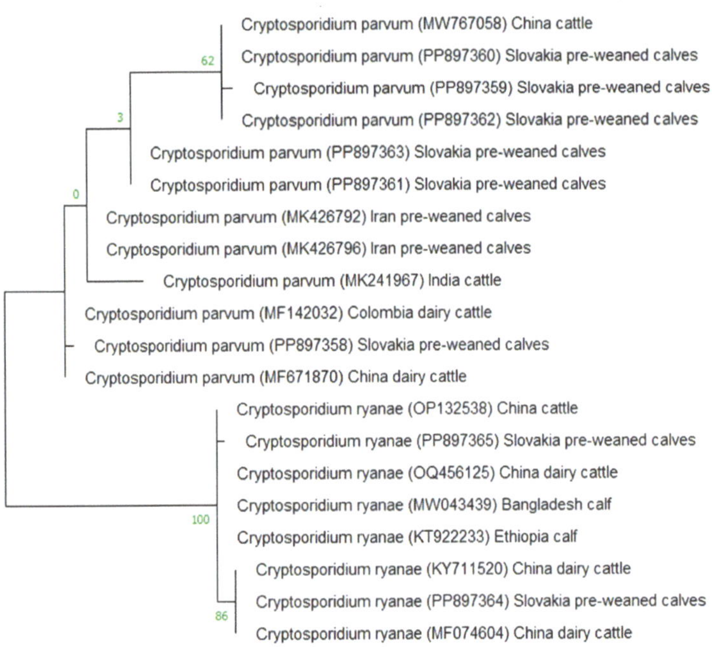

Figure 2. Phylogenetic tree constructed using the maximum likelihood method and Tamura–Nei model and depicting the relationships among *C. parvum* and *C. ryanae* based on the small subunit 18S rRNA gene sequence data available in the GenBank database.

4. Discussion

Ruminants are common hosts of *Cryptosporidium* spp., and contact with them represents a major risk factor for cryptosporidiosis infection in other animals or humans [3]. In this study, the dynamics of *Cryptosporidium* species was monitored and it was confirmed that the highest detection rate of infection occurred in the second week of the calves' lives. This study found that the dominant species in calves up to five weeks old was the zoonotic *C. parvum*, which has also been confirmed by studies from various countries with intensive animal husbandry, in which these species were dominant in pre-weaned calves [31–36]. PCR amplification targeting the 18S rRNA gene also demonstrated the presence of *C. ryanae* on a farm in the eastern part of Slovakia in calves in their second week of life. Studies have indicated that *C. ryanae* is the dominant species primarily in post-weaned calves and young cattle; however, in certain regions of Sweden, China, and Sudan, *C. ryanae*, along with *C. bovis*, has been isolated from pre-weaned diarrheic and healthy calves [37–42]. This study confirms the first occurrence of *C. ryanae* in newborn calves in Slovakia.

Since the identification of *C. parvum* at the subtype level is essential from an epidemiological perspective with regard to cryptosporidiosis, all positive isolates at the gp60 locus were characterized and the presence of subtype IIaA17G1R1 was confirmed in all samples. This subtype had previously been identified in calves in Slovakia [43,44]. Additionally, subtype IIaA17G1R1 was diagnosed in Slovakia in hemato-oncological patients, further confirming its zoonotic potential [45]. The first report of the occurrence of the IIaA10G1R1, IIaA11G2R1, IIaA12G2R1, IIaA13G1R1, and IIaA14G1R1 animal subtypes of *C. parvum* in humans in Slovakia was confirmed by Hatalová et al., 2019 [46]. The occurrence of the IIaA17G1R1 subtype has been reported in various European countries, including Hungary, Sweden, the United Kingdom, Poland, and Germany [32,37,47–49]. The IIaA17G1R1 subtype, along with the IIaA16G1R1 subtype, has not only been identified in calves but also confirmed in sheep, pigs, lambs, dogs, horses, and donkeys [50–55]. The fact that these zoonotic subtypes have been isolated from various hosts increases the risk of *C. parvum* transmission from cattle to other animals and humans.

In the neighboring Czech Republic, according to the study by Ondráčková et al., 2009, the highest prevalence of *Cryptosporidium* spp. in cattle was found in those aged 12 to 18 months. A total of 995 samples were examined, with the predominant species being *C. andersoni* (4.1%), *C. bovis* (0.2%), and *C. parvum* (0.1%). The subtyping of the *C. parvum* species revealed the IIaA16G1R1 subtype [56]. Another study examining the prevalence of species in calves up to 2 months old confirmed the same species' occurrence: *C. parvum* in 137 samples, *C. andersoni* in 21 samples, and *C. bovis* in 3 samples out of a total of 750 samples. The species *C. ryanae* was not confirmed in the calves. The subtyping of *C. parvum* confirmed the subtypes IIaA15G2R1, A16G1R1, A22G1R1, A18G1R1, and A15G1R1 [57]. An Austrian study in calves up to 180 days old confirmed the species *C. parvum* in 69 calves, *C. ryanae* in 11, and *C. bovis* in 7 out of a total of 177. The sequencing of the gp60 locus for *C. parvum* confirmed the subtypes IIaA15G2R1, IIaA21G2R1, IIaA19G2R1, and IIaA14G1R1 [58]. A Polish study reports the prevalence of cryptosporidiosis in cattle of 45.3% out of a total of 1601 tested animals, sampled from calves up to 4 months old, between the years 2014 and 2018. The dominant species were *C. bovis* and *C. ryanae*, with *C. parvum* being the third most common species [59].

The percentage of positive samples that were confirmed by commercially available ELISA tests, compared to the modified Ziehl–Neelsen staining method, corresponds to findings from other studies that have indicated that ELISA methods are more sensitive than microscopic methods [60,61]. Felefel et al., 2023 reported the lower sensitivity of the modified Ziehl–Neelsen method (19.23%) compared to coproantigen confirmation by the ELISA test (32.5%). Conversely, the study by Khurana et al., 2012 indicated that Auramine 'O'-phenol staining (fluorescent staining, AP) is a highly sensitive method compared to the ELISA method [62].

Since calves can shed extremely large amounts of resistant and immediately infectious cysts into the surrounding environment, there is a high probability of infecting other calves. Delafosse et al., 2015 identified the risk factors affecting infection in calves. Multivariate analysis confirmed that the walls of the buckets were highly contaminated with oocysts, which were difficult to remove by applying standard washing procedures. This led to the subsequent contamination of the food placed in the buckets. Similarly, the early separation of calves from their mothers, due to which calves suckle each other (cross-suckling) or suckle other objects in the housing area, may cause possible oocyst infection [63]. The prevention of cryptosporidiosis is influenced by several factors, including the age of the calves, their overall health status, the colostrum intake, the feed quality, the water sources, the housing conditions, and the presence of other diarrheal diseases [64,65].

The control of cryptosporidiosis on farms can only be managed by combining good hygiene management with efficient preventive medications. Although the licensed drug that has been approved for the treatment of cryptosporidiosis in calves is halofuginone lactate, paromomycin has also proven to be efficient in the treatment of acute infections in cattle. For newborn calves, treatment with halofuginone lactate is recommended orally during the first seven days of life at a dose of 100 µg per kg of body weight [66–68].

In the present study, the treatment of cryptosporidiosis was not initiated due to the good clinical condition of the calves. After the second week, in which the incidence of the disease was the highest, the disease gradually self-resolved, as indicated by the declining infection rate observed in the following weeks. According to the farm's veterinarian, hygiene measures were implemented to reduce the risk of infection spread. Thomson et al., 2017 noted that, in newborn calves, the disease may self-resolve due to the sufficient absorption of colostral antibodies, while maintaining the animals in a warm and dry environment, with supportive treatments and without other possible co-infections with other gastrointestinal pathogens [69].

Calves are one of the main reservoirs of zoonotic *Cryptosporidium* spp. species and contact with them can pose a possible risk of infection, especially for farm workers, animal technicians, veterinarians, and veterinary medicine students. Immunodeficient individuals, children, the elderly, and AIDS patients are particularly at risk [70,71].

The disease is of great importance from a public health perspective, highlighting the need for broader research and the identification of infection sources of *Cryptosporidium* spp. species and their genotypes.

Author Contributions: M.K, G.Š. and A.S.: laboratory examination of samples and evaluation of results; M.K., G.Š. and A.S.: methodology; M.K. and G.Š.: conceptualization; M.K. and A.S.: molecular identification; C.T., P.M. and P.G.: collection of samples and data; M.K., G.Š. and A.S.: writing—original draft preparation; M.K., G.Š. and A.S.: writing—review and editing; C.T. and P.M.: visualization and supervision. All authors have read and agreed to the published version of the manuscript.

Funding: This research was funded by the projects of the Scientific Grant Agency of the Ministry of Education of the SR and the Slovak Academy of Sciences, VEGA 1/0709/23.

Data Availability Statement: The original contributions presented in the study are included in the article, further inquiries can be directed to the corresponding author.

Conflicts of Interest: The authors declare no conflicts of interest.

References

1. Shafieyan, H.; Alborzi, A.; Ahmadinejad, H.; Tabandeh, M.R.; Hajikolaei, M.R.H. Prevalence of *Cryptosporidium* species in ruminants of Lorestan province, Iran. *J. Parasit. Dis.* **2014**, *12*, 63–94. [CrossRef]
2. Feng, Y.; Ryan, U.M.; Xiao, L. Genetic diversity and population structure of *Cryptosporidium*. *Trends Parasitol.* **2018**, *34*, 997–1011. [CrossRef] [PubMed]
3. Ryan, U.M.; Feng, Y.; Fayer, R.; Xiao, L. Taxonomy and molecular epidemiology of *Cryptosporidium* and *Giardia*—A 50 year perspective (1971–2021). *Int. J. Parasitol.* **2021**, *51*, 1099–1119. [CrossRef] [PubMed]
4. Holubová, N.; Zikmundová, V.; Kicia, M.; Zajaczkowska, Z.; Rajský, M.; Konečný, R.; Rost, M.; Mravcová, K.; Sak, B.; Kváč, M. Genetic diversity of *Cryptosporidium* spp., *Encephalitozoon* spp. and *Enterocytozoon bieneusi* in feral and captive pigeons in Central Europe. *Parasitol. Res.* **2024**, *123*, 128. [CrossRef] [PubMed]
5. Panciera, R.J.; Thomassen, R.W.; Garner, F.M. Cryptosporidial infection in a calf. *Vet. Pathol.* **1971**, *8*, 479–484. [CrossRef]
6. Santín, M.; Trout, J.M.; Fayer, R. A longitudinal study of cryptosporidiosis in dairy cattle from birth to 2 years of age. *Vet. Parasitol.* **2008**, *155*, 15–23. [CrossRef] [PubMed]
7. Fayer, R.; Santin, M.; Macarisin, D. *Cryptosporidium ubiquitum* n. sp. in animals and humans. *Vet. Parasitol.* **2010**, *172*, 23–32. [CrossRef] [PubMed]
8. Xiao, L. Molecular epidemiology of cryptosporidiosis: An update. *Exp. Parasitol.* **2010**, *124*, 80–89. [CrossRef]
9. Ebiyo, A.; Haile, G. Prevalence and Factors Associated with *Cryptosporidium* Infection in Calves in and around Nekemte Town, East Wollega Zone of Ethiopia. *Vet. Med. Int.* **2022**, *2022*, 1468242. [CrossRef]
10. Klein, P.; Kleinová, T.; Volek, Z.; Šimunek, J. Effect of *Cryptosporidium parvum* infection on the absorptive capacity and paracellular permeability of the small intestine in neonatal calves. *Vet. Parasitol.* **2008**, *152*, 53–59. [CrossRef]
11. Thompson, S. Cryptosporidiosis in Farm Livestock. Ph.D. Thesis, University of Glasgow, Glasgow, UK, 2016.
12. Robertson, L.J.; Bjorkman, C.; Axen, C.; Fayer, R. Cryptosporidiosis in Farmed Animals. In *Cryptosporidium: Parasite and Disease*; Caccio, S.M., Widmer, G., Eds.; Springer: Wien, Austria, 2014; pp. 149–180.
13. Gulliksen, S.M.; Jor, E.; Lie, K.I.; Hamnes, I.S.; Løken, T.; Åkerstedt, J.; Østerås, O. Enteropathogens and risk factors for diarrhea in Norwegian dairy calves. *J. Dairy Sci.* **2009**, *92*, 5057–5066. [CrossRef] [PubMed]
14. Bartels, C.J.M.; Holzhauer, M.; Jorritsma, R.; Swart, W.A.J.M.; Lam, T.J.G.M. Prevalence, prediction and risk factors of enteropathogens in normal and non-normal faeces of young Dutch dairy calves. *Prev. Vet. Med.* **2010**, *93*, 162–169. [CrossRef] [PubMed]
15. Mullusew, G.; Negesse, W.; Dinka, A.; Waktole, H. Study on *eimeria* and *cryptosporidium* infection in dairy cattle farms of holeta, west shoa zone, Oromia, Ethiopia. *Am. J. Sci.* **2020**, *16*, 44–60.
16. Manyazewal, A.; Francesca, S.; Mahenda, P.; Gezahegn, M.; Tesfaye, M.; Lucy, M.; Teklu, W.; Getachew, T. Prevalence, risk factors and molecular characterization of *Cryptosporidium* infection in cattle in Addis Ababa and its environs, Ethiopia. *Vet. Parasitol. Reg. Stud. Rep.* **2018**, *13*, 79–84. [CrossRef]
17. Fayer, R.; Nerad, T. Effects of low temperatures on viability of *Cryptosporidium parvum* oocysts. *Appl. Environ. Microbiol.* **1996**, *62*, 1431–1433. [CrossRef] [PubMed]
18. Trotz-Williams, L.A.; Wayne, S.; Martin, K.E.; Leslie, T.; Duffield, T.; Nydam, D.V.; Peregrine, A.S. Calf level risk factors for neonatal diarrhea and shedding of *Cryptosporidium parvum* in Ontario dairy calves. *Prev. Vet. Med.* **2007**, *82*, 12–28. [CrossRef]
19. Yang, X.; Guo, Y.; Xiao, L.; Feng, Y. Molecular epidemiology of human cryptosporidiosis in low-and middle-income countries. *Clin. Microbiol. Rew.* **2021**, *34*, 10. [CrossRef] [PubMed]

20. Ahmed, S.A.; Karanis, P. Comparison of current methods used to detect *Cryptosporidium* oocysts in stools. *Int. J. Hyg. Environ. Health* **2018**, *221*, 743–763. [CrossRef]
21. Florin, M.F.; Schnitther, L.; Bastos, R.G.; Rathinasamy, V.A.; Cooke, B.M.; Alzan, H.F.; Suarez, C.E. Pursuing effective vaccines against cattle diseases caused by apicomplexan protozoa. *CABI Rev.* **2021**, *16*, 1–23. [CrossRef]
22. Mravcová, K.; Štrkolcová, G.; Mucha, R.; Barbušinová, E.; Goldová, M.; Kačírová, J.; Mad'ar, M. *Cryptosporidium parvum*–zoonotic subtype IIdA15G1 in a Slovakian patient. *Ann. Agric. Environ. Med.* **2020**, *27*, 485–488. [CrossRef]
23. Wielinga, P.R.; de Vries, A.; Goot, T.H.; Mank, T.; Mars, M.H.; Kortbeek, L.M.; Giessen, J.W.B. Molecular epidemiology of *Cryptosporidium* in humans and cattle in The Netherlands. *Int. J. Parasitol.* **2007**, *38*, 809–817. [CrossRef] [PubMed]
24. Soba, B.; Logar, J. Genetic classification of *Cryptosporidium* isolates from humans and calves in Slovenia. *Parasitology* **2008**, *135*, 1263–1270. [CrossRef] [PubMed]
25. Garcia, L.S.; Bruckner, D.A. (Eds.) Collection, Preservation, and Shipment of Fecal Specimens. In *Diagnostic Medical Parasitology*, 3rd ed.; ASM Press: Washington, DC, USA, 1997; pp. 593–607.
26. Garcia, L.S. *Diagnostic Medical Parasitology*, 6th ed.; ASM Press: Washington, DC, USA, 2016.
27. Henriksen, S.A.; Pohlenz, J.F.L. Staining of cryptosporidia by a modified Ziehl-Neelsen technique. *Acta Vet. Scand.* **1981**, *22*, 594. [CrossRef] [PubMed]
28. Xiao, L.; Morgan, U.M.; Limor, J.; Escalante, A.; Arrowood, M.; Shulaw, W.; Thompson, R.C.; Fayer, R.; Lal, A.A. Genetic diversity within *Cryptosporidium parvum* and related *Cryptosporidium* species. *Appl. Environ. Microbiol.* **1999**, *65*, 3386–3391. [CrossRef] [PubMed]
29. Glaberman, S.; Moore, J.E.; Lowery, C.J.; Chalmers, R.M.; Sulaiman, I.; Elwin, K.; Rooney, P.J.; Millar, B.C.; Dooley, J.S.; Lal, A.A.; et al. Three drinking-water-associated cryptosporidiosis outbreaks, Northern Ireland. *Emerg. Infect. Dis.* **2002**, *8*, 631–633. [CrossRef] [PubMed]
30. Kumar, S.; Stecher, G.; Li, M.; Knyaz, C.; Tamura, K. MEGA X: Molecular Evolutionary Genetics Analysis across Computing Platforms. *Mol. Biol. Evol.* **2018**, *35*, 1547–1549. [CrossRef] [PubMed]
31. Santin, M.; Trout, J.M.; Xiao, L.; Zhou, L.; Greiner, E.; Fayer, R. Prevalence and age-related variation of *Cryptosporidium* species and genotypes in dairy calves. *Vet. Parasitol.* **2004**, *122*, 10–117. [CrossRef] [PubMed]
32. Fayer, R.; Santin, M.; Trout, J.M. Prevalence of *Cryptosporidium* species and genotypes in mature dairy cattle on farms in eastern United States compared with younger cattle from the same locations. *Vet. Parasitol.* **2007**, *145*, 260–266. [CrossRef]
33. Broglia, A.; Reckinger, S.; Cacció, S.M.; Nöckler, K. Distribution of *Cryptosporidium parvum* subtypes in calves in Germany. *Vet. Parasitol.* **2008**, *154*, 8–13. [CrossRef]
34. Kaupke, A.; Rzeżutka, A. Emergence of novel subtypes of *Cryptosporidium parvum* in calves in Poland. *Parasitol. Res.* **2015**, *114*, 4709–4716. [CrossRef]
35. Yildirim, A.; Adanir, R.; Inci, A.; Yukari, B.A.; Duzlu, O.; Onder, Z.; Ciloglu, A.; Simsek, E. Prevalence and genotyping of bovine *Cryptosporidium* species in the Mediterranean and Central Anatolia region of Turkey. *Comp. Immunol. Microbiol. Infect. Dis.* **2020**, *69*, 101425. [CrossRef] [PubMed]
36. Chen, Y.; Huang, J.; Qin, J.; Wang, L.; Li, J.; Zhang, L. *Cryptosporidium parvum* and gp60 genotype prevalence in dairy calves worldwide: A systematic review and meta-analysis. *Acta Trop.* **2023**, *240*, 106843. [CrossRef] [PubMed]
37. Silverlås, C.; Bosaeus-Reineck, H.; Näslund, K.; Björkman, C. Is there a need for improved *Cryptosporidium* diagnostics in Swedish calves? *Int. J. Parasitol.* **2013**, *43*, 55–161. [CrossRef] [PubMed]
38. Wang, R.; Wang, H.; Sun, Y.; Zhang, L.; Jian, F.; Qi, M.; Ning, C.; Xiao, L. Characteristics of *Cryptosporidium* transmission in preweaned dairy cattle in Henan, China. *J. Clin. Microbiol.* **2011**, *49*, 1077–1082. [CrossRef] [PubMed]
39. Silverlas, C.; Blanco-Penedo, I. *Cryptosporidium* spp. in calves and cows from organic and conventional dairy herds. *Epidemiol. Infect.* **2013**, *141*, 529–539. [CrossRef] [PubMed]
40. Taha, S.; Elmalik, K.; Bangoura, B.; Lendner, M.; Mossaad, E.; Daugschies, A. Molecular characterization of bovine *Cryptosporidium* isolated from diarrheic calves in the Sudan. *Parasitol. Res.* **2017**, *116*, 2971–2979. [CrossRef] [PubMed]
41. Aberg, M.; Emanuelson, U.; Troell, K.; Bjorkman, C. Infection dynamics of *Cryptosporidium bovis* and *Cryptosporidium ryanae* in a Swedish dairy herd. *Vet. Parasitol.* **2019**, *27*, 100010. [CrossRef] [PubMed]
42. Guo, Y.; Ryan, U.; Feng, Y.; Xiao, L. Emergence of zoonotic *Cryptosporidium parvum* in China. *Trends Parasitol.* **2022**, *38*, 335–343. [CrossRef] [PubMed]
43. Hatalová, E.; Valenčáková, A.; Kalinova, J.; Luptáková, L.; Danišová, O.; Húska, M. Comparison of PCR primers for the identification of *Cryptosporidium* species and genotypes in calves. *Bulg. J. Vet. Med.* **2017**, *20*, 154–157.
44. Mravcová, K.; Štrkolcová, G.; Goldová, M.; Mucha, R. *Cryptosporidium parvum* infection in Calves from animal farm in Slovakia. *J. Vet. Med. Res.* **2019**, *6*, 1–4.
45. Hatalová, E.; Guman, T.; Bednárová, V.; Turčok Simonová, V.; Logoida, M.; Halánová, M. Occurrence of *cryptosporidium parvum* IIaA17G1R1 in hospitalized hemato-oncological patients in Slovakia. *Parasitol. Res.* **2022**, *121*, 471–476. [CrossRef] [PubMed]
46. Hatalová, E.; Valenčáková, A.; Luptáková, L.; Špalková, M.; Kalinová, J.; Halánová, M.; Bednárová, V.; Gabzdilová, J.; Dedinská, K.; Ondriska, F.; et al. The first report of animal genotypes of *Cryptosporidium parvum* in immunosuppressed and immunocompetent humans in Slovakia. *Transbound. Emerg. Dis.* **2019**, *66*, 243–249. [CrossRef] [PubMed]
47. Plutzer, J.; Karanis, P. Genotype and subtype analyses of *Cryptosporidium* isolates from cattle in Hungary. *Vet. Parasitol.* **2007**, *146*, 357–362. [CrossRef]

48. Smith, R.P.; Clifton-Hadley, F.A.; Cheney, T.; Giles, M. Prevalence and molecular typing of *Cryptosporidium* in dairy cattle in England and Wales and examination of potential on-farm transmission routes. *Vet. Parasitol.* **2014**, *204*, 111–119. [CrossRef]
49. Holzhausen, I.; Lendner, M.; Daugschies, A. Bovine *Cryptosporidium parvum* field isolates differ in cytopathogenicity in HCT-8 monolayers. *Vet. Parasitol.* **2019**, *273*, 67–70. [CrossRef] [PubMed]
50. Kvác, M.; Hanzlíková, D.; Sak, B.; Kvetonová, D. Prevalence and age-related infection of *Cryptosporidium suis*, *C. muris* and *Cryptosporidium* pig genotype II in pigs on a farm complex in the Czech Republic. *Vet. Parasitol.* **2009**, *160*, 319–322. [CrossRef] [PubMed]
51. Kváč, M.; Kvetoňová, D.; Sak, B.; Ditrich, O. *Cryptosporidium* pig genotype II in immunocompetent man. *Emer. Inf. Dis.* **2009**, *15*, 982. [CrossRef]
52. Imre, K.; Sala, C.; Morar, A.; Ilie, M.S.; Plutzer, J.; Imre, M.; Hora, F.S.; Badea, C.; Herbei, M.V.; Darabus, G. *Giardia duodenalis* and *Cryptosporidium* spp. as contaminant protozoa of the main rivers of western Romania: Genetic characterization and public health potential of the isolates. *Environ. Sci. Pollut. Res.* **2017**, *24*, 18672–18679. [CrossRef] [PubMed]
53. Laatamna, A.E.; Wagnerová, P.; Sak, B.; Květoňová, D.; Xiao, L.; Rost, M.; McEvoy, J.; Saadi, A.R.; Aissi, M.; Kváč, M. Microsporidia and *Cryptosporidium* in horses and donkeys in Algeria: Detection of a novel *Cryptosporidium hominis* subtype family (Ik) in a horse. *Vet. Parasitol.* **2015**, *208*, 135–142. [CrossRef]
54. Kaupke, A.; Michalski, M.M.; Rzeżutka, A. Diversity of *Cryptosporidium* species occurring in sheep and goat breeds reared in Poland. *Parasitol. Res.* **2017**, *116*, 871–879. [CrossRef]
55. Rosanowski, S.M.; Banica, M.; Ellis, E.; Farrow, E.; Harwood, C.; Jordan, B.; James, C.; McKenna, D.; Fox, M.; Blake, D.P. The molecular characterisation of *Cryptosporidium* species in relinquished dogs in Great Britain: A novel zoonotic risk? *Parasitol. Res.* **2018**, *117*, 1663–1667. [CrossRef]
56. Ondráčková, Z.; Kvác, M.; Sak, B.; Kvetonová, D.; Rost, M. Prevalence and molecular characterization of *Cryptosporidium* spp. in dairy cattle in South Bohemia, the Czech Republic. *Vet. Parasitol.* **2009**, *165*, 141–144. [CrossRef]
57. Kvác, M.; Hromadová, N.; Květoňová, D.; Rost, M.; Sak, B. Molecular characterization of *Cryptosporidium* spp. in pre-weaned dairy calves in the Czech Republic: Absence of *C. ryanae* and management-associated distribution of *C. andersoni*, *C. bovis* and *C. parvum* subtypes. *Vet. Parasitol.* **2011**, *177*, 378–382. [CrossRef]
58. Lichtmannsperger, K.; Harl, J.; Freudanthaler, K.; Hinney, B.; Wittek, T.; Joachim, A. *Cryptosporidium parvum*, *Cryptosporidium ryanae*, and in samples from calves in Austria. *Parasitol. Res.* **2020**, *119*, 4291–4295. [CrossRef]
59. Rzezutka, A.; Kaupke, A. *Cryptosporidium* infections in asymptomatic calves up to 4 months in Poland: A cross-sectional population study. *Sci. Rep.* **2023**, *13*, 20997. [CrossRef]
60. Elgun, G.; Koltas, I.S. Investigation of *Cryptosporidium* spp. antigen by ELISA method in stool specimens obtained from patients with diarrhea. *Parasitol. Res.* **2011**, *108*, 395–397. [CrossRef]
61. Khurana, S.; Sharma, P.; Sharma, A.; Malla, N. Evaluation of Ziehl-Neelsen staining, auramine phenol staining, antigen detection enzyme-linked immunosorbent assay and polymerase chain reaction, for the diagnosis of intestinal cryptosporidiosis. *Trop. Parasitol.* **2012**, *2*, 20. [CrossRef]
62. Felefel, W.; Abd El-Rady, A.; El Rahim, I.; Elkamshishi, M.M.; Mostafa, W. Detection of *Cryptosporidium parvum* in calf feces using microscopical, serological, and molecular methods. *Iraqi J. Veter Sci.* **2023**, *37*, 383–389. [CrossRef]
63. Delafosse, A.; Chartier, C.; Dupuy, M.C.; Dumoulin, M.; Pors, I.; Paraud, C. *Cryptosporidium parvum* infection and associated risk factors in dairy calves in western France. *Prev. Vet. Med.* **2015**, *118*, 406–412. [CrossRef]
64. Nasir, A.; Avais, M.; Khan, M.; Ahmad, N. Prevalence of *Cryptosporidium parvum* infection in Lahore (Pakistan) and its association with diarrhea in dairy calves. *Int. J. Agric. Biol.* **2009**, *11*, 221–224.
65. Ogendo, A.; Obonyo, M.; Wasswa, P.; Bitek, A.; Mbugua, A.; Thumbi, S.M. *Cryptosporidium* infection in calves and the environment in Asembo, Western Kenya: 2015. *Pan. Afr. Med. J.* **2017**, *28*, 9. [CrossRef]
66. Santin, M. Clinical and subclinical infections with *Cryptosporidium* in animals. *N. Z. Vet. J.* **2013**, *61*, 1–10. [CrossRef]
67. Deplazes, P.; Eckert, J.; Mathis, A.; Samson-Himmelstjerna, G.; Zahner, H. *Parasitology in Veterinary Medicine*, 1st ed.; Wageningen Academic: Wageningen, The Netherlands, 2016. [CrossRef]
68. Santin, M. *Cryptosporidium* and *Giardia* in ruminants. *Vet. Clin. N. Am. Food Anim. Pract.* **2020**, *36*, 223–238. [CrossRef]
69. Thomson, S.; Hamilton, C.A.; Hope, C.J.; Katzer, F.; Mabbott, N.A.; Morrison, L.J.; Innes, E.A. Bovine cryptosporidiosis: Impact, host-parasite interaction and control strategies. *Vet. Res.* **2017**, *48*, 1–16. [CrossRef]
70. Feng, Y.; Li, N.; Duan, N.; Xiao, L. *Cryptosporidium genotype* and subtype distribution in raw wastewater in Shanghai, China: Evidence for possible unique *Cryptosporidium hominis* transmission. *J. Clin. Microbiol.* **2009**, *47*, 153–157. [CrossRef]
71. Gharpure, R.; Perez, A.; Miller, A.D.; Wikswo, M.E.; Silver, R.; Hlavsa, C.M. Cryptosporidiosis outbreaks—United states, 2009–2017. *Am. J. Transplant.* **2019**, *19*, 2650–2654. [CrossRef]

microorganisms

Article

Parasitic Effects on the Congenital Transmission of *Trypanosoma cruzi* in Mother–Newborn Pairs

Ana Gabriela Herrera Choque [1], Washington R. Cuna [1], Simona Gabrielli [2], Simonetta Mattiucci [2], Roberto Passera [3] and Celeste Rodriguez [1,*]

1 Unidad de Inmunología Parasitaria, Facultad de Medicina, Universidad Mayor de San Andrés, La Paz, Bolivia; dra_anagherrera@yahoo.com (A.G.H.C.); washingtoncuna@gmail.com (W.R.C.)
2 Department of Public Health and Infectious Diseases, Sapienza University of Rome, 00185 Rome, Italy; simona.gabrielli@uniroma1.it (S.G.); simonetta.mattiucci@uniroma1.it (S.M.)
3 Department of Medical Sciences, University of Torino, 10126 Torino, Italy; passera.roberto@gmail.com
* Correspondence: celesterodriguez6@gmail.com

Abstract: Maternal parasitemia and placental parasite load were examined in mother–newborn pairs to determine their effect on the congenital transmission of *Trypanosoma cruzi*. Parasitemia was qualitatively assessed in mothers and newborns by the microhematocrit test; parasite load was determined in the placental tissues of transmitting and non-transmitting mothers by the detection of *T. cruzi* DNA and by histology. Compared to transmitter mothers, the frequency and prevalence of parasitemia were found to be increased in non-transmitter mothers; however, the frequency and prevalence of parasite load were higher among the transmitter mothers than among their non-transmitter counterparts. Additionally, serum levels of interferon (IFN)-γ were measured by an enzyme-linked immunosorbent assay (ELISA) in peripheral, placental, and cord blood samples. Median values of IFN-γ were significantly increased in the cord blood of uninfected newborns. The median IFN-γ values of transmitter and non-transmitter mothers were not significantly different; however, non-transmitter mothers had the highest total IFN-γ production among the group of mothers. Collectively, the results of this study suggest that the anti-*T. cruzi* immune response occurring in the placenta and cord is under the influence of the cytokines from the mother's blood and results in the control of parasitemia in uninfected newborns.

Keywords: *Trypanosoma cruzi*; congenital; placenta; human; parasitemia; DNA; IFN-γ

1. Introduction

The etiological agent of Chagas disease, *Trypanosoma cruzi*, is responsible for the endemicity of this disease in the Americas, together with risk factors strongly linked to socioeconomic factors, such as poverty and marginalization. The parasite is transmitted most often by blood-sucking triatomine bugs, but also through blood transfusion, organ transplant, congenital transmission, or oral transmission in the sylvatic cycle. Although control programs have made possible the reduction of the prevalence of *T. cruzi* infection [1,2], uncontrolled transmission from mothers infected with *T. cruzi* to their fetuses is a potential threat to spreading this infection over time, which can be recurrent at each pregnancy and transmitted from one generation to the next [3]. The incidence of *T. cruzi* infection in pregnant women varies from 6% to 54% [4–7]. Transmission from mothers chronically infected with *T. cruzi* to newborns, occurring in 1–10% of pregnancies [8–10], can be associated with severe disease and mortality [11,12].

Despite many experimental approaches, there is still no clear understanding of the factors associated with congenital Chagas disease. Among these factors, high maternal parasitemia and diminished IFN-γ production [8,13–16] have been postulated to account for transplacental transmission, but none of these issues has been unequivocally implicated as crucial in the transmission process.

Given the crucial role of IFN-γ in the context of this study, its concentration was measured in maternal, placenta, and cord blood samples of uninfected and infected mothers, and their newborns. In particular, IFN-γ production by whole blood cells upon specific stimulation observed in pregnant mothers limits the occurrence of newborn congenital infection [17]; depressed production of IFN-γ by blood cells of pregnant women was associated with congenital transmission of *T. cruzi* [8]; a study by Hermann and collaborators revealed a reduced production of IFN-γ by cord blood NK cells obtained from congenitally infected newborns [18]; and in our previous study, increased production of IFN-γ elicited by enhanced parasitemia was observed in maternal, placenta, and cord samples of chronically infected mothers delivering uninfected newborns [19]. These results point to a strong pro-inflammatory immune response in the absence of congenital transmission.

Within the scope of this study, the median and total median values from the mother, placenta, and cord were compared between samples from mothers of uninfected newborns (non-transmitters) and mothers of infected newborns (transmitters). Moreover, parasite load at the tissue level was measured through molecular analysis and microscopic observation of placental tissues from pregnant women chronically infected with *T. cruzi*. The ensuing question is to define which factors, such as parasitemia, parasitic load, and cytokine interactions, in peripheral, placental, and cord blood are involved in the outcome of parasite transmission to newborns.

2. Materials and Methods

2.1. Study Population and Sample Collection

The study was carried out at the maternity hospital Odón Ortega in Yacuiba, South Bolivia, an endemic area for *T. cruzi* infection. Informed consent for participation was obtained from women (14 transmitters, 15 non-transmitters) before their inclusion in the study. Maternal information (i.e., age, village of birth and village of residence, medical history, and present symptoms) was documented at enrollment. Women with concomitant infections (tuberculosis, toxoplasmosis) were excluded from the study. Peripheral blood samples were drawn by venipuncture from all pregnant women before delivery, and cord blood was collected after the umbilical section. Placental blood was obtained from the maternal side by rinsing several times with a sterile physiological solution (0.9% NaCl), as previously described [20]. Briefly, the excision of a 5-cm^3 block of tissue from the cleaned maternal side permitted aspiration of intervillous blood using a sterile Pasteur pipette. Thereafter, blood samples were centrifuged to separate the packed erythrocytes, and the serum was frozen at −80 °C until assayed for antibodies. *Trypanosoma cruzi* infection in pregnant women was determined by specific serological tests, that is, indirect hemagglutination (HAI, Chagas, Polychaco S.A.I.C., Buenos Aires, Argentina), with a sensitivity and specificity of 99%, followed by enzyme-linked immunosorbent assays (first- and second-generation, Wiener Laboratories, Buenos Aires, Argentina), with sensitivity and specificity between 98–99%, for confirmation of diagnosis. The maternal parasitemia in chronic chagasic mothers and infection in neonates were qualitatively assessed by detecting living parasites through microscopic examination of the buffy coat from the mother's peripheral and cord blood by the microhematocrit method because of its ease of performance, reliability, and high sensitivity (97.4%) in infants under 6 months old [21]. Blood was collected in four 75-μL microhematocrit tubes, for a volume of 300 μL in total, and the results were scored as "detectable" and "undetectable", denoting higher and lower parasitemia, respectively. Newborns' negative microhematocrit test results were confirmed at one month of age in case the peak of parasitemia was observed [9,22,23]. Hereinafter, detectable and undetectable parasitemia will be referred to as positive microhematocrit (+μHT) and −μHT, respectively. Samples from non-infected mothers (controls) were defined as samples from mothers with negative serology for *T. cruzi* and negative μHT test results in which there was no DNA amplification. The study received approval from the ethics committee of the Faculty of Medicine and the Medical College of Bolivia (Supplementary Materials).

2.2. Tissues

Placental tissue sections 1 cm long and adjacent to the umbilical cord insertion were obtained from each parturient woman immediately after delivery and washed in sterile physiologic saline solution. One tissue sample was kept frozen at −80 °C until molecular analysis, while another one was prepared for tissue staining.

2.3. Paraffin-Embedded Tissue Sections

Tissue sections were fixed in buffered formalin (3.7% formaldehyde in 10 mM phosphate buffer, pH 7.4), dehydrated through immersion in an alcohol series (70–100%) for 20 min each, and then 'cleared' by placing the tissues into a xylene bath for 20 min. Next, the tissues were embedded in molten paraffin in a mold. Sections (3–4 μm) were cut and affixed onto glass slides coated with aminoalkylsilane (Sigma-Aldrich, St. Louis, MO, USA), designed to enhance tissue adhesion.

2.4. Hematoxylin and Eosin Staining and Bright-Field Microscopy

Slides were stained with hematoxylin and eosin (H&E) solution (~30 s), thoroughly rinsed with distilled water then dehydrated to xylene and permanently mounted. Stained samples were examined microscopically using an Olympus BX53 microscope (Shinjuku-ku, Tokyo, Japan), provided with an Olympus DP73 high-performance Peltier-cooled digital camera (Olympus Corporation, Shinjuku-ku, Tokyo, Japan).

2.5. Molecular Analysis

DNA was extracted from placental tissue samples using a commercial kit (DNeasy Blood & Tissue Kit, Qiagen, GmbH, Hilden, Germany) according to the manufacturer's instructions. Extracted DNAs were analyzed using the NanoDrop 2000 spectrophotometer (Thermo Scientific, Middlesex, MA, USA) and electrophoresed on 1% agarose gel to determine the quantity and quality of the extracted DNA as a means to verify that DNA was being extracted correctly. Each DNA was submitted to a TaqMan Real-time (RT)-PCR assay to amplify a region of the 18S ribosomal RNA (rRNA) gene of *T. cruzi* (Genesig Primer-Design, Camberley, UK) according to the manufacturer's instructions. In addition, RT-PCR positive samples were subjected to a nested PCR (N-PCR) to perform a sequence analysis of amplified products. Briefly, nuclear DNA was first amplified using primers TCZ1 (5′-CGAGCTCTTGCCACACGGG-3′) and TCZ2 (5′-CCTCCAAGCAGCGGATAGTTCAGG-3′), to yield 188 base pairs (bp) product using the described protocol, followed by the N-PCR reaction with TCZ3 (5′-TGCTGCASTCGGCTGATCGTTTTCGA-3′) and TCZ4 (5′-CAR GSTTGTTTGGTGTCCAGTGTTGTGA-3′), which yield a product of 149 bp for sequence analysis [24]. The first PCR amplification was performed in 25 μL volumes under the following final conditions: 1× buffer including 1.5 mM $MgCl_2$, 0.2 mM of each deoxynucleoside triphosphate (dNTP), 1 μM each of forward and reverse primers, and 1 U of *Taq* polymerase (BIOTAQTM DNA Polymerase, Aurogene, Rome, Italy). The thermal profile used was as follows: 94 °C for 30 s, 60 °C for 30 s, and 72 °C for 1 min for 30 cycles, followed by a final extension for 7 min at 72 °C. One microliter of the reaction was used for the second amplification, in which primers TCZ3 and TCZ4 amplified a 149-nucleotide internal sequence of the same repetitive sequence. The N-PCR conditions and protocol were the same as for the first amplification. Amplicons were purified (SureClean Bioline, Aurogene) and then sequenced (Eurofins MWG Operon, Ebersberg, Germany). Sequences corrected by visual analysis of the electropherograms and aligned using ClustalW (http://www.genome.jp/tools/clustalw (accessed on 10 November 2018), were compared with those available in the GenBank (http://www.ncbi.nlm.nih.gov/genbank/ (accessed on 10 November 2018) dataset by BLAST analysis.

2.6. Measurement of IFN-γ in Peripheral, Placental, and Cord Serum

Peripheral, placental, and cord blood derived-serum samples were analyzed by sandwich ELISA, using pairs of capture and biotinylated-specific secondary antibodies

(BioSource Europe S.A., Nivelles, Belgium). A standard curve of recombinant human cytokine was run with each plate. The sensitivity of the IFN-γ assay is 0.03 IU/mL. For a more representative comparison of the cytokine response between uninfected, transmitter, and non-transmitter mothers, the total production of IFN-γ from the three sites—periphery, placenta, and cord—was compared between the groups of mothers.

2.7. Statistical Analysis

Continuous covariates were described as the median (IQR, or interquartile range), while categorical ones were described as absolute/relative frequencies. A non-parametric method for inferential tests on independent data was employed. The differences between continuous variables were assessed by the Kruskal–Wallis test, using the Mann–Whitney test for post hoc analysis. All reported p-values were obtained by the two-sided exact method at the conventional 5% significance level. Data were analyzed as of May 2024 using R 4.4.0 (R Foundation for Statistical Computing, Vienna, Austria).

3. Results

3.1. Characterization of *T. cruzi*-Infected Mothers and Newborns

The frequency (i.e., number of times) of maternal parasitemia as well as the parasite load of *T. cruzi*-infected pregnant women was determined by μHT test and RT-PCR, respectively. A higher frequency of patent parasitemia in non-transmitters (12/15) than in transmitter mothers (6/14) was associated with a lower frequency of parasite load in non-transmitters (5/15) than transmitter mothers (14/16). Congenital transmission determined by μHT findings was observed in 15 newborns of 14 infected mothers. Of note, a transmitter mother (Code 1072) had triplet newborns, two infected and one uninfected. (Table 1).

Table 1. Characteristics of mother/newborn pairs.

	Transmitter Mother		Newborn *	Placenta
Code	**Serology**	**μHT**	**μHT**	**RT-PCR 18S rRNA**
525	+	+	+	+
1072	+	+	+ (a)	+
		+	+ (b)	+
			− (c)	+
1226	+	+	+	+
2016	+	+	+	+
2097	+	+	+	−
1961	+	+	+	−
1056	+	−	+	+
1088	+	−	+	+
1089	+	−	+	+
1211	+	−	+	+
1229	+	−	+	+
1234	+	−	+	+
1977	+	−	+	+
2056	+	−	+	+
	Non-Transmitter Mother		**Newborn**	**Placenta**
Code	**Serology**	**μHT**	**μHT**	**RT-PCR 18S rRNA**
368	+	+	−	−
1968	+	+	−	−
322	+	+	−	−
323	+	+	−	−
527	+	+	−	−
298	+	+	−	−
461	+	+	−	−

Table 1. *Cont.*

	Non-Transmitter Mother		Newborn	Placenta
Code	Serology	μHT	μHT	RT-PCR 18S rRNA
496	+	+	−	−
2115	+	+	−	−
338	+	+	−	+
437	+	+	−	+
1246	+	+	−	+
1150	+	−	−	+
1157	+	−	−	+
1137	+	−	−	−
	Uninfected Mother		**Newborn**	**Placenta**
Code	**Serology**	**μHT**	**μHT**	**RT-PCR 18S rRNA**
1154	−	−	−	−
1159	−	−	−	−
1161	−	−	−	−
1151	−	−	−	−

* a, b, and c under "Newborn" indicate the triplets born to mother 1072.

3.2. RT-PCR Assay

Molecular analyses of tissue samples from the placenta of transmitter and non-transmitter mothers were positive for *T. cruzi* DNA in 87.5% (14/16) and 33.3% (5/15) of cases, respectively. Positivity amount of the target DNA ranged from 10^1 to 10^4 copies/μL. Two placental samples of transmitter mothers with +μHT test results (2097, 1961) were RT-PCR negative (Tables 1 and 2). All positive samples on RT-PCR assay were also confirmed with the N-PCR assay (Figure 1) and the amplicons of 149 base pairs were correctly sequenced. The sequences obtained showed high nucleotide identity (98–100%) with those from *T. cruzi* repetitive DNA sequences available in the GenBank database (accession number KX235537).

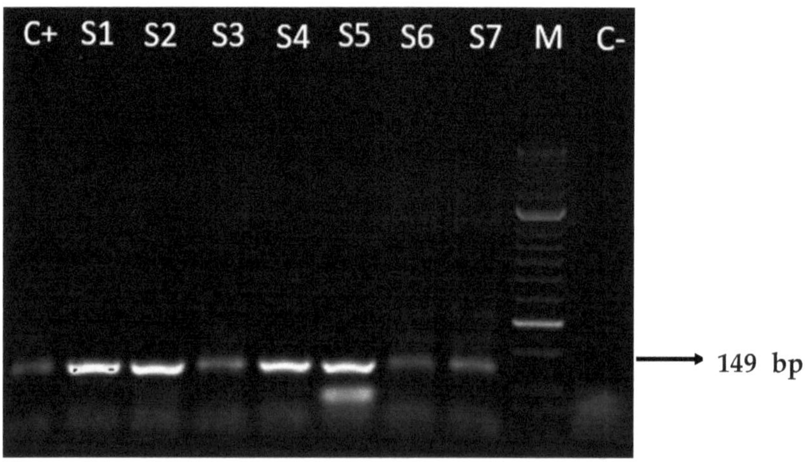

Figure 1. Results of N-PCR amplifications with *T. cruzi* DNA nuclear, electrophoresed on a 2% agarose gel and visualized by ethidium bromide staining. The 149 base pairs (bp) were amplified through N-PCR with primers TCZ3 and TCZ4. M: molecular weight marker (50 bp); C+: positive control (*T. cruzi* II of Y strain); S1–S7: representative amplicons of positive patients, from transmitter (S2–S4) and non-transmitter (S1, S5–S7) mothers; C−: negative control from a patient with negative serology for *T. cruzi*.

Table 2. Maternal IFN-γ production, parasitemia, parasite load, and newborn parasitemia in transmitter and non-transmitter mothers.

Mother	IFN-γ †	Blood	Placenta	Newborn
		μHT %	RT-PCR %	μHT %
Transmitters	0.59 (0.29–1.13)	42.8 (6/14)	87.5 (14/16) [a]	93.7 (15/16) [b]
Non-transmitters	1.30 (0.56–2.65)	80.0 (12/15)	33.3 (5/15)	0.0 (0/15)

† IU/mL, median (interquartile range). [a], [b], data include triplets born to one mother.

3.3. *Association between IFN-γ, Parasitic Factors, and Newborn Infection Status*

As can be seen in Table 2, the pregnant women who did not transmit the parasites to their newborns presented higher IFN-γ production than their transmitter counterparts. Likewise, parasitemia was proportionally higher (i.e., more prevalent) in the non-transmitter (80%) than in the transmitter mothers (42.8%), and this was associated with a reduced prevalence of parasite load (33.3%) compared to transmitter women (87.5%).

3.4. *Production of IFN-γ in Uninfected, Transmitter, and Non-Transmitter Mothers*

Median concentrations of IFN-γ were not significantly different between samples from transmitter and non-transmitter mothers ($p = 0.224$). However, total IFN-γ production was highest in non-transmitter mothers ($p = 0.002$), and uninfected newborns displayed a significantly higher level of IFN-γ than newborns of uninfected or transmitter mothers ($p = 0.037$) (Table 3).

Table 3. Serum concentrations of IFN-γ from peripheral, placental, and cord blood samples in uninfected, transmitter, and non-transmitter mothers.

Mother	IFN-γ †					
	Periphery	Placenta	Cord	p *	Total	p *
Uninfected	0.81 (0.37–1.26)	0.24 (0.04–0.86)	−0.08 (−0.17–0.53)		0.33 (−0.3–1.09)	
Transmitter	0.59 (0.29–1.13)	0.65 (0.27–2.37)	0.62 (0.24–1.12)	0.224 [a]	0.62 (0.24–1.35)	
Non-transmitter	1.30 (0.56–2.65)	1.19 (0.22–1.88)	1.56 (0.12–2.18)	0.037 [b]	1.39 (0.22–2.47)	0.002 [c]

† IU/mL, median (interquartile range). * p value. [a] Comparison between transmitter and non-transmitter mothers. [b] Comparisons of values between newborns of uninfected, transmitter, and non-transmitter mothers. [c] Comparison of total values between uninfected, transmitter, and non-transmitter mothers.

3.5. *Histological Studies*

Histological preparations of placental tissue sections representative of non-transmitter and transmitter mothers are shown in Figures 2 and 3. Different forms of the parasite were detected through the microscopic observation of chorionic villous human placenta from these mothers. Amastigote nests were observed in preparations from non-transmitter (Figure 2A,B) and transmitter (Figure 3A) mothers, while released parasites were found in placentas from non-transmitter (Figure 2C) and transmitter (Figure 3B,C) mothers.

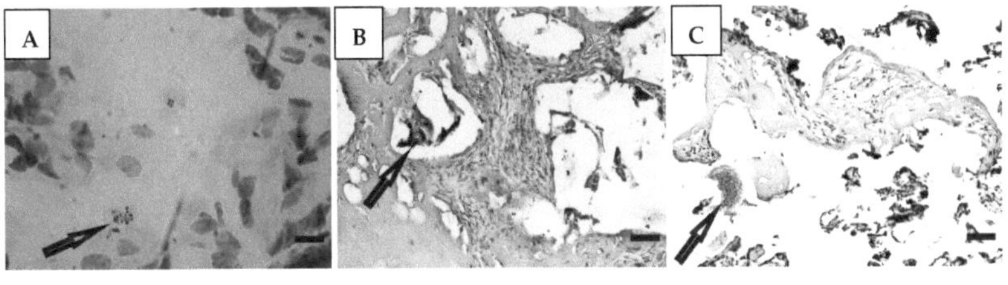

Figure 2. Histological section of the placenta from a non-transmitter mother. Arrows point to (**A**,**B**) amastigote nests; and (**C**) released parasites (H&E). Scale bar: 25 μm.

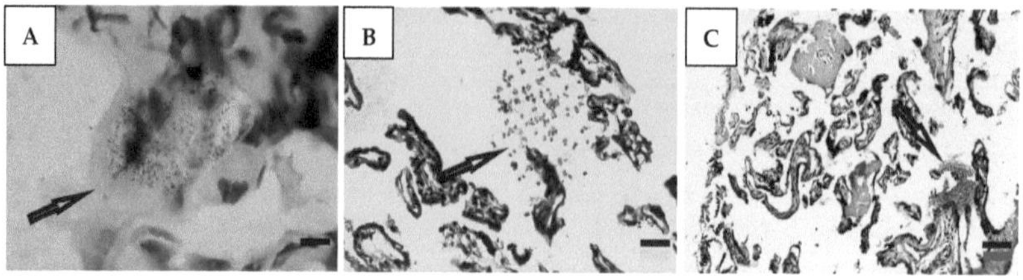

Figure 3. Histological section of the placenta from a transmitter mother. Arrows point to (**A**) amastigote nest; and (**B**,**C**) released parasites (H&E). Scale bar: 25 μm.

4. Discussion

The main feature of this study was the evaluation of parasitic factors and IFN-γ production in chronically infected pregnant women associated with the occurrence or absence of congenital *T. cruzi* infection. We were able to show an inverse correlation between maternal parasitemia and placental parasite load in the two groups of pregnant mothers. A higher frequency of detectable parasitemia was observed in non-transmitter mothers than in transmitter mothers, but the contrary was observed in transmitter mothers, suggesting that the higher parasitemia was not a transmission risk in our study. This negative correlation was ascribed to higher levels of circulating IFN-γ in non-transmitter than transmitter mothers. It is well known that IFN-γ confers protection from *T. cruzi* infection [25–27] through macrophage nitric oxide production to kill parasites in synergy with TNF-α [28,29].

Protection from vertical transmission likely requires the coordination of different components of the immune system. Controlling the parasite load in non-transmitter mothers through the induction of IFN-γ tempts us to speculate that maternal IFN-γ influences the immune response at the level of the placenta and cord and protects the newborn from infection in utero. Relevant to that point, previous studies reported higher levels of cytokines, including IFN-γ, in the periphery and placenta of malaria uninfected compared to infected women, suggesting that these cytokines are involved in the control of parasitemia in peripheral blood and the placenta [30]. Along the same line of reasoning, it was determined that the total IFN-γ concentration was highest in non-transmitter mothers, suggesting a synergistic effect of IFN-γ between the mother, placenta, and cord in the control of congenital infection. In addition, the increased level of this cytokine in the cord of uninfected newborns compared to newborns of uninfected and transmitter mothers probably results from the combined effect of maternal and cord IFN-γ contributing to the prevention of congenital infection. Taking into account the protective role of the placenta, IFN-γ might be transported from the placenta to the cord. Evidence suggests that IFN-γ involved in maternal immune activation can cross the placenta and predispose to neuropsychiatric disorders [31]. Another study points to the induction of the mother's immune system to release proinflammatory cytokines after infection during pregnancy which can cross the placenta and enter the fetal circulation [32].

Part of the focus of this study was the evaluation of parasite load in the congenital transmission process, through the histological and molecular analyses of placental tissue samples. Patterns of amastigote nests and free parasites were observed in histological samples of infected mothers. However, as determined by RT-PCR analyses suggesting live parasites in the placenta, samples with a higher and lower prevalence of parasite load corresponded with infected and non-infected newborns, respectively. We observed that the interplay between the level of parasitemia and the production of IFN-γ determines the occurrence or absence of congenital infection. Therefore, when parasitemia diminishes the stimulus for the production of this cytokine diminishes, there is less control of parasite load, and congenital infection occurs. Conversely, when parasitemia rises, production of IFN-γ increases, and parasite load and congenital infection are controlled.

Regarding the placenta, the case of triplets born to a *T. cruzi*-infected mother (Table 1, sample 1072) from a triamniotic trichorionic delivery of which two newborns were infected and one uninfected is indicative of the protection offered by the placenta. Previous reports on the placenta´s role in protecting the fetus from infection point to the effect of lysosomal subfractions of the placenta on parasite viability and infectivity [33], a decreased viability of parasites in villous explants coincident with high levels of nitric oxide production by placental cells [34], and our previous study in chronically infected mothers showing that higher levels of IFN-γ in their placentas prevented *T. cruzi* infection of neonates [19]. Interestingly, placental samples 2097 and 1961 despite being RT-PCR negatives were associated with vertical transmission of *T. cruzi*, suggesting low levels of parasite load in these samples, undetected in the RT-PCR assay.

Our results demonstrate a potential relation between maternal immune profile during pregnancy and vertical *T. cruzi* transmission. However, more in-depth laboratory testing to identify likely mechanisms of congenital transmission, such as the study of genes differentially expressed in the placenta of transmitter and non-transmitter mothers related to the immune system as well as the polymorphism of some genes expressed in the placenta in association with congenital *T. cruzi* transmission [35–37], which was not possible to undertake, could have contributed to improving understanding to our observations.

A limited number of cases due to budget limitations hindered our ability to have stronger statistical support. Budget support was secured for two years and due to the low incidence of congenital transmission access to these cases is limited.

Supplementary Materials: The following supporting information can be downloaded at: https://www.mdpi.com/article/10.3390/microorganisms12061243/s1, ethical approval materials.

Author Contributions: Conceptualization, W.R.C. and C.R.; data curation, W.R.C.; formal analysis, R.P.; funding acquisition, C.R.; investigation, W.R.C. and C.R.; methodology, A.G.H.C., W.R.C., S.G., S.M., and C.R.; project administration, C.R.; resources, S.M. and C.R.; supervision, C.R.; writing—original draft, W.R.C. and C.R.; writing—review and editing, R.P. and C.R. All authors have read and agreed to the published version of the manuscript.

Funding: This work was supported by the "Institut de Recherche pour le Développement" (IRD) through its program of "Jeunes Equipes Associées à l´ IRD" and the Swedish International Developing Cooperation Agency (SIDA) through the Program SIDA-Universidad Mayor de San Andrés.

Institutional Review Board Statement: The study was conducted in accordance with the Declaration of Helsinki, and approved by the Research Ethics Committee of the Universidad Mayor de San Andrés (UMSA), CEI-UMSA; Approval Code, None; Approval Date, 14 September 2008.

Informed Consent Statement: Informed consent was obtained from all subjects included in the study.

Data Availability Statement: The data that support the findings of this study are included within the article.

Acknowledgments: We thank the medical doctors and technical staff of the Maternity Clinic Odón Ortega in Yacuiba for their invaluable support and help in the collection of biological samples.

Conflicts of Interest: The authors declare no conflicts of interest.

References

1. Moncayo, A. Chagas disease: Current epidemiological trends after the interruption of vectorial and transfusional transmission in the southern cone countries. *Mem. Inst. Oswaldo* **2003**, *98*, 577–591. [CrossRef] [PubMed]
2. Schofield, C.J.; Jannin, J.; Salvatella, R. The future of Chagas disease control. *Trends Parasitol.* **2006**, *22*, 583–588. [CrossRef] [PubMed]
3. Sanchez, N.O.; Mora, M.C.; Basombrio, M.A. High prevalence of congenital *Trypa nosoma cruzi* infection and family clustering in Salta, Argentina. *Pediatrics* **2005**, *115*, e668–e672.
4. Azogue, E.; Darras, C. Prospective study of Chagas disease in newborn children with placental infection caused *by Trypanosoma cruzi* (Santa Cruz-Bolivia). *Rev. Soc. Bras. Med. Trop.* **1991**, *24*, 105–109. [CrossRef] [PubMed]

5. Contreras, S.; Fernandez, M.R.; Agüero, F.; Desse Desse, J.D.; Orduna, T.; Martino, O. Congenital Chagas-Mazza disease in Salta, Argentina. *Rev. Soc. Bras. Med. Trop.* **1999**, *32*, 633–636. [CrossRef] [PubMed]
6. Blanco, S.B.; Segura, E.L.; Cura, E.N.; Chuit, R.; Tulián, I.; Flores, I.; Garbarino, G.; Villalonga, J.F. Congenital transmission of *Trypanosoma cruzi*: An operational outline for detecting and treating infected infants in north–western Argentina. *Trop. Med. Int. Health* **2000**, *6*, 293–301. [CrossRef] [PubMed]
7. Russomando, G.; Almirón, M.; Candia, N.; Franco, I.; Sánchez, Z.; de Guillen, I. Implementation and evaluation of a locally sustainable system of prenatal diagnosis to detect cases of congenital Chagas disease in endemic areas of Paraguay. *Rev. Soc. Bras. Med. Trop.* **2005**, *38* (Suppl. 2), 49–54.
8. Hermann, R.; Truyens, C.; Alonso-vega, C.; Rodriguez, P.; Berthe, S.; Torrico, F.; Carlier, Y. Congenital transmission of *Trypanosoma cruzi* is associated with maternal enhanced parasitemia and decreased production of interferon-γ in response to parasite antigens. *J. Infect. Dis.* **2004**, *189*, 1274–1281. [CrossRef] [PubMed]
9. Bern, C.; Verastegui, M.; Gilman, R.H.; La Fuente, C.; Galdos-Cardenas, G.; Calderon, M.; Pacori, J.; del Carmen Abastoflor, M.; Aparicio, H.; Brady, M.F.; et al. Congenital Trypanosoma cruzi transmission in Santa Cruz, Bolivia. *Clin. Infect. Dis.* **2009**, *49*, 1667–1674. [CrossRef]
10. Howard, E.J.; Xiong, X.; Carlier, Y.; Sosa-Estani, S.; Buekens, P. Frequency of the congenital transmission of *Trypanosoma cruzi*: A systematic review and meta-analysis. *Int. J. Obstet. Gynaecol.* **2014**, *121*, 22–23. [CrossRef]
11. Azogue, E.; La Fuente, C.; Darras, C. Congenital Chagas disease in Bolivia: Epidemiological aspects and pathological findings. *Trans. R. Soc. Trop. Med. Hyg.* **1985**, *79*, 176–180. [CrossRef]
12. Bittencourt, A.L. *American trypanosomiasis* (Chagas diseases). In *Parasitic Infections in Pregnancy and the Newborn*; McLeod, C., Ed.; Oxford University Press: Oxford, UK, 1988; pp. 62–86.
13. Alonso-Vega, C.; Hermann, E.; Truyens, C.; Rodríguez, P.; Torrico, M.C.; Torrico, F.; Carlier, Y. Immunological status of mothers infected with *Trypanosoma cruzi*. *Rev. Soc. Bras. Med. Trop.* **2005**, *38* (Suppl. 2), 101–104.
14. Virreira, M.; Truyens, C.; Alonso Cega, C.; Brutus, L.; Jijena, J.; Torrico, F.; Carlier, Y.; Svoboda, M. Comparison of *Trypanosoma cruzi* lineages and levels of parasitic DNA in infected mothers and their newborns. *Am. J. Trop. Med. Hyg.* **2007**, *77*, 102–106. [CrossRef] [PubMed]
15. Brutus, L.; Castillo, H.; Bernal, C.; Salas, N.A.; Schneider, D.; Santalla, J.A.; Chippaux, J.P. Detectable *Trypanosoma cruzi* parasitemia during pregnancy and delivery as a risk factor for congenital Chagas disease. *Am. J. Trop. Med. Hyg.* **2010**, *83*, 1044–1047. [CrossRef] [PubMed]
16. Bua, J.; Volta, B.J.; Velázquez, E.B.; Ruiz, A.M.; Rissio, A.M.; Cardoni, R.L. Vertical transmission of *Trypanosoma cruzi* infection: Quantification of parasite burden in mothers and their children by parasite DNA amplification. *Trans. R. Soc. Trop. Med. Hyg.* **2012**, *106*, 623–628. [CrossRef]
17. Vekemans, J.; Truyens, C.; Ttorrico, F.; Solano, M.M.; Torrico, M.C.; Rodriguez, P.; Alonso Vega, C.; Carlier, Y. Maternal *Trypanosoma cruzi* infection upregulates capacity of uninfected neonate cells to produce pro-and anti-inflammatory cytokines. *Infect. Immun.* **2000**, *68*, 5430–5434. [CrossRef] [PubMed]
18. Hermann, E.; Alonso-Vega, C.; Berthe, A.; Truyens, C.; Flores, A.; Cordova, M.; Moretta, I.; Torrico, F.; Braud, V.; Carlier, Y. Human congenital infection with Trypanosoma cruzi induces phenotypic and functional modifications of cord blood NK cells. *Pediatr. Res.* **2006**, *60*, 38–43. [CrossRef]
19. Cuna, W.R.; Choque, A.G.; Passera, R.; Rodriguez, C. Pro-inflammatory cytokine production in chagasic mothers and their uninfected newborns. *J. Parasitol.* **2009**, *95*, 891–894. [CrossRef]
20. Bouyou-Akotet, M.K.; Kombila, M.; Kremsner, P.G.; Mavoungou, E. Cytokine profiles in peripheral, placental and cord blood in pregnant women from an area endemic for *Plasmodium falciparum*. *Eur. Cytokine Netw.* **2004**, *15*, 120–125.
21. Freilij, H.; Muller, L.; Gonzalez Cappa, S.M. Direct micromethod for diagnosis of acute and congenital Chagas' disease. *J. Clin. Microbiol.* **1983**, *18*, 327–330. [CrossRef]
22. Alonso-Vega, C.; Billot, C.; Torrico, F. Achievements and challenges upon the implementation of a program for national control of congenital Chagas in Bolivia: Results 2004–2009. *PLoS Neglected Trop. Dis.* **2013**, *7*, e2304. [CrossRef]
23. Bua, J.; Volta, B.J.; Perrone, A.E.; Scollo, K.; Velázquez, E.B.; Ruiz, A.M.; De Rissio, A.M.; Cardoni, R.L. How to improve the early diagnosis of Trypanosoma cruzi infection: Relationship between validated conventional diagnosis and quantitative DNA amplification in congenitally infected children. *PLoS Neglected Trop. Dis.* **2013**, *7*, e2476. [CrossRef] [PubMed]
24. Marcon, G.E.; Andrade, P.D.; de Albuquerque, D.M.; Wanderley, J.S.; de Almeida, E.A.; Guariento, M.E.; Costa, S.C. Use of a nested polymerase chain reaction (N-PCR) to detect *Trypanosoma cruzi* in blood samples from chronic chagasic patients and patients with doubtful serologies. *Diagn. Microbiol. Infect. Dis.* **2002**, *43*, 39–43. [CrossRef]
25. Plata, F.; Wietzerbin, J.; Pons, F.G.; Falcoff, E.; Eisen, H. Synergistic protection by specific antibodies and interferon against infection by *Trypanosoma cruzi* in vitro. *Eur. J. Immunol.* **1984**, *14*, 930–935. [CrossRef]
26. Reed, S.G. In vivo administration of recombinant IFN-gamma induces macrophage activation, and prevents acute disease, immune suppression, and death in experimental *Trypanosoma cruzi* infections. *J. Immunol.* **1988**, *140*, 4342–4347. [CrossRef]
27. Torrico, F.; Heremans, H.; Rivera, M.T.; Van Marck, E.; Billiau, A.; Carlier, Y. Endogenous IFN-gamma is required for resistance to acute *Trypanosoma cruzi* infection in mice. *J. Immunol.* **1991**, *146*, 3626–3632. [CrossRef]
28. Vespa, G.N.; Cunha, F.Q.; Silva, J.S. Nitric oxide is involved in control of Trypanosoma cruzi–induced parasitemia and directly kills the parasite in vitro. *Infect. Immun.* **1994**, *62*, 5177–5182. [CrossRef]

29. Muñoz-Fernández, M.A.; Fernández, M.A.; Fresno, M. Synergism between tumor necrosis factor-alpha and interferon-gamma on macrophage activation for the killing of intracellular Trypanosoma cruzi through a nitric oxide-dependent mechanism. *Eur. J. Immunology.* **1992**, *22*, 301–307. [CrossRef] [PubMed]
30. Bayoumi, N.K.; Bakhet, K.H.; Mohmmed, A.A.; Eltom, A.M.; Elbashir, M.I.; Mavoungou, E.; Adam, I. Cytokine profiles in peripheral, placental and cord blood in an area of unstable malaria transmission in eastern Sudan. *J. Trop. Pediatr.* **2009**, *55*, 233–237. [CrossRef]
31. Kathuria, A.; Lopez-Lengowski, K.; Roffman, J.L.; Karmacharya, R. Distinct effects of interleukin-6 and interferon-γ on differentiating human cortical neurons. *Brain Behav. Immun.* **2022**, *103*, 97–108. [CrossRef]
32. Zawadzka, A.; Cieślik, M.; Adamczyk, A. The Role of Maternal Immune Activation in the Pathogenesis of Autism: A Review of the Evidence, Proposed Mechanisms and Implications for Treatment. *Int. J. Mol. Sci.* **2021**, *22*, 11516. [CrossRef] [PubMed]
33. Frank, F.; Sartori, M.J.; Asteggiano, C.; Lin, S.; de Fabro, S.P.; Fretes, R.E. The effect of placental subfractions on *Trypanosoma cruzi*. *Exp. Mol. Pathol.* **2000**, *69*, 144–151. [CrossRef] [PubMed]
34. Díaz-Luján, C.; Friquell, M.F.; Schijman, A.; Paglini, P.; Fretes, R.E. Differential susceptibility of isolated human trophoblasts to infection by *Trypanosoma cruzi*. *Placenta* **2012**, *33*, 264–270. [CrossRef] [PubMed]
35. Juiz, N.A.; Cayo, N.M.; Burgos, M.; Salvo, M.E.; Nasser, J.R.; Búa, J.; Silvia, A.; Longhi, S.A.; Schijman, A.G. Human polymorphisms in placentally expressed genes and their association with susceptibility to congenital *Trypanosoma cruzi* infection. *J. Infect. Dis.* **2016**, *213*, 1299–1306. [CrossRef] [PubMed]
36. Juiz, N.A.; Solana, M.E.; Acevedo, G.R.; Benatar, A.F.; Ramirez, J.C.; da Costa, P.A.; Macedo, A.M.; Longhi, S.A.; Schijman, A.G. Different genotypes of Trypanosoma cruzi produce distinctive placental environment genetic response in chronic experimental infection. *PLoS Neglected Trop. Dis.* **2017**, *8*, e0005436. [CrossRef]
37. Juiz, N.A.; Torrejón, I.; Burgos, M.; Torres, A.M.F.; Duffy, T.; Cayo, N.M.; Tabasco, A.; Salvo, M.; Longhi, S.A.; Schijman, A.G. Alterations in placental gene expression of pregnant women with chronic Chagas disease. *Am. J. Pathol.* **2018**, *188*, 1345–1353. [CrossRef]

microorganisms

MDPI

Article

Subtype Distribution of *Blastocystis* spp. in Patients with Gastrointestinal Symptoms in Northern Spain

Cristina Matovelle [1,2], Joaquín Quílez [3,4], María Teresa Tejedor [5,6], Antonio Beltrán [2,7], Patricia Chueca [5] and Luis Vicente Monteagudo [3,5,*]

1 Faculty of Medicine, University of Zaragoza, 50009 Zaragoza, Spain; crismatovelle@gmail.com
2 Environmental Sciences Institute (IUCA), University of Zaragoza, 50009 Zaragoza, Spain; abeltranros@salud.aragon.es
3 AgriFood Institute of Aragon (IA2), 50013 Zaragoza, Spain; jquilez@unizar.es
4 Department of Animal Pathology, Faculty of Veterinary Sciences, University of Zaragoza, 50013 Zaragoza, Spain
5 Department of Anatomy, Embryology and Animal Genetics, Faculty of Veterinary Sciences, University of Zaragoza, 50013 Zaragoza, Spain; ttejedor@unizar.es (M.T.T.); pachueca@unizar.es (P.C.)
6 Aragon Institute of Health Sciences (IACS), Centro de Investigación Biomédica en Red-Enfermedades Cardiovasculares (CIBERCV), 50009 Zaragoza, Spain
7 Service of Microbiology and Parasitology, Hospital Clínico Universitario Lozano Blesa, 50009 Zaragoza, Spain
* Correspondence: monteagu@unizar.es

Abstract: Limited molecular data exist on the prevalence and subtype distribution of *Blastocystis* spp., the most prevalent parasite in human and animal feces worldwide. A total of 44 different subtypes (STs) of *Blastocystis* are currently recognized based on the sequence of the small subunit ribosomal RNA (*SSU-rRNA*) gene. This is a molecular study of *Blastocystis* spp. in hospitalized patients with gastrointestinal symptoms in northern Spain. We analyzed 173 *Blastocystis*-positive patients with gastrointestinal symptoms by using nested PCR for molecular detection, subtype identification, phylogenetic analyses, and genetic diversity assessment. ST2 (34.1%) and ST3 (34.7%) predominated, followed by ST1 (15.6%) and ST4 (15.6%). Mixed infections with different subtypes were observed in some patients. Sequence analysis revealed for the first time in European humans the allele 88 (a variant of ST1). In other cases, alleles commonly found in animal samples were detected (allele 9 in ST2, allele 34 in ST3, and allele 42 in ST4). Phylogenetic analysis showed high variability in ST1 and ST2, suggesting a polyphyletic origin, while both ST3 and ST4 exhibited higher genetic homogeneity, indicating a possible monophyletic origin and recent transmission to humans. These data confirm *Blastocystis* spp. subtype diversity and may help in understanding the evolutionary processes and potential zoonotic transmission of this parasite.

Keywords: *Blastocystis* spp.; genetic diversity; phylogenetic analysis; subtypes

1. Introduction

Blastocystis spp. is the most prevalent intestinal protozoon detected in humans and animals worldwide. This enteric protist is known to be associated with gastrointestinal symptoms in people across both industrialized and developing countries [1]. Various factors account for the high prevalence of *Blastocystis* spp. Its transmission occurs through the fecal–oral route, with several sources of infection, including person-to-person, zoonotic, and waterborne transmission. In developing countries, it is associated with socio-economic factors leading to poor sanitation [2]. This protozoon could have implications for public health since it can be transmitted to humans from animals, suggesting its potential zoonotic nature. The "One Health" strategy, promoted by the World Health Organization, encourages interdisciplinary collaboration to achieve optimal health for humans, animals, and the environment, effectively addressing zoonotic infections like *Blastocystis* spp. This approach enhances the understanding of the disease and facilitates the adoption of specific control measures for the benefit of both human and animal populations [3].

Over the past decade, the scientific community has increasingly focused its attention on unraveling the genetic diversity of *Blastocystis* spp. by using the small subunit ribosomal RNA (*SSU-rRNA*) gene as a molecular marker. This genetic tool has yielded invaluable insights into its taxonomy, population structure and potential pathogenicity, significantly enhancing our comprehension of *Blastocystis* spp. [4]. A total of 44 different subtypes and numerous subtype subgroups have been reported to date based on variations in the *SSU-rRNA* gene in humans and animals [5]. However, not all strains of a specific subtype have confirmed clinical significance and the potential relationship between different subtypes and their ability to cause disease is still a topic of active debate [6].

Infection with *Blastocystis* spp. in humans has been reported across the globe [7,8]; in Europe, the reported prevalence of this protist in humans ranges from 3% to 7% in France, Italy and the United Kingdom (UK) by using optical microscopy, but higher prevalence levels (14.5–24.2%) were found when PCR-based studies were conducted in France, the Netherlands and Denmark [9–14]. Subtype ST3 exhibits the highest global distribution, with subtypes ST1 and ST2 following closely in prevalence [15]. In Europe, subtype ST3 is also the most frequent [16–18], followed by ST4 [16,19], then ST2 [20,21] and lastly ST1 [11,22]. Additional *Blastocystis* subtypes that are rare in the human population have been identified in Europe, such as ST5 [18], ST6 [23], ST7 [24], ST8 [21], and ST9 [25].

A long-debated topic is related to the pathogenicity of *Blastocystis* spp. and the ongoing challenge to determine whether this protozoon is genuinely pathogenic, a commensal or only pathogenic in specific situations such as immunosuppression, malnutrition or recurrent infections [26,27]. Some studies suggest that *Blastocystis* spp. could be part of a healthy intestinal microbiota, potentially mitigating inflammation and autoimmune disorders; it prompts interleukin-22 release, thus assisting in intestinal mucosal secretion to relieve colitis symptoms and it may also contribute to host metabolism by breaking down cellulose [28]. Researchers have also explored pathogenicity variations among *Blastocystis* subtypes (ST), yet conclusive findings remain elusive. ST1, ST2, and ST4 are implicated as potential sources of gastrointestinal symptoms, with studies indicating their higher prevalence in symptomatic patients compared to controls. ST1 has been linked to irritable bowel syndrome [29], while ST2 is associated with gastrointestinal issues and urticaria, and is particularly prevalent in patients with diarrhea in Colombia, whereas asymptomatic individuals carry ST1 [30]. However, some studies present inconsistent support for ST2's pathogenicity [31]. ST3 is predominantly found in patients with urticaria and gastrointestinal symptoms [32] while ST4 shows high prevalence in severe diarrhea cases [33]. Subtypes ST5, ST6 and ST7 also exhibit potential pathogenicity [34]. Although rare in humans, ST8 has been linked to severe symptoms in two studies [35].

In Spain, the reported prevalence of *Blastocystis* in human populations is highly variable, and it could be grossly underestimated in several studies due to the low diagnostic sensitivity of some detection techniques; specifically, molecular analyses are much more sensitive than microscopy and in vitro xenic culture for detecting *Blastocystis* spp. in fecal samples from humans and animals [19,36]. A recent study conducted in Zaragoza (Spain) using conventional microscopy found a prevalence of 9.2% [37]. This result is consistent with other research using conventional microscopy in the central region of Spain, where a prevalence value of 9.6% was documented in HIV-positive children, and 5.3–19.4% in children attending nurseries and primary schools [38,39]. Nevertheless, the figure rises to 13% among asymptomatic schoolchildren using PCR-based methods in the same geographical area of central Spain [40]. Higher values (27.8%) have been reported in adult patients in northeastern Spain using microscopic examination and PCR [41], while a similar procedure reported a prevalence of 35.2% among humans cohabiting with dogs and cats in northern Spain [42]. Studies investigating the subtype distribution of *Blastocystis* spp. are limited in Spain and predominantly focused on specific population groups. Subtypes ST1–ST4 and ST8 have been identified in both asymptomatic and symptomatic schoolchildren in Madrid. Subtype ST4 was the most prevalent in a human population in Valencia, while subtype ST2 was the most frequently detected in Alava [20,21,33,37,42]. The aim of the current study was to analyze the genetic diversity of *Blastocystis* spp. subtypes circulating in infected patients in an area of northern Spain and to investigate any differences in clinical significance among the various subtypes.

2. Materials and Methods

2.1. Ethics Approval Statement

This study was conducted in accordance with the guidelines of the Declaration of Helsinki (1975, revised in 2013) to ensure ethical considerations in human research. Approval for this study was obtained from the Ethics Committee of Aragón (ref 18/081) before commencing the research, ensuring compliance with national and international guidelines. All participating patients were anonymized and provided signed informed consent. This study also adheres to the requirements of the Health Insurance Portability and Accountability Act (HIPAA, 1996). Throughout the research, mandatory health and safety procedures were adhered to.

2.2. Sampling of Fecal Specimens

A total of 6807 stool samples from 3682 patients showing gastrointestinal symptoms in the year 2018 were analyzed as described in a previous report [37]. Among the 338 *Blastocystis*-positive (by microscopy) patients detected in that previous report, 173 fecal samples providing good DNA sequences (following the procedure described in the next sections) were included in the present study. *Blastocystis* positivity and sufficient quality of the genetic sequence were the inclusion criteria. The categorical variables analyzed for association with *Blastocystis* spp. infection were: demographic origin (Spain, rest of Europe, Africa, American continent and Asia); age group (16 years or younger and >16 years); gender (male and female) and *Blastocystis* subtypes (ST1, ST2, ST3 and ST4 subtypes).

2.3. Molecular Detection of Blastocystis spp.

According to the manufacturer's instructions, DNA extraction was performed on the 173 samples using a DNA Stool Kit (NORGEN BIOTEK CORP., Thorold, ON, Canada). For the molecular detection of *Blastocystis*, a nested PCR was performed. The primary PCR amplified the conserved eukaryotic region of the 18S rRNA gene with universal primers EUK-F and EUK-R in a 50 µL final volume [43]. For the secondary PCR, a specific *SSU-rRNA* gene fragment of *Blastocystis* spp. was amplified following the protocol by Santín et al. [44] using the primary PCR product as a template. The primers Blast 505–532 and Blast 998–1017 amplify a ~479 bp fragment, including a variable region of the *SSU-rRNA* gene that enables the subtyping of *Blastocystis* spp. The reaction mix for the secondary PCR was prepared in a final volume of 50 µL. Both PCR reactions were performed using a MJ Research MINICYCLER-PCR-THERMAL CYCLER and Applied Biosystems™ 2720 Thermal Cycler (Applied Biosystems, Whaltman, MA, USA). In order to verify that the PCR generated amplicons were of the desired size, agarose gel electrophoresis was performed using the products from the secondary PCR.

2.4. Subtype Identification, Phylogenetic Analyses and Genetic Diversity

The PCR products were purified using the Speedtools PCR Clean Up Kit (Biotools, Madrid, Spain) and sequenced on both strands by the Sanger method. The sequences obtained were edited and assembled in BioEdit software version 7.0.0 URL https://bioedit.software.informer.com/7.0/ (accessed on 24 May 2024). To confirm the identity of the sequences as *Blastocystis* spp., they were compared with the reference sequences of the different *Blastocystis* spp. subtypes available in the GenBank® database using the nucleotide BLAST program provided by the National Center for Biotechnology Information (NCBI) [45,46]. Subsequently, the sequences were assembled in FASTA format and submitted to the *Blastocystis* Subtype database (18S), which is a multilocus sequence typing (MLST) database available at http://pubmlst.org/blastocystis/ (accessed on 24 May 2024), and the ST and corresponding alleles were determined through sequence comparison. As of 15 January 2024, the database contained 357 alleles for the gene investigated in the present study [47,48].

Following alignment using ClustalW [49] in BioEdit 7.0 [50], the *SSU-rRNA* gene sequences of *Blastocystis* spp. were analyzed. Finally, phylogenetic analysis was performed using the Neighbor Joining (NJ) method based on genetic distances calculated using the 2-parameter or Kimura 2 model [51] with MEGA5.10 software [52]. A sequence from *Proteromonas lacertae* (GenBank® accession number U37108) was used as an outgroup. The resulting trees were exported in Newick format (which allows tree representation using parentheses and commas) [53]. For the graphical representation of the obtained phylogenetic trees, the online software iTOL v5 (https://itol.embl.de/about.cgi, accessed on 24 May 2024) [54] was used.

To assess the genetic diversity of the sequences, the DNAsp v6.12.01 software [55] available at http://www.ub.edu/dnasp/ (last accessed on 15 January 2024) was used. The parameters used to measure genetic diversity among the sequences included the number of polymorphic sites (S) [a site is considered polymorphic when different sequences yield distinct nucleotides at that site or position], number of haplotypes (h) [each specific combination of nucleotides in the sequence is considered a different haplotype], haplotype diversity (Hd) [the likelihood of two randomly sampled haplotypes being distinct], and nucleotide diversity (π) [the average number of nucleotide differences per site between two sequences].

3. Results

3.1. Subtypes

A total of four different subtypes were identified among the 173 *Blastocystis*-positive samples that were previously analyzed in the GenBank® database. The predominant subtypes were *Blastocystis* ST2 (34.1%) and ST3 (34.7%), which were found in a similar proportion of patients, followed by subtypes ST1 (15.6%) and ST4 (15.6%), which were both identified in an equal number of infected patients. Only 111 among the 173 *Blastocystis*-positive samples met the requirements of length and quality to be deposited in the GenBank® database with the accession numbers OP495227–OP495337 and to be compared with sequences previously deposited in the database by means of BLASTn [45,46]. Nevertheless, after aligning the *Blastocystis* spp. sequences and identifying a partial length bias in some sequences, the shortest ones were discarded, leaving only 83 sequences that met the length requirements (minimum 352 bp) to be selected for genetic diversity study and phylogenetic analysis. Genetic diversity analysis performed in the 352 aligned nucleotide sites provided nucleotide diversity per site (Pi = 0.14808) and an average of 50.05260 nucleotide differences between haplotypes.

Out of the 173 *Blastocystis* samples, 11 (6.3%) exhibited mixed *Blastocystis* sequences, which were identified by double peaks in the *SSU-rRNA* gene chromatograms. These peaks occurred within a 30-nucleotide segment, enabling specific variant identification within the rest of the sequence. Four out of these eleven sequences belonged to the subtype ST1 (36.4%) and seven to the subtype ST2 (63.6%). Another significant finding was the subtype variation observed in four of the patients who submitted repeated samples. Namely, one patient switched from ST1 to ST3 in nine days, another from ST3 to ST1 in just one day. Notably, one patient showed ST1, then ST2 after three days, and reverted to ST1 three days later (Table 1).

Table 1. Patients with variation of subtypes in repeated samples.

Patient	Sample Date	Subtype	Age (years)	Country of Birth
1	17 September 2018 26 September 2018	ST1 ST3	2	Ukraine

Table 1. *Cont.*

Patient	Sample Date	Subtype	Age (years)	Country of Birth
2	3 October 2018 4 October 2018	ST3 ST1	70	Spain
3	9 January 2018 12 January 2018 15 January 2018	ST1 ST2 ST1	8	Gambia
4	19 February 2018 22 February 2018	ST4 ST2	62	Spain

3.2. Alleles

Analysis of allele distribution according to the age, gender and geographic origin of the patients is indicated in Tables S1–S4. Allele 88 was the most common within ST1-positive patients (9/12; 75%) followed by alleles 2 (2/12: 16.6%) and 4 (1/12, 8.33%); allele 2 was found in both Spanish and European patients. ST2 sequences featured only two alleles: allele 9 (15/23; 65.2%) and allele 13 (8/23; 34.8%), with patients from different geographic distributions (Table S2). ST3 sequences were mainly associated with allele 34 (29/32; 90.6%), while three samples showed allele 36 (3/32, 9.4%), and most of them (31/32; 97%) originating from patients born in Europe and Africa (Table S3). The 16 ST4 sequences uniformly exhibited allele 42, with all patients being of Spanish origin, and a majority (10/16; 62.5%) being under 16 years of age (Table S4).

3.3. Blastn Alignments with Sequences Existing in GenBank

The 111 *Blastocystis* spp. sequences were subjected to BLASTn comparisons with reference sequences deposited in GenBank [45,46] (Tables S5–S8). Notably, some ST1 and ST4 sequences showed 100% similarity with *Blastocystis* sequences from animals (*Bos taurus*, *Sus scrofa domesticus*) but lower similarity with strains from humans, suggesting a potential zoonotic origin. In contrast, most ST2 and ST3 sequences displayed the highest similarity with human-origin sequences, suggesting a potential human-to-human transmission pattern.

3.4. Genetic Diversity and Phylogenetic Analysis

Nucleotide comparisons among partial *SSU-rRNA* gene sequences of each *Blastocystis* subtype are shown in Tables S9–S12. Comparison of the ST1 subtype revealed that all 17 sequences had a nucleotide similarity higher than 94.3% to each other but only eight sequences were 100% identical. The number of sequences exhibiting 100% identity to each other was greater within subtypes 2 and 3. Notably, all but two of the twenty sequences belonging to the ST4 subtype showed 100% similarity to each other.

Among 83 *Blastocystis* spp. sequences that met the requirements to be selected for genetic diversity study and phylogenetic analysis, 26 distinct haplotypes were identified, yielding a joint haplotype diversity index of 0.927. Genetic diversity was assessed for each *Blastocystis* spp. subtype (Table 2). ST1, ST2, ST3, and ST4 displayed eight, seven, nine and two haplotypes, respectively. Notably, ST1 exhibited the highest haplotype diversity (Hd: 0.924), while ST4 had the lowest (Hd: 0.233), with 30 and 1 polymorphic sites, respectively. ST2 showed the highest nucleotide diversity (π: 0.03945), indicating substantial variability in the *SSU-rRNA* gene fragment.

Figure 1 presents a phylogenetic tree of the *SSU-rRNA* gene sequences of *Blastocystis* rooted with *Proteromonas lacertae* (U37108) as the outgroup. Four well-supported branches (green) with 1000 bootstrap replicates (support value of one) are depicted. In the ST1 branch, observed in the vertical representation, two well-supported sub-branches with an absolute support value of one are apparent. The first sub-branch holds two sequences associated with allele 2, while the second sub-branch comprises nine sequences displaying allele 88, along with one sequence with allele 4. In ST2, three well-supported sub-branches emerge from the bootstrap replicates. The first branch includes seven sequences linked to allele 9 and two to allele 13. The second group comprises solely allele 13 sequences, while the third contains allele 9 sequences, except for one linked to allele 13. Within the ST3 branch, two distinct alleles are observed: allele 34 dominates most sequences, while only three sequences correspond to allele 36. On the opposite side, in the ST4 branch, all sequences exhibit allele 42, and no sub-branches are observed, indicating a possible recent monophyletic pattern.

Table 2. Genetic variability of *Blastocystis* spp. subtypes in patients in the present study.

Subtypes	Frequency	Monomorphic Sites	Polymorphic Sites	h*	Hd≠	$\pi^{\#}$
ST1	12	315	30	8	0.924	0.02429
ST2	23	316	28	7	0.842	0.03945
ST3	31	342	7	9	0.774	0.00438
ST4	16	346	1	2	0.233	0.00067
Total	83	234	104	26	0.927	0.14808

h*: Number of haplotypes, Hd≠: Haplotype diversity, $\pi^{\#}$: Nucleotide diversity.

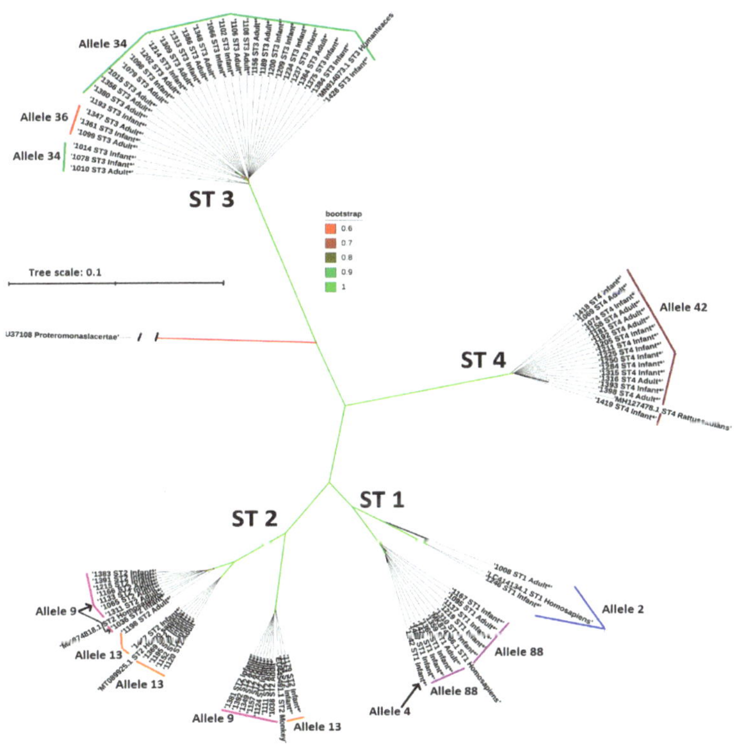

Figure 1. Neighbor joining analysis of the partial sequences of the *SSU-rRNA* gene of *Blastocystis* and reference sequences representative of different subtypes. A sequence of *Proteromonas lacertae* was used as the outgroup and sequences of other *Blastocystis* subtypes were obtained from GenBank® (LC414134.1 and MK874786.1: ST1 *Homo sapiens*, EU445491.1: ST2 Monkey (sic.), M25.1 and MK874818.1: ST2 *Homo sapiens*, MN914073.1: ST3 *Homo sapiens*, and MH127478.1: ST4 *Rattus exulans*). Genetic distances were calculated using the Kimura 2 model (own image). The length of the branch connecting the outgroup sample was reduced by 61% to simplify the image. The symbol * indicates the sequences obtained in the present work.

4. Discussion

The present report is based on the analysis of fecal samples obtained from patients exhibiting gastrointestinal symptoms in a limited geographical area of Spain (not from healthy individuals). In spite of the limited sample size, marked genetic variability was observed in the set of samples. The distribution of *Blastocystis* spp. subtypes varies across countries and continents, as summarized in Tables S13 and S14 [56]. The laboratory protocol used in the present study, based on PCR amplification of a 479 bp fragment of the *Blastocystis SSU-rRNA* gene and proposed by Santín et al. [44], is widely utilized in specialized literature, even if other methodologies are used too.

The current study identified ST3 and ST2 as the most prevalent subtypes. Subtype ST3 is the predominant *Blastocystis* subtype globally [23,57–60]. In Spain, this subtype was not detected in symptomatic individuals in Valencia but ranked second among pets and their owners in Álava and was less common in the Madrid region [20,21,33,42].

In warm climates like those in Senegal, Lebanon, Saudi Arabia and Bolivia, ST2 is among the most common subtypes, suggesting adaptation to such environments, even if it was also the most prevalent one in a study performed in Ireland [9,13,23,24,59–61]. The high frequency of ST2 and ST3 in South America has been associated with poverty, sanitation issues, civil conflicts, biodiversity and limited access to clean water, promoting *Blastocystis* spp. transmission [8]. ST2 was also the most frequent subtype reported in previous Spanish studies in Alava, Leganés (Madrid) and Central Spain, being mostly associated with children, even when they were asymptomatic [20,21,42]. Conversely, ST2 was less common in Valencia, where most fecal specimens were from adult patients [33]. These observations suggest a potential fecal–oral transmission cycle among school-age children for ST2 [62].

In our study, both ST4 and ST1 were the less-prevalent *Blastocystis* subtypes, each detected in over 15% of patients. The ST4 subtype is commonly found in temperate European countries like Denmark and France, while it is infrequent in Japan, Malaysia, China, Latin American and African countries [1,16,58,60,63–66]. In Spain, ST4 was previously found to be highly prevalent in Valencia (over 94% of *Blastocystis*-infected patients) [33]. ST1 is among the less common subtypes in Europe, except in some studies in Germany, Greece, France, Denmark and Italy [22,24,61,67,68]. Conversely, ST1 is the most common *Blastocystis* variant in several countries across the rest of the continents, including Colombia, Brazil, Libya, Nigeria, Tanzania, Iran, Turkey, Saudi Arabia, the Philippines and Australia [18,44,69–73].

A high proportion (42%) of the patients studied in the present report are immigrant residents from various countries [37]. This could be a possible explanation for the different *Blastocystis* subtype distribution when compared with that found in other points of Spain, highlighting the need for molecular studies in order to understand *Blastocystis* spp. genetic diversity globally.

A search in GenBank® via Blastn revealed that some of the sequences obtained in this research exhibit 100% identity with previous isolates obtained from animals and human beings all over the world. For instance, the ST1 sequences closely match those from humans in several countries such as Colombia, Mexico, Laos and Malaysia [44,74]. Such coincidences among samples from distant origins are not uncommon: in a Chinese study, two ST1 sequences were identical with a *Blastocystis* sequence from humans in Turkey [75].

Our ST2 sequences matched those from human samples globally and from various animal species in Germany, Spain, China and Iran [9,13,44,74,76–79].

The ST3 sequences showed similarities with sequences from human samples from Mexico, Colombia, Germany, South Korea, the Philippines, Malaysia, Libya, Egypt and Turkey, as well as high similarity with animal sequences from the USA, Spain, Iran, Malaysia, China and Japan [44,80–84]. This observation is consistent across multiple studies owing to the widespread prevalence of ST3 globally [44,77,81,85–87].

The ST4 sequences also showed high similarity with sequences of human and animal origin from various regions [77,88–90]. Numerous ST4 sequences in GenBank® are linked to various rodent species, supporting their role as primary hosts for this subtype [91–94].

In Spain, diverse investigations explored *Blastocystis* spp. isolates in animals and the environment. A study on fecal samples from free-living carnivores in various regions identified ST2 and ST4 *Blastocystis* subtypes in red fox, exhibiting significant sequence similarity (98.5–100%) to those in the present study [95]. Similar patterns were observed in urban wastewater in Valencia, where ST2 was the predominant subtype [96]. Additionally, 100% identity was found with ST3 sequences from *Blastocystis* isolates from cattle from Álava [83] and a 99.5% similarity was found with ST4 sequences from *Blastocystis* in *Rattus* spp. from a zoo in Córdoba [97]. The substantial similarity among *Blastocystis* spp. isolates across different regions and animals implies a potential zoonotic connection, raising public health concerns. Widespread international trade and travel could explain the identical sequences observed between our study and those from distant countries.

This study found mixed *Blastocystis* spp. infections with diverse subtypes, in agreement with worldwide observations. In contrast to previous Spanish studies, a French investigation utilized PCR product cloning to identify three subtypes (ST2, ST3, and ST4) within the same host [20,21,33,42,98]. However, the practical use of this technique in larger cohorts is hindered by its labor-intensive nature and its high cost. Various protocols, including the one in this study, have detected mixed infections using DNA from in vitro cultures or fecal samples. Sanger sequencing of PCR-amplified 18S rRNA gene products, with universal or *Blastocystis*-specific primers, has been used widely. Overlapping peaks in chromatograms indicate mixed infections, with PCR often amplifying DNA from the predominant subtype. In cases of similar subtype concentrations, the amplifications of both of them are equally efficient and result in double peaks [16,77,99]. Next-Generation Sequencing (NGS) has revolutionized the detection of mixed *Blastocystis* spp. infections, providing precise quantification even at low levels (as low as 5%). However, its application is restricted by high costs and the need for highly skilled personnel [84]. Mixed *Blastocystis* infections are frequently overlooked in research due to detection limitations. In a meta-analysis, fewer than half of the studies (24/55) reported mixed *Blastocystis* infections, with a prevalence of around 6% [18,100]. Our study mirrors this prevalence, identifying mixed infections in 6.3% (11 out of 173) of subtyped samples. Four of these eleven samples belonged to ST1 and seven to ST2.

The finding of different subtypes in repeated samples of the same patient is also an unexpected observation in this study. PCR amplifies the predominant subtype, raising uncertainty about whether undetected coexisting subtypes were present initially due to technical limitations. In one case involving three samples, the detected subtype in the first and last samples was the same. Another explanation could be reinfection with a different strain, but this is less plausible because of the short interval between the two samples (1–9 days, depending on the patient). It is noteworthy that three of the four individuals showing this pattern were infected with ST1, which is a minor subtype in this study but commonly found in mixed infections [101].

The DnaSP results indicate varying levels of diversity, with ST1 sequences exhibiting the highest number of haplotypes and polymorphic sites and ST4 sequences showing the lowest diversity. In a study in Iran, ST2 had the highest haplotype diversity (Hd: 0.934) and ST1 the lowest one (Hd: 0.564) [102]. Conversely, ST1 was the most variable subtype in a study in Saudi Arabia [71]. In our study, ST2 also showed the highest nucleotide diversity (π: 0.03945). Overall, these findings confirm significant variability in the amplified *SSU-rRNA* gene fragment among *Blastocystis* subtypes, suggesting a longer evolutionary history for ST1 and ST2, while ST3 and ST4 are more recent. In fact, the low variability observed in ST4 could also be due to a recent transmission of this subtype to humans. The sequences for ST4 obtained in rats, Guinea pigs and opossum and most human patients show a high similarity. Rodents are considered a possible reservoir for *Blastocystis* transmission to humans [30,103].

Various authors highlight the value of distinguishing *Blastocystis* alleles for insights into host specificity, geographic distribution and clinical manifestations [30,104,105]. In our study, allele diversity within each subtype was limited. ST1 exhibited the highest diversity with three identified alleles (2, 4, and 88), consistent with findings by other authors. While allele 4 is commonly reported in European studies, allele 88, previously identified in the Middle East and South America, especially in immunocompromised patients, has been detected for the first time in European human populations in the present report [106–108].

Two alleles (9 and 13) were identified within ST2 sequences, with 9 being the most frequent. Previous research [1,30,105] revealed that allele 9 is the most prevalent in humans and animals in South America, too [20]. In contrast, other studies identified allele 12, which was not observed in this study, as the most common in humans and animals, and did not find allele 9 [21,42,97]. Among the fifteen sequences with allele 9 in our study, over a third (7/15) were from individuals from Africa (five) and from South America (two). Furthermore, allele 9 has been observed in studies involving dogs and rats, indicating potential zoonotic transmission [1,109].

In ST3, we identified two alleles (34 and 36), with a clear predominance of the former (29/32 ST3 isolates), consistent with the higher frequency of allele 34 in the European human population [21,42,97]. In contrast, allele 36 is dominant in African children, where some studies do not detect allele [15,34,110]. Sequences of ST4 exhibited minimal variability in our study, with only allele 42 being observed, consistent with other research [21,110]. ST4 is prevalent in rodents and the previous detection of allele 42 in stray cats suggests their potential role as a source of *Blastocystis* spp. infections in humans [111].

Neighbor joining analysis showed more than one branch in the phylogenetic tree for both ST1 and ST2 sequences. This suggests possible diverse sources of infection from humans or animals with these subtypes and potential ancient origins and extensive evolutionary processes [112]. In contrast, the dominant ST3 exhibited a limited variability, which could be explained by a higher infectivity, by a potential origin from limited sources or by a shorter evolutionary timeline. The greatest genetic homogeneity was observed in ST4 sequences, suggesting that its transmission to humans is more recent than that of the other subtypes.

Furthermore, it is noticeable that ST4 sequences from rats, guinea pigs, opossums and most humans are highly conserved. This supports the theory that rodents can be a reservoir for human infections with *Blastocystis* ST4, as suggested in other works [30,103]. In the present dataset, patients showing this subtype were of Spanish origin; the only two sequences presenting similarity slightly below 100% came from patients living in a rural area.

5. Conclusions

In summary, this molecular study revealed that the distribution of *Blastocystis* spp. subtypes infecting humans showed variations compared to other geographical areas in Spain, a circumstance that could be related to the high percentage of immigrants residing in the population investigated. Two predominant subtypes (ST2 and ST3) and two minor subtypes (ST1 and ST4) have been identified. Also, in some patients, mixed infections with different subtypes of *Blastocystis* spp. have been detected by the presence of double peaks in chromatograms following Sanger sequencing of the *SSU-rRNA* gene. The analysis of the sequences of the different subtypes has revealed an allele previously undescribed in European human samples (allele 88 in ST1) and other alleles coincident with those found in animal samples (9 in ST2, 34 in ST3 and 42 in ST4). In fact, high genetic similarity has been found with isolates of *Blastocystis* spp. from samples obtained in both human and animal species obtained in geographically distant regions; these findings could support the potential zoonotic nature of this parasite. The highest genetic variability was observed in ST1 and ST2, suggesting a polyphyletic origin of these variants, indicative of diverse origins or a longer evolutionary process. The greatest similarity was observed among the

ST3 and especially ST4 sequences, indicating a probable monophyletic origin and/or a more recent transmission for these subtypes.

Supplementary Materials: The following supporting information can be downloaded at: https://www.mdpi.com/article/10.3390/microorganisms12061084/s1, Table S1. Distribution of *Blastocystis* sp. alleles, country of origin of the patients, age and sex in sequences belonging to ST1. Table S2. Distribution of *Blastocystis* sp. alleles, country of origin of the patients, age and sex in sequences belonging to ST2. Table S3. Distribution of *Blastocystis* sp. alleles, country of origin of the patients, age and sex in sequences belonging to ST3. Table S4. Distribution of *Blastocystis* sp. alleles, country of origin of the patients, age and sex in sequences belonging to ST4. Table S5. Comparisons of the genetic sequences obtained from ST1 with the genetic sequences stored in the GenBank® database. Table S6. Comparisons of the genetic sequences obtained from ST2 with the genetic sequences stored in the GenBank® database. Table S7. Comparisons of the genetic sequences obtained from ST3 with the genetic sequences stored in the GenBank® database. Table S8. Comparisons of the genetic sequences obtained from ST4 with the genetic sequences stored in the GenBank® database. between partial SSU-rRNA gene sequences identified as *Blastocystis* ST1. There was a total of 136 combinations in the final data set. The pink background indicates the maximum range (100%) and the green background the minimum range (94.26%). Table S9 Homology between partial SSU-rRNA gene sequences identified as *Blastocystis* ST1. There was a total of 136 combinations in the final data set. The pink background indicates the maximum range (100%) and the green background the minimum range (94.26%). Table S10 Homology between partial SSU-rRNA gene sequences identified as ST2 of *Blastocystis*. There was a total of 465 combinations in the final data set. The pink background indicates the maximum range (100%) and the green background the minimum range (94.72%). Table S11. Homology between partial SSU-rRNA gene sequences identified as ST3 of *Blastocystis*. There was a total of 903 combinations in the final data set. The pink background indicates the maximum range (100%) and the green background the minimum range (91.38%). Table S12 Homology between partial SSU-rRNA gene sequences identified as ST4 of *Blastocystis*. There was a total of 190 combinations in the final data set. The pink background indicates the maximum range (100%) and the green background the minimum range (99.72%). Table S13. Distribution of *Blastocystis* sp. subtypes by continents. Table S14. Frequency of subtypes (%) of *Blastocystis* sp. identified in human infections in various studies in Spain.

Author Contributions: Conceptualization, C.M., J.Q., M.T.T., A.B. and L.V.M.; Methodology, C.M., J.Q., A.B. and L.V.M.; Software, M.T.T.; Validation, C.M., A.B. and P.C.; Formal Analysis, J.Q. and L.V.M.; Investigation, C.M., J.Q., A.B., P.C. and L.V.M.; Resources, L.V.M. and J.Q., Data Curation, M.T.T., Writing—Original Draft Preparation, C.M., J.Q., M.T.T., A.B. and L.V.M.; Writing—Review and Editing, C.M., J.Q., M.T.T., A.B. and L.V.M.; Visualization, C.M., J.Q. and L.V.M.; Supervision, J.Q. and L.V.M.; Project Administration, J.Q. and L.V.M.; Funding Acquisition, J.Q. and L.V.M. All authors have read and agreed to the published version of the manuscript.

Funding: Funding for this research and for the article processing charges was provided by the Government of Aragón (Spain) via the Research Group A16_23R (Zoonosis and Emerging Diseases of Interest to Public Health) and the Research Group B43_23R ("Water and Environmental Health").

Institutional Review Board Statement: Approval for this study was obtained from the Ethics Committee of Aragón (Spain) (ref 18/081) before commencing the research, ensuring compliance with national and international guidelines.

Informed Consent Statement: All participating patients were anonymized and provided signed informed consent. As stated before, the complete procedure was supervised and authorized by the Ethics Committee of Aragón (Spain).

Data Availability Statement: The sequences used for analysis in the present manuscript are available in the GenBank repository [www.ncbi.nlm.nih.gov/genbank/] [OP495227-OP495337]. The other datasets are available on request from the authors.

Acknowledgments: The authors thank Demetris Savva for his help in the improvement of English grammar and style. They also acknowledge the assistance of the General Support Service for Research-(SAI) of the University of Zaragoza regarding the molecular study of this work.

Conflicts of Interest: The authors declare no conflicts of interest.

References

1. Sinniah, B.; Hassan, A.K.R.; Sabaridah, I.; Soe, M.M.; Ibrahim, Z.; Ali, O. Prevalence of intestinal parasitic infections among communities living in different habitats and its comparison with one hundred and one studies conducted over the past 42 years (1970 to 2013) in Malaysia. *Trop. Biomed.* **2014**, *31*, 190–206.
2. Javanmard, E.; Niyyati, M.; Ghasemi, E.; Mirjalali, H.; Asadzadeh Aghdaei, H.; Zali, M.R. Impacts of Human Development Index and Climate Conditions on Prevalence of *Blastocystis*: A Systematic Review and Meta-Analysis. *Acta Trop.* **2018**, *185*, 193–203. [CrossRef] [PubMed]
3. Jinatham, V.; Maxamhud, S.; Popluechai, S.; Tsaousis, A.D.; Gentekaki, E. *Blastocystis* One Health Approach in a Rural Community of Northern Thailand: Prevalence, Subtypes and Novel Transmission Routes. *Front. Microbiol.* **2021**, *12*, 746340. [CrossRef] [PubMed]
4. Maloney, J.G.; Jang, Y.; Molokin, A.; George, N.S.; Santin, M. Wide Genetic Diversity of *Blastocystis* in White-tailed Deer (*Odocoileus Virginianus*) from Maryland, USA. *Microorganisms* **2021**, *9*, 1343. [CrossRef] [PubMed]
5. Santin, M.; Figueiredo, A.; Molokin, A.; George, N.S.; Köster, P.C.; Dashti, A.; González-Barrio, D.; Carmena, D.; Maloney, J.G. Division of *Blastocystis* ST10 into Three New Subtypes: ST42-ST44. *J. Eukaryot. Microbiol.* **2023**, *71*, e12998. [CrossRef] [PubMed]
6. Kosik-Bogacka, D.; Lepczyńska, M.; Kot, K.; Szkup, M.; Łanocha-Arendarczyk, N.; Dzika, E.; Grochans, E. Prevalence, Subtypes and Risk Factors of *Blastocystis* spp. Infection among Pre- and Perimenopausal Women. *BMC Infect. Dis.* **2021**, *21*, 1125. [CrossRef] [PubMed]
7. Yowang, A.; Tsaousis, A.D.; Chumphonsuk, T.; Thongsin, N.; Kullawong, N.; Popluechai, S.; Gentekaki, E. High Diversity of *Blastocystis* Subtypes Isolated from Asymptomatic Adults Living in Chiang Rai, Thailand. *Infect. Genet. Evol.* **2018**, *65*, 270–275. [CrossRef] [PubMed]
8. Jiménez, P.A.; Jaimes, J.E.; Ramírez, J.D. A Summary of *Blastocystis* Subtypes in North and South America. *Parasites Vectors* **2019**, *12*, 376. [CrossRef]
9. Stensvold, C.R.; Traub, R.J.; von Samson-Himmelstjerna, G.; Jespersgaard, C.; Nielsen, H.V.; Thompson, R.C.A. *Blastocystis*: Subtyping Isolates Using Pyrosequencing^{TM} Technology. *Exp. Parasitol.* **2007**, *116*, 111–119. [CrossRef]
10. Masucci, L.; Graffeo, R.; Bani, S.; Bugli, F.; Boccia, S.; Nicolotti, N.; Fiori, B.; Fadda, G.; Spanu, T. Intestinal Parasites Isolated in a Large Teaching Hospital, Italy, 1 May 2006 to 31 December 2008. *Eurosurveillance* **2011**, *16*, 19891. [CrossRef]
11. El Safadi, D.; Cian, A.; Nourrisson, C.; Pereira, B.; Morelle, C.; Bastien, P.; Bellanger, A.; Botterel, F.; Candolfi, E.; Desoubeaux, G.; et al. Prevalence, Risk Factors for Infection and Subtype Distribution of the Intestinal Parasite *Blastocystis* spp. from a Large-Scale Multi-Center Study in France. *BMC Infect. Dis.* **2016**, *16*, 451. [CrossRef] [PubMed]
12. González-Moreno, O.; Domingo, L.; Teixidor, J.; Gracenea, M. Prevalence and Associated Factors of Intestinal Parasitisation: A Cross-Sectional Study among Outpatients with Gastrointestinal Symptoms in Catalonia, Spain. *Parasitol. Res.* **2011**, *108*, 87–93. [CrossRef]
13. Bart, A.; Wentink-Bonnema, E.M.S.; Gilis, H.; Verhaar, N.; Wassenaar, C.J.A.; van Vugt, M.; Goorhuis, A.; van Gool, T. Diagnosis and Subtype Analysis of *Blastocystis* spp. in 442 Patients in a Hospital Setting in the Netherlands. *BMC Infect. Dis.* **2013**, *13*, 389. [CrossRef]
14. Windsor, J.J.; Macfarlane, L.; Hughes-Thapa, G.; Jones, S.K.A.; Whiteside, T.M. Incidence of *Blastocystis hominis* in Faecal Samples Submitted for Routine Microbiological Analysis. *Br. J. Biomed. Sci.* **2002**, *59*, 154–157. [CrossRef] [PubMed]
15. Poulsen, C.S.; Efunshile, A.M.; Nelson, J.A.; Stensvold, C.R. Epidemiological Aspects of *Blastocystis* Colonization in Children in Ilero, Nigeria. *Am. J. Trop. Med. Hyg.* **2016**, *6*, 175–179. [CrossRef] [PubMed]
16. Scicluna, S.M.; Tawari, B.; Clark, C.G. DNA Barcoding of Blastocystis. *Protist* **2006**, *157*, 77–85. [CrossRef]
17. Stensvold, C.R.; Arendrup, M.C.; Jespersgaard, C.; Mølbak, K.; Nielsen, H.V. Detecting *Blastocystis* Using Parasitologic and DNA-Based Methods: A Comparative Study. *Diagn. Microbiol. Infect. Dis.* **2007**, *59*, 303–307. [CrossRef] [PubMed]
18. Alfellani, M.A.; Stensvold, C.R.; Vidal-Lapiedra, A.; Onuoha, E.S.U.; Fagbenro-Beyioku, A.F.; Clark, C.G. Variable Geographic Distribution of *Blastocystis* Subtypes and Its Potential Implications. *Acta Trop.* **2013**, *126*, 11–18. [CrossRef] [PubMed]
19. Poirier, P.; Wawrzyniak, I.; Albert, A.; El Alaoui, A.H.; Delbac, F.; Livrelli, V. Development and Evaluation of a Real-Time PCR Assay for Detection and Quantification of *Blastocystis* Parasites in Human Stool Samples: Prospective Study of Patients with Hematological Malignancies. *J. Clin. Microbiol.* **2011**, *49*, 975–983. [CrossRef]
20. Hernández-Castro, C.; Dashti, A.; Vusirikala, A.; Balasegaram, S.; Köster, P.C.; Bailo, B.; Imaña, E.; López, A.; Llorente, M.T.; González-Barrio, D.; et al. Prevalence and Temporal Dynamics of Cryptosporidium spp., Giardia Duodenalis, and *Blastocystis* spp. among Toddlers Attending Day-Care Centres in Spain. A Prospective Molecular-Based Longitudinal Study. *Eur. J. Pediatr.* **2023**, *182*, 213–223. [CrossRef]
21. Muadica, A.S.; Köster, P.C.; Dashti, A.; Bailo, B.; Hernández-de-Mingo, M.; Reh, L.; Balasegaram, S.; Verlander, N.Q.; Chércoles, E.R.; Carmena, D. Molecular Diversity of *Giardia duodenalis*, *Cryptosporidium* spp. and *Blastocystis* spp. in Asymptomatic School Children in Leganés, Madrid (Spain). *Microorganisms.* **2020**, *8*, 446. [CrossRef] [PubMed]
22. Gabrielli, S.; Furzi, F.; Fontanelli Sulekova, L.; Taliani, G.; Mattiucci, S. Occurrence of *Blastocystis* -Subtypes in Patients from Italy Revealed Association of ST3 with a Healthy Gut Microbiota. *Parasite Epidemiol. Control* **2020**, *9*, e00134. [CrossRef] [PubMed]
23. Rudzińska, M.; Kowalewska, B.; Wąż, P.; Sikorska, K.; Szostakowska, B. *Blastocystis* Subtypes Isolated from Travelers and Non-Travelers from the North of Poland—A Single Center Study. *Infect. Genet. Evol.* **2019**, *75*, 103926. [CrossRef] [PubMed]

24. Souppart, L.; Sanciu, G.; Cian, A.; Wawrzyniak, I.; Delbac, F.; Capron, M.; Dei-Cas, E.; Boorom, K.; Delhaes, L.; Viscogliosi, E. Molecular Epidemiology of Human *Blastocystis* Isolates in France. *Parasitol. Res.* **2009**, *105*, 413–421. [CrossRef] [PubMed]
25. Stensvold, C.R.; Lewis, H.C.; Hammerum, A.M.; Porsbo, L.J.; Nielsen, S.S.; Olsen, K.E.P.; Arendrup, M.C.; Nielsen, H.; Mølbak, K. *Blastocystis*: Unravelling Potential Risk Factors and Clinical Significance of a Common but Neglected Parasite. *Epidemiol. Infect.* **2009**, *137*, 1655–1663. [CrossRef] [PubMed]
26. Shirvani, G.; Fasihi-Harandi, M.; Raiesi, O.; Bazargan, N.; Zahedi, M.J.; Sharifi, I.; Kalantari-Khandani, B.; Nooshadokht, M.; Shabandoust, H.; Mohammadi, M.A.; et al. Prevalence and Molecular Subtyping of *Blastocystis* from Patients with Irritable Bowel Syndrome, Inflammatory Bowel Disease and Chronic Urticaria in Iran. *Acta Parasitol.* **2020**, *65*, 90–96. [CrossRef] [PubMed]
27. Scanlan, P.D.; Stensvold, C.R.; Rajilić-Stojanović, M.; Heilig, H.G.H.J.; De Vos, W.M.; O'Toole, P.W.; Cotter, P.D. The Microbial Eukaryote *Blastocystis* Is a Prevalent and Diverse Member of the Healthy Human Gut Microbiota. *FEMS Microbiol. Ecol.* **2014**, *90*, 326–330. [CrossRef]
28. Lepczyńska, M.; Białkowska, J.; Dzika, E.; Piskorz-Ogórek, K.; Korycińska, J. Blastocystis: How Do Specific Diets and Human Gut Microbiota Affect Its Development and Pathogenicity? *Eur. J. Clin. Microbiol. Infect. Dis.* **2017**, *36*, 1531–1540. [CrossRef]
29. Fouad, S.A.; Basyoni, M.M.; Fahmy, R.A.; Kobaisi, M.H. The Pathogenic Role of Different *Blastocystis* Hominis Genotypes Isolated from Patients with Irritable Bowel Syndrome. *Arab J. Gastroenterol.* **2011**, *12*, 194–200. [CrossRef]
30. Ramírez, J.D.; Sánchez, L.V.; Bautista, D.C.; Corredor, A.F.; Flórez, A.C.; Stensvold, C.R. *Blastocystis* Subtypes Detected in Humans and Animals from Colombia. *Infect. Genet. Evol.* **2014**, *22*, 223–228. [CrossRef]
31. Dogruman-Al, F.; Dagci, H.; Yoshikawa, H.; Kurt, Ö.; Demirel, M. A Possible Link between Subtype 2 and Asymptomatic Infections of *Blastocystis* Hominis. *Parasitol. Res.* **2008**, *103*, 685–689. [CrossRef] [PubMed]
32. Tan, T.C.; Suresh, K.G.; Smith, H.V. Phenotypic and Genotypic Characterisation of *Blastocystis Hominis* Isolates Implicates Subtype 3 as a Subtype with Pathogenic Potential. *Parasitol. Res.* **2008**, *104*, 85–93. [CrossRef] [PubMed]
33. Domínguez-Márquez, M.V.; Guna, R.; Muñoz, C.; Gómez-Muñoz, M.T.; Borrás, R. High Prevalence of Subtype 4 among Isolates of *Blastocystis Hominis* from Symptomatic Patients of a Health District of Valencia (Spain). *Parasitol. Res.* **2009**, *105*, 949–955. [CrossRef] [PubMed]
34. Koltas, I.S.; Eroglu, F. Subtype Analysis of *Blastocystis* Isolates Using SSU RRNA-DNA Sequencing in Rural and Urban Population in Southern Turkey. *Exp. Parasitol.* **2016**, *170*, 247–251. [CrossRef] [PubMed]
35. Roberts, T.; Stark, D.; Harkness, J.; Ellis, J. Subtype Distribution of *Blastocystis* Isolates from a Variety of Animals from South Wales, Australia. *Vet. Parasitol.* **2013**, *196*, 85–89. [CrossRef]
36. Stensvold, C.R.; Ahmed, U.N.; Andersen, L.O.B.; Nielsen, H.V. Development and Evaluation of a Genus-Specific, Probe-Based, Internal-Process-Controlled Real-Time PCR Assay for Sensitive and Specific Detection of *Blastocystis* spp. *J. Clin. Microbiol.* **2012**, *50*, 1847–1851. [CrossRef] [PubMed]
37. Matovelle, C.; Tejedor, M.T.; Monteagudo, L.V.; Beltrán, A.; Quílez, J. Prevalence and Associated Factors of *Blastocystis* spp. Infection in Patients with Gastrointestinal Symptoms in Spain: A Case-Control Study. *Trop. Med. Infect. Dis.* **2022**, *7*, 226. [CrossRef] [PubMed]
38. Martín-Sánchez, A.M.; Canut-Blasco, A.; Rodriguez-Hernandez, J.; Montes-Martínez, I.; García-Rodríguez, J.A. Epidemiology and Clinical Significance of *Blastocystis* Hominis in Different Population Groups in Salamanca (Spain). *Eur. J. Epidemiol.* **1992**, *8*, 553–559. [CrossRef]
39. Aguila, C.; Navajas, R.; Gurbindo, D.; Ramos, J.; Mellado, M.; Fenoy, S.; Muñoz Fernandez, M.; Subirats, M.; Ruiz, J.; Pieniazek, N. Microsporidiosis in HIV-Positive Children in Madrid (Spain). *J. Eukaryot. Microbiol.* **1997**, *44*, 84–85. [CrossRef]
40. Reh, L.; Muadica, A.S.; Köster, P.C.; Balasegaram, S.; Verlander, N.Q.; Chércoles, E.R.; Carmena, D. Substantial Prevalence of Enteroparasites *Cryptosporidium* spp., *Giardia Duodenalis* and *Blastocystis* spp. in Asymptomatic Schoolchildren in Madrid, Spain, November 2017 to June 2018. *Eurosurveillance* **2019**, *24*, 1900241. [CrossRef]
41. Salvador, F.; Lobo, B.; Goterris, L.; Alonso-Cotoner, C.; Santos, J.; Sulleiro, E.; Bailo, B.; Carmena, D.; Sánchez-Montalvá, A.; Bosch-Nicolau, P.; et al. *Blastocystis* spp. Carriage and Irritable Bowel Syndrome: Is the Association Already Established? *Biology* **2021**, *10*, 340. [CrossRef] [PubMed]
42. Paulos, S.; Köster, P.C.; de Lucio, A.; Hernández-de-Mingo, M.; Cardona, G.A.; Fernández-Crespo, J.C.; Stensvold, C.R.; Carmena, D. Occurrence and Subtype Distribution of *Blastocystis* spp. in Humans, Dogs and Cats Sharing Household in Northern Spain and Assessment of Zoonotic Transmission Risk. *Zoonoses Public Health* **2018**, *65*, 993–1002. [CrossRef]
43. DeLong, E.F. Archaea in Coastal Marine Environments. *Proc. Natl. Acad. Sci. USA* **1992**, *89*, 5685–5689. [CrossRef] [PubMed]
44. Santín, M.; Gómez-Muñoz, M.T.; Solano-Aguilar, G.; Fayer, R. Development of a New PCR Protocol to Detect and Subtype *Blastocystis* spp. from Humans and Animals. *Parasitol. Res.* **2011**, *109*, 205–212. [CrossRef] [PubMed]
45. Altschul, S.F.; Gish, W.; Miller, W.; Myers, E.W.; Lipman, D.J. Basic Local Alignment Search Tool. *J. Mol. Biol.* **1990**, *215*, 403–410. [CrossRef] [PubMed]
46. Altschul, S.F.; Madden, T.L.; Schäffer, A.A.; Zhang, J.; Zhang, Z.; Miller, W.; Lipman, D.J. Gapped BLAST and PSI-BLAST: A New Generation of Protein Database Search Programs. *Nucleic Acids Res.* **1997**, *25*, 3389–3402. [CrossRef]
47. Jolley, K.A.; Bray, J.E.; Maiden, M.C.J. Open-Access Bacterial Population Genomics: BIGSdb Software, the PubMLST.Org Website and Their Applications. *Wellcome Open Res.* **2018**, *3*, 124. [CrossRef] [PubMed]
48. Engsbro, A.L.; Stensvold, C.R. Blastocystis: To Treat or Not to Treat . . . but How? *Clin. Infect. Dis.* **2012**, *55*, 1431–1432. [CrossRef]

49. Thompson, J.D.; Higgins, D.G.; Gibson, T.J. CLUSTAL W: Improving the Sensitivity of Progressive Multiple Sequence Alignment through Sequence Weighting, Position-Specific Gap Penalties and Weight Matrix Choice. *Nucleic Acids Res.* **1994**, *22*, 4673–4680. [CrossRef]
50. Hall, T.A. BioEdit: A User-Friendly Biological Sequence Alignment Editor and Analysis Program for Windows 95/98/NT. *Nucleic Acids Symp. Ser.* **1999**, *41*, 95–98.
51. Kimura, M. A Simple Method for Estimating Evolutionary Rates of Base Substitutions through Comparative Studies of Nucleotide Sequences. *J. Mol. Evol.* **1980**, *16*, 111–120. [CrossRef] [PubMed]
52. Tamura, K.; Peterson, D.; Peterson, N.; Stecher, G.; Nei, M.; Kumar, S. MEGA5: Molecular Evolutionary Genetics Analysis Using Maximum Likelihood, Evolutionary Distance, and Maximum Parsimony Methods. *Mol. Biol. Evol.* **2011**, *28*, 2731–2739. [CrossRef] [PubMed]
53. Olsen, G. Gary Olsen's Interpretation of the "Newick's 8:45" Tree Format Standard. Available online: https://phylipweb.github.io/phylip/newick_doc.html (accessed on 24 May 2024).
54. Letunic, I.; Bork, P. Interactive Tree of Life (ITOL) v5: An Online Tool for Phylogenetic Tree Display and Annotation. *Nucleic Acids Res.* **2021**, *49*, W293–W296. [CrossRef] [PubMed]
55. Rozas, J.; Ferrer-Mata, A.; Sánchez-DelBarrio, J.C.; Guirao-Rico, S.; Librado, P.; Ramos-Onsins, S.; Sánchez-Gracia, A. Dna SP v6: DNA Sequence Polymorphism Analysis of Large Datasets. *Mol. Biol. Evol.* **2017**, *34*, 3299–3302. [CrossRef]
56. Stensvold, C.R.; Alfellani, M.A.; Nørskov-Lauritsen, S.; Prip, K.; Victory, E.L.; Maddox, C.; Nielsen, H.V.; Clark, C.G. Subtype Distribution of *Blastocystis* Isolates from Synanthropic and Zoo Animals and Identification of a New Subtype. *Int. J. Parasitol.* **2009**, *39*, 473–479. [CrossRef] [PubMed]
57. Clark, C.G. Extensive Genetic Diversity in *Blastocystis Hominis*. *Mol. Biochem. Parasitol.* **1997**, *87*, 79–83. [CrossRef] [PubMed]
58. Stensvold, R.; Brillowska-Dabrowska, A.; Nielsen, H.V.; Arendrup, M.C. Detection of *Blastocystis Hominis* in Unpreserved Stool Specimens by Using Polymerase Chain Reaction. *J. Parasitol.* **2006**, *92*, 1081–1087. [CrossRef] [PubMed]
59. Meloni, D.; Sanciu, G.; Poirier, P.; El Alaoui, H.; Chabé, M.; Delhaes, L.; Dei-Cas, E.; Delbac, F.; Luigi Fiori, P.; Di Cave, D.; et al. Molecular Subtyping of *Blastocystis* spp. Isolates from Symptomatic Patients in Italy. *Parasitol. Res.* **2011**, *109*, 613–619. [CrossRef] [PubMed]
60. Forsell, J.; Granlund, M.; Stensvold, C.R.; Clark, G.C.; Evengård, B. Subtype Analysis of *Blastocystis* Isolates in Swedish Patients. *Eur. J. Clin. Microbiol. Infect. Dis.* **2012**, *31*, 1689–1696. [CrossRef]
61. Menounos, P.G.; Spanakos, G.; Tegos, N.; Vassalos, C.M.; Papadopoulou, C.; Vakalis, N.C. Direct Detection of *Blastocystis* spp. in Human Faecal Samples and Subtype Assignment Using Single Strand Conformational Polymorphism and Sequencing. *Mol. Cell. Probes* **2008**, *22*, 24–29. [CrossRef]
62. El Safadi, D.; Gaayeb, L.; Meloni, D.; Cian, A.; Poirier, P.; Wawrzyniak, I.; Delbac, F.; Dabboussi, F.; Delhaes, L.; Seck, M.; et al. Children of Senegal River Basin Show the Highest Prevalence of *Blastocystis* spp. Ever Observed Worldwide. *BMC Infect. Dis.* **2014**, *14*, 164. [CrossRef] [PubMed]
63. Kaneda, Y.; Horiki, N.; Cheng, X.J.; Fujita, Y.; Maruyama, M.; Tachibana, H. Ribodemes of *Blastocystis Hominis* Isolated in Japan. *Am. J. Trop. Med. Hyg.* **2001**, *65*, 393–396. [CrossRef]
64. Noradilah, S.A.; Lee, I.L.; Anuar, T.S.; Salleh, F.M.; Abdul Manap, S.N.A.; Husnie, N.S.; Azrul, S.M.; Moktar, N. *Blastocystis* spp. Contaminated Water Sources in Aboriginal Settlements. *Trop. Biomed.* **2017**, *34*, 110–117.
65. Li, L.H.; Zhou, X.N.; Du, Z.W.; Wang, X.Z.; Wang, L.B.; Jiang, J.Y.; Yoshikawa, H.; Steinmann, P.; Utzinger, J.; Wu, Z.; et al. Molecular Epidemiology of Human *Blastocystis* in a Village in Yunnan Province, China. *Parasitol. Int.* **2007**, *56*, 281–286. [CrossRef] [PubMed]
66. Stensvold, C.R.; Christiansen, D.B.; Olsen, K.E.P.; Nielsen, H.V. *Blastocystis* spp. Subtype 4 Is Common in Danish *Blastocystis*-Positive Patients Presenting with Acute Diarrhea. *Am. J. Trop. Med. Hyg.* **2011**, *84*, 883. [CrossRef] [PubMed]
67. Böhm-Gloning, B.; Knobloch, J.; Walderich, B. Five Subgroups of *Blastocystis* Hominis Isolates from Symptomatic and Asymptomatic Patients Revealed by Restriction Site Analysis of PCR-Amplified 16S-like RDNA. *Trop. Med. Int. Health* **1997**, *2*, 771–778. [CrossRef] [PubMed]
68. Rene, B.; Stensvold, C.; Badsberg, J.; Nielsen, H. Subtype Analysis of *Blastocystis* Isolates from *Blastocystis* Cyst Excreting Patients. *Am. J. Trop. Med. Hyg.* **2009**, *80*, 588–592. [CrossRef] [PubMed]
69. Malheiros, A.F.; Stensvold, C.R.; Clark, C.G.; Braga, G.B.; Shaw, J.J. Short Report: Molecular Characterization of *Blastocystis* Obtained from Members of the Indigenous Tapirapé Ethnic Group from the Brazilian Amazon Region, Brazil. *Am. J. Trop. Med. Hyg.* **2011**, *85*, 1050–1053. [CrossRef]
70. Eroglu, F.; Gene, A.; Elgun, K.; Koltas, I.S. Identification of *Blastocystis hominis* Isolates from Asymptomatic and Symptomatic Patients by PCR. *Parasitol. Res.* **2009**, *105*, 1589–1592. [CrossRef]
71. Wakid, M.H.; Aldahhasi, W.T.; Alsulami, M.N.; El-Kady, A.M.; Elshabrawy, H.A. Identification and Genetic Characterization of *Blastocystis* Species in Patients from Makkah, Saudi Arabia. *Infect. Drug Resist.* **2022**, *15*, 491–501. [CrossRef]
72. Rivera, W.L.; Tan, M.A.V. Molecular Characterization of *Blastocystis* Isolates in the Philippines by Riboprinting. *Parasitol. Res.* **2005**, *96*, 253–257. [CrossRef] [PubMed]
73. Nagel, R.; Cuttell, L.; Stensvold, C.R.; Mills, P.C.; Bielefeldt-Ohmann, H.; Traub, R.J. *Blastocystis* Subtypes in Symptomatic and Asymptomatic Family Members and Pets and Response to Therapy. *Intern. Med. J.* **2012**, *42*, 1187–1195. [CrossRef] [PubMed]

74. Rojas-Velázquez, L.; Maloney, J.G.; Molokin, A.; Morán, P.; Serrano-Vázquez, A.; González, E.; Pérez-Juárez, H.; Ximénez, C.; Santin, M. Use of Next-Generation Amplicon Sequencing to Study *Blastocystis* Genetic Diversity in a Rural Human Population from Mexico. *Parasites Vectors* **2019**, *12*, 566. [CrossRef] [PubMed]
75. Zhou, Y.L.; Zhao, N.; Yang, Y.; Li, Y.; Zhang, X.; Chen, J.; Peng, X.; Zhao, W. Molecular Identification and Subtype Analysis of *Blastocystis* in Captive Asiatic Black Bears (Ursus Thibetanus) in China's Heilongjiang and Fujian Provinces. *Front. Cell. Infect. Microbiol.* **2022**, *12*, 993312. [CrossRef] [PubMed]
76. Kim, M.J.; Won, E.J.; Kim, S.H.; Shin, J.H.; Chai, J.Y. Molecular Detection and Subtyping of Human *Blastocystis* and the Clinical Implications: Comparisons between Diarrheal and Non-Diarrheal Groups in Korean Populations. *Korean J. Parasitol.* **2020**, *58*, 321–326. [CrossRef]
77. Sarzhanov, F.; Dogruman-Al, F.; Santin, M.; Maloney, J.G.; Gureser, A.S.; Karasartova, D.; Taylan-Ozkan, A. Investigation of Neglected Protists *Blastocystis* spp. and *Dientamoeba fragilis* in Immunocompetent and Immunodeficient Diarrheal Patients Using Both Conventional and Molecular Methods. *PLoS Neglected Trop. Dis.* **2021**, *15*, e0009779. [CrossRef] [PubMed]
78. Barbosa, C.V.; Barreto, M.M.; de Jesus Andrade, R.; Sodré, F.; D'Avila-Levy, C.M.; Peralta, J.M.; Igreja, R.P.; de Macedo, H.W.; Santos, H.L.C. Intestinal Parasite Infections in a Rural Community of Rio de Janeiro (Brazil): Prevalence and Genetic Diversity of *Blastocystis* Subtypes. *PLoS ONE* **2018**, *13*, e0193860. [CrossRef] [PubMed]
79. Abdulsalam, A.M.; Ithoi, I.; Al-Mekhlafi, H.M.; Khan, A.H.; Ahmed, A.; Surin, J.; Mak, J.W. Prevalence, Predictors and Clinical Significance of Blastocystis spp. in Sebha, Libya. *Parasites Vectors* **2013**, *6*, 86. [CrossRef]
80. Abe, N. Molecular and Phylogenetic Analysis of *Blastocystis* Isolates from Various Hosts. *Vet. Parasitol.* **2004**, *120*, 235–242. [CrossRef]
81. Noradilah, S.A.; Moktar, N.; Anuar, T.S.; Lee, I.L.; Salleh, F.M.; Manap, S.N.A.A.; Mohtar, N.S.H.M.; Azrul, S.M.; Abdullah, W.O.; Nordin, A.; et al. Molecular Epidemiology of Blastocystosis in Malaysia: Does Seasonal Variation Play an Important Role in Determining the Distribution and Risk Factors of *Blastocystis* Subtype Infections in the Aboriginal Community? *Parasites Vectors* **2017**, *10*, 360. [CrossRef]
82. Maloney, J.G.; Lombard, J.E.; Urie, N.J.; Shivley, C.B.; Santin, M. Zoonotic and Genetically Diverse Subtypes of *Blastocystis* in US Pre-Weaned Dairy Heifer Calves. *Parasitol. Res.* **2019**, *118*, 575–582. [CrossRef] [PubMed]
83. Abarca, N.; Santín, M.; Ortega, S.; Maloney, J.G.; George, N.S.; Molokin, A.; Cardona, G.A.; Dashti, A.; Köster, P.C.; Bailo, B.; et al. Molecular Detection and Characterization of *Blastocystis* spp. and *Enterocytozoon bieneusi* in Cattle in Northern Spain. *Vet. Sci.* **2021**, *8*, 191. [CrossRef] [PubMed]
84. Maloney, J.G.; Molokin, A.; Santin, M. Next Generation Amplicon Sequencing Improves Detection of *Blastocystis* Mixed Subtype Infections. *Infect. Genet. Evol.* **2019**, *73*, 119–125. [CrossRef] [PubMed]
85. Rivera, W.L. Phylogenetic Analysis of *Blastocystis* Isolates from Animal and Human Hosts in the Philippines. *Vet. Parasitol.* **2008**, *156*, 178–182. [CrossRef] [PubMed]
86. Abdulsalam, A.M.; Ithoi, I.; Al-Mekhlafi, H.M.; Al-Mekhlafi, A.M.; Ahmed, A.; Surin, J. Subtype Distribution of *Blastocystis* Isolates in Sebha, Libya. *PLoS ONE* **2013**, *8*, e84372. [CrossRef] [PubMed]
87. Wylezich, C.; Caccio, S.M.; Walochnik, J.; Beer, M.; Höper, D. Untargeted Metagenomics Shows a Reliable Performance for Synchronous Detection of Parasites. *Parasitol. Res.* **2020**, *119*, 2623–2629. [CrossRef] [PubMed]
88. Maloney, J.G.; Molokin, A.; da Cunha, M.J.R.; Cury, M.C.; Santin, M. *Blastocystis* Subtype Distribution in Domestic and Captive Wild Bird Species from Brazil Using next Generation Amplicon Sequencing. *Parasite Epidemiol. Control* **2020**, *9*, e00138. [CrossRef] [PubMed]
89. Yoshikawa, H.; Yoshida, K.; Nakajima, A.; Yamanari, K.; Iwatani, S.; Kimata, I. Faecal-Oral Transmission of the Cyst Form of *Blastocystis hominis* in Rats. *Parasitol. Res.* **2004**, *94*, 391–396. [CrossRef]
90. Zhang, S.X.; Carmena, D.; Ballesteros, C.; Yang, C.L.; Chen, J.X.; Chu, Y.H.; Yu, Y.F.; Wu, X.P.; Tian, L.G.; Serrano, E. Symptomatic and Asymptomatic Protist Infections in Hospital in Patients in Southwestern China. *Pathogens* **2021**, *10*, 684. [CrossRef]
91. Noël, C.; Dufernez, F.; Gerbod, D.; Edgcomb, V.P.; Delgado-Viscogliosi, P.; Ho, L.C.; Singh, M.; Wintjens, R.; Sogin, M.L.; Capron, M.; et al. Molecular Phylogenies of *Blastocystis* Isolates from Different Hosts: Implications for Genetic Diversity, Identification of Species, and Zoonosis. *J. Clin. Microbiol.* **2005**, *43*, 348–355. [CrossRef]
92. Noël, C.; Peyronnet, C.; Gerbod, D.; Edgcomb, V.P.; Delgado-Viscogliosi, P.; Sogin, M.L.; Capron, M.; Viscogliosi, E.; Zenner, L. Phylogenetic Analysis of *Blastocystis* Isolates from Different Hosts Based on the Comparison of Small-Subunit RRNA Gene Sequences. *Mol. Biochem. Parasitol.* **2003**, *126*, 119–123. [CrossRef] [PubMed]
93. Silberman, J.D.; Sogin, M.L.; Leipe, D.D.; Clark, C.G. Human Parasite Finds Taxonomic Home. *Nature* **1996**, *380*, 398. [CrossRef] [PubMed]
94. Yoshikawa, H.; Nagano, I.; Wu, Z.; Yap, E.H.; Singh, M.; Takahashi, Y. Genomic Polymorphism among *Blastocystis hominis* Strains and Development of Subtype-Specific Diagnostic Primers. *Mol. Cell. Probes* **1998**, *12*, 153–159. [CrossRef] [PubMed]
95. Calero-Bernal, R.; Santín, M.; Maloney, J.G.; Martín-Pérez, M.; Habela, M.A.; Fernández-García, J.L.; Figueiredo, A.; Nájera, F.; Palacios, M.J.; Mateo, M.; et al. *Blastocystis* spp. Subtype Diversity in Wild Carnivore Species from Spain. *J. Eukaryot. Microbiol.* **2020**, *67*, 273–278. [CrossRef] [PubMed]
96. Moreno-Mesonero, L.; Amorós, I.; Moreno, Y.; Alonso, J.L. Simultaneous Detection of Less Frequent Waterborne Parasitic Protozoa in Reused Wastewater Using Amplicon Sequencing and QPCR Techniques. *J. Environ. Manag.* **2022**, *314*, 115029. [CrossRef] [PubMed]

97. Köster, P.C.; Dashti, A.; Bailo, B.; Muadica, A.S.; Maloney, J.G.; Santín, M.; Chicharro, C.; Miguel, S.; Nieto, F.J.; Cano-Terriza, D.; et al. Occurrence and Genetic Diversity of Protist Parasites in Captive Non-Human Primates, Zookeepers, and Free-Living Sympatric Rats in the Córdoba Zoo Conservation Centre, Southern Spain. *Animals* **2021**, *11*, 700. [CrossRef] [PubMed]
98. Meloni, D.; Poirier, P.; Mantini, C.; Noël, C.; Gantois, N.; Wawrzyniak, I.; Delbac, F.; Chabé, M.; Delhaes, L.; Dei-Cas, E.; et al. Mixed Human Intra- and Inter-Subtype Infections with the Parasite *Blastocystis* spp. *Parasitol. Int.* **2012**, *61*, 719–722. [CrossRef] [PubMed]
99. Stensvold, C.R. Blastocystis: Genetic Diversity and Molecular Methods for Diagnosis and Epidemiology. *Trop. Parasitol.* **2013**, *3*, 26. [CrossRef]
100. Alfellani, M.A.; Jacob, A.S.; Perea, N.O.; Krecek, R.C.; Taner-Mulla, D.; Verweij, J.J.; Levecke, B.; Tannich, E.; Clark, C.G.; Stensvold, C.R. Genetic Diversity of *Blastocystis* in Livestock and Zoo Animals. *Protist* **2013**, *164*, 497–509. [CrossRef]
101. Scanlan, P.D.; Stensvold, C.R.; Cotter, P.D. Development and Application of a *Blastocystis* Subtype-Specific PCR Assay Reveals That Mixed-Subtype Infections Are Common in a Healthy Human Population. *Appl. Environ. Microbiol.* **2015**, *81*, 4071–4076. [CrossRef]
102. Rezaei Riabi, T.; Mirjalali, H.; Haghighi, A.; Rostami Nejad, M.; Pourhoseingholi, M.A.; Poirier, P.; Delbac, F.; Wawrzyniak, I.; Zali, M.R. Genetic Diversity Analysis of *Blastocystis* Subtypes from Both Symptomatic and Asymptomatic Subjects Using a Barcoding Region from the 18S RRNA Gene. *Infect. Genet. Evol.* **2018**, *61*, 119–126. [CrossRef] [PubMed]
103. Stensvold, C.R.; Alfellani, M.; Clark, C.G. Levels of Genetic Diversity Vary Dramatically between *Blastocystis* Subtypes. *Infect. Genet. Evol.* **2012**, *12*, 263–273. [CrossRef]
104. Alfellani, M.A.; Jacob, A.S.; Perea, N.O.; Krecek, R.C.; Taner-Mulla, D.; Verweij, J.J.; Levecke, B.; Tannich, E.; Clark, C.G.; Stensvold, C.R. Diversity and Distribution of *Blastocystis* spp. Subtypes in Non-Human Primates. *Parasitology* **2013**, *140*, 966–971. [CrossRef] [PubMed]
105. Ramírez, J.D.; Sánchez, A.; Hernández, C.; Flórez, C.; Bernal, M.C.; Giraldo, J.C.; Reyes, P.; López, M.C.; García, L.; Cooper, P.J.; et al. Geographic Distribution of Human *Blastocystis* Subtypes in South America. *Infect. Genet. Evol.* **2016**, *41*, 32–35. [CrossRef] [PubMed]
106. Rahimi, H.M.; Karamati, S.A.; Nemati, S.; Mirjalali, H.; Zali, M.R. Molecular Identification, Subtypes Distribution, and Alleles Discrimination of *Blastocystis* spp., Isolated from Immunocompromised Subjects in Iran. *Iran. J. Parasitol.* **2022**, *17*, 184–193. [CrossRef]
107. Candela, E.; Goizueta, C.; Periago, M.V.; Antoli, C.M. Prevalence of Intestinal Parasites and Molecular Characterization of *Giardia intestinalis*, *Blastocystis* spp. and *Entamoeba histolytica* in the Village of Fortín Mbororé (Puerto Iguazú, Misiones, Argentina). *Parasites Vectors* **2021**, *14*, 510. [CrossRef] [PubMed]
108. Öncü Öner, T.; Karabey, M.; Can, H.; Değirmenci Döşkaya, A.; Karakavuk, M.; Gül, A.; Köseoğlu, A.E.; Döşkaya, M.; Ün, C.; Gürüz, A.Y.; et al. Molecular Investigation of *Blastocystis* spp. and Its Subtypes in Cancer Patients under Chemotherapy in Aegean Region, Turkey. *Acta Trop.* **2022**, *233*, 106577. [CrossRef]
109. Mohammadpour, I.; Bozorg-Ghalati, F.; Gazzonis, A.L.; Manfredi, M.T.; Motazedian, M.H.; Mohammadpour, N. First Molecular Subtyping and Phylogeny of *Blastocystis* spp. Isolated from Domestic and Synanthropic Animals (Dogs, Cats and Brown Rats) in Southern Iran. *Parasites Vectors* **2020**, *13*, 365. [CrossRef] [PubMed]
110. Muadica, A.S.; Köster, P.C.; Dashti, A.; Bailo, B.; Hernández-De-mingo, M.; Balasegaram, S.; Carmena, D. Molecular Diversity of *Giardia duodenalis*, *Cryptosporidium* spp., and *Blastocystis* spp. in Symptomatic and Asymptomatic Schoolchildren in Zambézia Province (Mozambique). *Pathogens* **2021**, *10*, 255. [CrossRef]
111. Mahdavi, F.; Asghari, A.; Shahabi, S.; Shamsi, L.; Soltani-Jazi, F.; Sadrebazzaz, A.; Shams, M. Distribution, Genetic Diversity, and Zoonotic Significance of *Blastocystis* Subtypes in Pet Dogs. *Comp. Immunol. Microbiol. Infect. Dis.* **2022**, *88*, 101848. [CrossRef]
112. Wibowo, M.C.; Yang, Z.; Borry, M.; Hübner, A.; Huang, K.D.; Tierney, B.T.; Zimmerman, S.; Barajas-Olmos, F.; Contreras-Cubas, C.; García-Ortiz, H.; et al. Reconstruction of Ancient Microbial Genomes from the Human Gut. *Nature* **2021**, *594*, 234–239. [CrossRef] [PubMed]

microorganisms

MDPI

Article

Synthetic Peptides Selected by Immunoinformatics as Potential Tools for the Specific Diagnosis of Canine Visceral Leishmaniasis

Gabriel Moreira [1,†], Rodrigo Maia [1,†], Nathália Soares [1], Thais Ostolin [1], Wendel Coura-Vital [2,3], Rodrigo Aguiar-Soares [1,4], Jeronimo Ruiz [5], Daniela Resende [5], Rory de Brito [1], Alexandre Reis [1,2,3,6] and Bruno Roatt [1,3,7,*]

[1] Laboratório de Imunopatologia, Núcleo de Pesquisas em Ciências Biológicas/NUPEB, Universidade Federal de Ouro Preto, Ouro Preto 35400-000, MG, Brazil; gabriel.lucas@aluno.ufop.edu.br (G.M.); rodrigodacostamaia@hotmail.com (R.M.); nathalias.vet@gmail.com (N.S.); thais.ostolin@gmail.com (T.O.); rodrigo.soares@ufop.edu.br (R.A.-S.); rorybrito@gmail.com (R.d.B.); alexreis@ufop.edu.br (A.R.)

[2] Departamento de Análises Clínicas, Escola de Farmácia, Universidade Federal de Ouro Preto, Ouro Preto 35400-000, MG, Brazil; wendelcoura@ufop.edu.br

[3] Programa de Pós-Graduação em Ciências Biológicas, Núcleo de Pesquisas em Ciências Biológicas/NUPEB, Universidade Federal de Ouro Preto, Ouro Preto 35400-000, MG, Brazil

[4] Programa de Pós-Graduação em Biotecnologia, Núcleo de Pesquisas em Ciências Biológicas/NUPEB, Universidade Federal de Ouro Preto, Ouro Preto 35400-000, MG, Brazil

[5] Grupo de Informática de Biossistemas e Genômica, Programa de Pós-Graduação em Ciências da Saúde, Instituto René Rachou, Fiocruz Minas, Belo Horizonte 30190-002, MG, Brazil; jeronimo@cpqrr.fiocruz.br (J.R.); dani.melo.resende@gmail.com (D.R.)

[6] Instituto Nacional de Ciência e Tecnologia em Doenças Tropicais, INCT-DT, Salvador 40296-710, BA, Brazil

[7] Departamento de Ciências Biológicas, Instituto de Ciências Exatas e Biológicas, Universidade Federal de Ouro Preto, Ouro Preto 35400-000, MG, Brazil

* Correspondence: bmroatt@gmail.com or roatt@ufop.edu.br; Tel.: +55-21-31-3559-1694; Fax: +55-21-31-3559-1680

† These authors contributed equally to this work.

Abstract: Diagnosing canine visceral leishmaniasis (CVL) in Brazil faces challenges due to the limitations regarding the sensitivity and specificity of the current diagnostic protocol. Therefore, it is urgent to map new antigens or enhance the existing ones for future diagnostic techniques. Immunoinformatic tools are promising in the identification of new potential epitopes or antigen candidates. In this study, we evaluated peptides selected by epitope prediction for CVL serodiagnosis in ELISA assays. Ten B-cell epitopes were immunogenic in silico, but two peptides (peptides No. 45 and No. 48) showed the best performance in vitro. The selected peptides, both individually and in combination, were highly diagnostically accurate, with sensitivities ranging from 86.4% to 100% and with a specificity of approximately 90%. We observed that the combination of peptides showed better performance when compared to peptide alone, by detecting all asymptomatic dogs, showing lower cross-reactivity in sera from dogs with other canine infections, and did not detect vaccinated animals. Moreover, our data indicate the potential use of immunoinformatic tools associated with ELISA assays for the selection and evaluation of potential new targets, such as peptides, applied to the diagnosis of CVL.

Keywords: canine visceral leishmaniasis; serological diagnosis; peptides-based ELISA; immunoinformatic

1. Introduction

Leishmaniasis belongs to a complex group of diseases caused by a parasite from the *Leishmania* genus, leading to a broad spectrum of clinical presentations [1]. *Leishmania infantum* is one of the primary causes of visceral leishmaniasis (VL), which is fatal in 95% of untreated cases [2]. It is estimated that 50,000–90,000 new VL cases occur annually worldwide, with only 25–45% of these cases being reported to the WHO. In Latin America, VL has been declared endemic in 12 countries, with 63,331 cases being registered between 2001 and 2018 [3]. In Brazil, VL remains a serious public health problem, with 3.466 new human cases reported in 2018, which corresponds to 97% of the total cases in America [3].

Particularly in Brazil, the Leishmaniasis Control and Surveillance Program includes the early diagnosis and treatment of human cases, control of insect vectors, and the identification and culling of seropositive-infected dogs [4]. The management of infected dogs is considered an essential component in the control of VL, given the zoonotic profile of the disease in Brazil [5], where dogs are regarded as urban hosts of the disease, representing an incidence of 5.4 per 1000 dogs-months and a prevalence of 8.1% in endemic areas [6].

Regarding the serodiagnosis of canine visceral leishmaniasis (CVL) in Brazil, the standard protocol recommended by the Brazilian Ministry of Health involves two sequential tests, an initial dual path platform (DPP®) test, which is an immunochromatographic test employed in the screening of seropositive dogs, followed by a sequential test, which is an enzyme-linked immunosorbent assay (ELISA) that is considered as the confirmatory test (Brazil. 2014). However, several studies have indicated that this current protocol is inefficient, demonstrating a lower diagnostic accuracy in asymptomatic dogs and those in the initial stages of the infection [7,8]. This represents a serious issue because up to 85% of *L. infantum*-infected dogs may be asymptomatic in endemic areas, acting as a reservoir for the transmission of *L. infantum* between sand flies and humans [9]. Moreover, the tests have yielded false-positive results for dogs infected with other pathogens such as *Ehrlichia canis* and *Babesia canis* [10,11], which could lead to unnecessary euthanasia of noninfected dogs. In addition, it has been reported that the vaccination of dogs might lead to seroconversion, which can directly impact the diagnosis of canine disease.

Despite these limitations, serological methods represent the most practical and flexible tools for epidemiological research and CVL diagnosis. Thus, reinforcing the importance of developing innovative alternatives and identifying new antigens. In this context, immunoinformatics presents itself as a rational tool, given its potential use as a catalyst in the prospecting processes of candidate components of diagnostic tests and vaccines [12,13]. In this study, we propose the use of an immunoinformatic approach to select B-cell epitopes as antigen candidates to be used in CVL serodiagnosis.

2. Materials and Methods

2.1. In Silico Strategies and Peptide Selection

For a better understanding, a flow chart with an experimental design is shown in Figure 1. The strategy is detailed and explained in the next topics. The peptide sequences were retrieved from the previously established *L. infantum* proteome relational database proposed by Brito et al. 2017 [14]. Using this database, the search simultaneously considered the highest values of 3 B-cell epitope predictor algorithms (AAP12, BCPred12, and BepiPred) [15–18]; in addition, all intracellular peptides were discarded after considering results obtained from cell location predictor algorithms (WoLF PSORT, Sigcleave, TargetP, and TMHMM) [19–22]. After that, to elucidate the interactions and metabolic pathways of proteins containing the peptides in the previous step, we used the STRING v.11 algorithms [23,24]; *Leishmania* was established as a reference organism, and text prospecting, experiments, database, co-expression, neighborhood, genetic fusion, and co-occurrence were used as active sources of interaction, with a mean confidence interval of 0.400, and the design of PPI networks was made with the aid of the Cytoscape program, with the addition of information obtained from KEGG [25]. Then, selected peptide sequences were compared against *Leishmania donovani, Leishmania braziliensis, Leishmania major, Trypanosoma cruzi, Ehrlichia,* and *Babesia* organisms using the Basic Local Alignment Search Tool (BLASTp) [26]. Furthermore, this same algorithm was used to eliminate proteins that, despite fulfilling the previous criteria, presented a similarity greater than 60% to humans, dogs, and mice, thus decreasing the possibility of obtaining proteins already present in these three organisms.

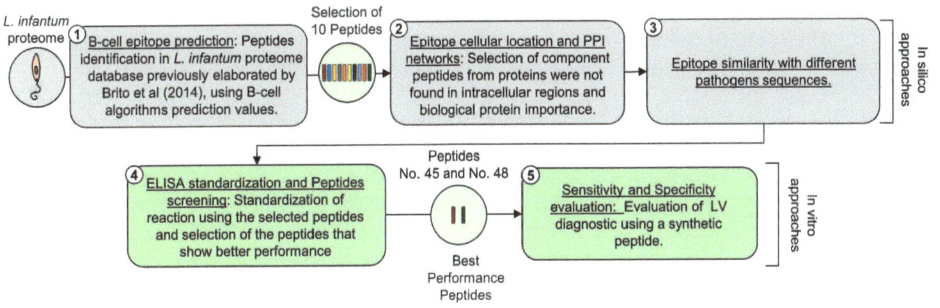

Figure 1. Flow chart with an experimental design for the selection and application of synthetic peptides, based on the epitope prediction approach, in the serological diagnosis of CVL [14].

2.2. Peptide Synthesis

Linear peptides from *L. infantum* (15–17 mer) were synthesized with a purity higher than 95% obtained through purification by high-performance liquid chromatography by the Genscript Co., Ltd. (Piscataway, NJ, USA). Once the peptides arrived at the Immunopathology Laboratory, they were stored in an ultra-freezer at −80 °C until use, when they were resuspended (1 mg/mL) in dimethyl sulfoxide (DMSO).

2.3. Enzyme-Linked Immunosorbent Assay

To identify the synthetic peptides that offered the highest performance, a peptide-based ELISA was performed, and the conditions were standardized for all 10 peptides. Sera of 10 animals infected with *L. infantum* and 10 healthy animals were used individually for standardization and peptide screening. Flat-bottom polystyrene plates (Nunc MaxiSorp®, San Diego, CA, USA) were sensitized with each peptide, at a concentration of 0.25 µg per well, each well containing 100 µL of carbonate-bicarbonate buffer (pH 9.6), after which they were incubated at 4 °C overnight. After incubation, 4 consecutive washes were carried out with a wash solution composed of PBS (pH 7.2) added with 0.05% Tween 20, to remove antigen excess. Then, using a wash solution added with 5% BSA, possible free sites were blocked. In this step, each well was filled with 100 µL of this blocking solution for 2 h at 37 °C. This was followed by another step of 4 consecutive washes, and, after that, the diluted samples (1:600) were added after dilution in blocking solution (100 µL/well) and incubated again at 37 °C for 1 h. After incubation and 4 consecutive washes, the plates were incubated for 1 h at 37 °C with peroxidase-conjugated sheep anti-dog IgG (Bethyl Laboratories, Inc., Montgomery, TX, USA) diluted 1:16,000 in a wash solution (100 µL/well). The reactions were carried out using 3,3′,5,5′-Tetramethylbenzidine as the substrate for 20 min in the dark, with subsequent interruption of the reaction using 30 µL of 2.5 M H_2SO_4, followed by analysis in a spectrophotometer (ELX800 Biotek Instruments, Winooski, VT, USA) at 450 nm. After the screening, two peptides were selected as presenting the best results after the assay with the serum samples. The sequence of the peptides No. 45 (Pep45) and No. 48 (Pep48), described in the current work, is registered at the Instituto Nacional da Propriedade Industrial (Brazil) under patent number BR 1020230118887, deposited on 15 June 2023.

2.4. Preparation of L. infantum Soluble Antigenic Extract

The *L. infantum* strain MCAN/BR/2008/OP46 was used for the preparation of the *L. infantum* soluble antigenic extract. Stationary phase promastigotes of *L. braziliensis* were grown at 24 °C in liver infusion tryptose (LIT) medium supplemented with 10% fetal bovine serum (FBS, Sigma-Aldrich, Saint Louis, MO, USA), 100 U/mL penicillin, and 100 µg/mL streptomycin, at pH 7.4. The soluble *Leishmania infantum* antigen (SLiA) was prepared as described previously by Reis et al. (2006) [27].

2.5. Animal Samples

The study protocol number 083/2007 was approved by the Universidade Federal de Ouro Preto Committees of Ethics in Animal Experimentation. Serum samples from 113 dogs were selected from a serum bank at the Laboratório de Imunopatologia from the Universidade Federal de Ouro Preto, where they were stored at −20 °C. The samples were categorized into distinct groups (Figure 2). Twenty (20) samples from noninfected dogs were included as the control group (CNI; n = 20). This group was composed of five (05) sera from control dogs born in a kennel of the Federal University of Ouro Preto (Minas Gerais, Brazil) and fifteen (15) sera from control dogs from an endemic area in Brazil (Governador Valadares, Minas Gerais, Brazil). The control dogs were characterized by negative parasitological and PCR-restriction fragment length polymorphism (RFLP) results for *L. infantum* in the bone marrow and seronegative results for *Leishmania* spp. using DPP® and BioManguinhos ELISA®. The *L. infantum*-infected group of dogs (CVL; n = 37) was divided into three groups based on their clinical status: asymptomatic dogs (AD; n = 18) with no clinical signs of CVL; oligosymptomatic dogs (OD; n = 9) presenting one to three signs; and symptomatic dogs (SD; n = 10) with more than three characteristic clinical signs of VL. The characteristic signs include opaque bristles, a severe loss of weight, onychogryphosis, cutaneous lesions, apathy, and keratoconjunctivitis (Reis 2006) [27]. The CVL group was determined based on the serological reactivity of dogs in the BioManguinhos ELISA®, DPP®, and PCR-RFLP in the bone marrow results. Sera from dogs infected with *E. canis* (n = 15), *B. canis* (n = 9), or *Trypanosoma cruzi* (n = 15) were used for cross-reactivity analyses. Each infection was previously characterized using specific serology (ELISA) and PCR-positive results, and samples were confirmed to be PCR-negative in the bone marrow for *L. infantum*. In addition to the groups described above, the study also used 17 samples from dogs vaccinated with Leish-Tec® (n = 7), a commercial vaccine against CVL available in Brazil, and a potential candidate vaccine, LBSap (n = 10) (Aguiar Soares 2020) [28]. All dogs presented negative serology using DPP® and ELISA and were PCR-negative in the bone marrow for *Leishmania* spp.

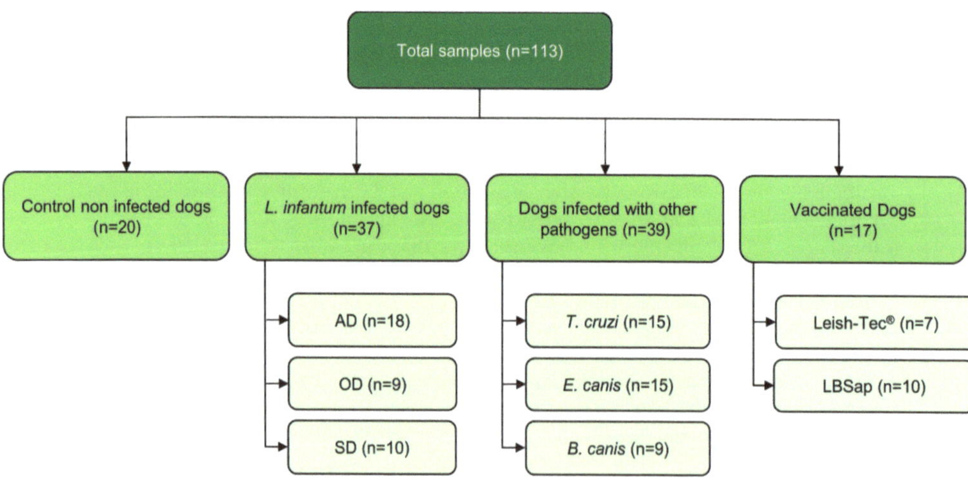

Figure 2. The experimental design employed in the serological test for peptides No. 45, No. 48, and mix. The control noninfected group (CNI), dogs from the endemic area and nonendemic area. The *L. infantum*-infected dogs (CVL) were stratified according to their statuses as asymptomatic dogs (AD), oligosymptomatic dogs (OD), and symptomatic dogs (SD). Vaccinated dogs (LBSap and Leish-Tec®) and dogs infected with other pathogens (*Trypanosoma cruzi*, *Ehrlichia canis*, and *Babesia canis*) constitute the other two groups evaluated.

2.6. Statistical Analysis

The OD cutoff assays were calculated using the receiver operating characteristic curve (ROC curve), by considering the point that yielded the highest combined value of sensitivity and specificity for each antigen. GraphPad Prism software (version 8.0 for Windows) was used to provide the area under the curve (AUC) and the ROC curve. The sensitivity, specificity, negative predictive value (NPV), positive predictive value (PPV), and accuracy were calculated according to Greenhalgh (1997).

3. Results

3.1. The Database Provides Potential Epitopes for Use in CVL Diagnosis

The database used had 8241 predicted proteins, which when submitted to the B-cell predictor algorithms, returned 47,482 epitopes according to BepiPred, 957,493 according to BCPred12, and 2,361,313 according to AAP12. A search script was then used to find the higher scores for all algorithms, resulting in peptides belonging to five proteins, which were predicted by all three predictive algorithms, and which were secreted/excreted and are not intracellular: LinJ.18.1500, LinJ.32.0970, LinJ.36.2160, LinJ.28.1850, and LinJ. 20.0350. After selection, these proteins have their biological importance analyzed using PPI networks, and all five proteins seem to relate to important cellular functions, especially in metabolic pathways. The peptide sequence, position, prediction scores, and protein ID of each peptide selected are shown in Table 1. The BLASTp analysis was performed to evaluate the identity and similarity between the selected peptides and non-redundant protein sequences from *L. donovani*, *L. braziliensis*, *L. major*, *T. cruzi*, *Ehrlichia*, and *Babesia* organisms (Table 2). It is possible to note that for peptides No. 50 and No. 52 to No. 55, no similarity with *Ehrlichia* was found. Furthermore, all peptides showed 100% identity and a low e-value for *L. donovani*. For *L. braziliensis*, the identity values were lower, not reaching 100% for any peptide. The identity for *L. major*, in general, was higher when compared to *L. braziliensis*, and it can still be observed that peptide No. 48 had the highest e-value among all *Leishmania* species and an identity lower than 90%. Regarding sequence similarity with other genera of pathogens, it was observed that peptides No. 46, No. 47, No. 48, No. 50, No. 51, and No. 52 had a high e-value and/or low identity.

Table 1. B-cell epitopes selected with the highest prediction scores in the databank. The protein identification, peptide sequence, position in protein sequence, prediction scores, and reactivity index are shown.

Protein ID	Peptide ID	Peptide Sequence	Position	Prediction Scores			Reactivity Index
				AAP12	BepiPred	BCPred12	
LinJ.18.1500 (XP_001464963)	45	VDPNFQFFHLPVLMF	692–706	0.9	0.8	0.8	7.94
	46	EGYSSQYYENSWFHRL	763–777	0.9	0.97	0.8	7.80
LinJ.32.0970 (XP_001467777)	47	WAPISEQKGTTYPTTPNGLPV	493–507	1	1.28	1	6.17
LinJ.36.2160 (XP_001469796)	48	FALIRQGFESFPPTPKT	374–390	1	1.67	0.98	7.51
LinJ.28.1850 (XP_001470212)	50	LAVQPAPSTSDAAGA	288–302	0.9	1.47	0.98	6.26
	51	AYQETPESERAELPP	115–129	0.9	1.25	0.98	5.36
	52	LPKGPSVPTLPYQEA	443–457	0.9	1.1	0.99	7.19
LinJ.19.0350 (XP_001464998)	53	SRRPPPLDPEEPEKV	171–185	1	1.74	1	3.93
	54	GLGEEEKEVRQTLRDLR	304–320	1	1	0.96	4.83
	55	CVERITPRVRDRRASYKQS	262–276	1	0.61	1	7.91

Table 2. Similarity values obtained by BLASTp for each selected B-cell epitope when confronted with the proteome of other species of parasites. Asterisks (*) indicate the absence of a similar sequence; identity values are indicated in percentages (%).

Peptide ID	Similarity											
	L. donovani		*L. braziliensis*		*L. major*		*T. cruzi*		*Ehrlichia*		*Babesia*	
	E-Value	Identity	E-Value	Identity	E-Value	Identity	E-Value	Identity	E-Value	Identity	E-Value	Identity
45	8×10^{-11}	100	1×10^{-8}	92.86	1×10^{-9}	93.33	2×10^{-8}	86.67	23	85.71	14	77.78
46	6×10^{-12}	100	18	62.5	9×10^{-9}	87.5	19	66.67	83	85.71	13	100
47	1×10^{-14}	100	2×10^{-11}	85	1×10^{-14}	100	48	61.11	69	76.92	38	71.43
48	2×10^{-11}	100	2×10^{-7}	87.5	44	85.71	22	75	11	80	44	100
50	2×10^{-7}	100	28	69.23	3×10^{-5}	92.86	61	78.57	*	*	57	90
51	3×10^{-9}	100	3×10^{-7}	92.86	3×10^{-9}	100	16	77.78	33	75	11	53.33
52	7×10^{-9}	100	2×10^{-7}	93.33	7×10^{-9}	100	40	80	*	*	66	64.29
53	2×10^{-9}	100	1×10^{-8}	93.33	2×10^{-9}	100	2×10^{-5}	80	*	*	16	85.71
54	6×10^{-11}	100	3×10^{-9}	94.12	8×10^{-9}	94.12	2×10^{-7}	88.24	*	*	39	70
55	9×10^{-14}	100	2×10^{-11}	89.47	1×10^{-12}	94.74	1×10^{-10}	89.47	*	*	21	77.78

3.2. All Peptides Were Able to Differentiate Infected and Uninfected Animals during the Standardization Step of the ELISA Reaction

After standardization, and to evaluate the peptides as immunodiagnostic using individual serum samples, ELISA reactions were performed using the 20 previously characterized samples from 10 infected animals and 10 uninfected animals. All peptides showed the ability to distinguish animals infected by *L. infantum*. Using the average of the values obtained in the individual ELISA, the reactivity index showed that each peptide was able to differentiate infected from uninfected animals by 7.94, 7.80, 6.17, 7.51, 6.26, 5.36, 7.19, 3.93, 4.83, and 7.91 times for peptides No. 45–No. 55, respectively (Table 1). By observing this index, it is possible to notice that all peptides significantly separated infected animals from animals not infected by *L. infantum*, with peptide No. 45 being the highest and peptide No. 53 having the smallest difference between optical density values.

3.3. The Peptides No. 45, No. 48, and the Combination of Both Showed the Best Capacity to Distinguish L. infantum-Infected Dogs with Different Clinical Forms from the Noninfected Dogs

Considering the results obtained in the ELISA standardization, our selection indicators for the next assays were a peptide with the highest reactivity level and another peptide that showed the lowest OD mean of the negative control. In this sense, the peptides No. 45 (VDPNFQFFHLPVLMF) and No. 48 (FALIRQGFESFPPTPKT) were selected for the next step. We observed that the peptides No. 45 (Pep45) and No. 48 (Pep48), in an isolated form as well as in combination (mix), demonstrated an increased efficiency in differentiating infected dogs from the noninfected, as shown in Figure 3a. Thus, SLiA, No. Pep45, No. Pep48, and mix presented 35/37 (94.6%), 32/37 (86.5%), 37/37 (100%), and 36/37 (97.3%) of the true-positive results, respectively. Therefore, peptide No. 48 and peptide mix were capable of detecting all the asymptomatic dogs (18/18; 100%), oligosymptomatic dogs (9/9; 100% using peptide No. 48, and 8/9; 88.8% using peptide mix), and all symptomatic dogs (10/10) (Figure 3b). The general performance of peptide No. 48 and the peptide mix was superior compared to that of peptide No. 45 and SLiA, demonstrating higher positive (94.8% and 97.3%) and negative predictive values (100% and 95%) and accuracy (both assays presented 96.5%).

3.4. The Selected Peptides Present Low Cross-Reactivity with Immunoglobulins from Dogs Infected with Other Canine Pathogens

The ELISA assays showed low reactivity when serum samples from *T. cruzi-*, *E. canis-*, and *B. canis*-infected dogs were tested (Figure 4). No cross-reactivity was observed in *T. cruzi*-infected animals in all tests, in contrast to that with SLiA, which presented 57% positive results for these samples. We observed that the peptide mix showed better results when compared to the results shown by isolated peptides and SLiA. As shown in Figure 4c, the peptide mix presented superior sensitivity (97.2%), specificity (84.7%), NPV

(98%), PPV (80%), and AUC (0.9794). Moreover, an accuracy analysis of this approach indicated an excellent performance (89.5%).

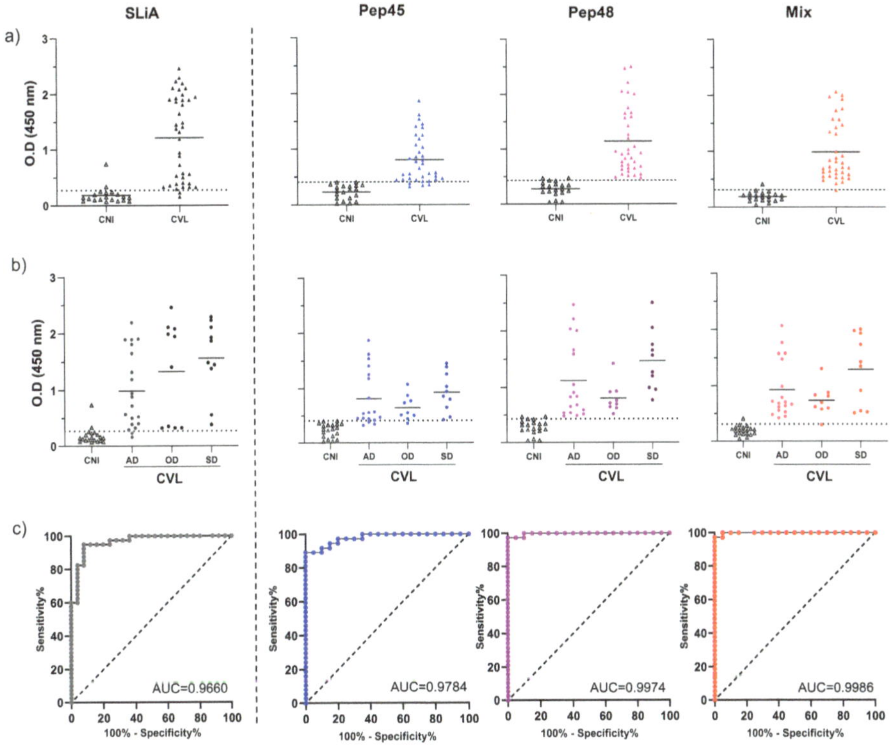

d)

Antigen	Results (% [95% CI])[α] Sensitivity	Specificity	NPV	PPV	Accuracy
SLiA	94.6 (82.3 - 98.5)	90 (69.9 - 97.2)	90 (69.9 - 97.2)	94.5 (82.3 - 98.5)	92.8 (83.3 - 97.2)
Pep 45	86.4 (72 - 94.0)	100 (83.9 - 100)	80 (60.8- 91.1)	100 (89.2 - 100)	91.2 (81 - 96.2)
Pep 48	100 (90.6 - 100)	90 (69.9 - 97.2)	100 (82.4 - 100)	94.8 (83.1 - 98.5)	96.5 (88.1 - 99)
Mix	97.2 (86.1 - 99.5)	95 (76.3 - 99.1)	95 (76.3 - 99.1)	97.3 (86.1 - 99.5)	96.5 (88.1 - 99)

[α] CI, confidence interval.

Figure 3. Performance of the peptides No. 45, No. 48, and mix in the discrimination of noninfected and *L. infantum*-infected dogs presenting different clinical forms. (**a**) Distribution of the individual optical density results of the control noninfected dogs (CNI) and *L. infantum*-infected dogs (CVL) tested by the peptides No. 45, No. 48, mix, and soluble *Leishmania infantum* antigen (SLiA). (**b**) Distribution of the individual optical density results of the control noninfected dogs (CNI) and *L. infantum*-infected dogs according to clinical forms. The CVL dogs were stratified according to their clinical statuses as asymptomatic dogs (AD), oligosymptomatic dogs (OD), and symptomatic dogs (SD). The dotted lines within the graphs represent the cut-offs calculated by the ROC curve between the negative and positive results of SLiA (OD = 0.273), Pep45 (OD = 0.41), Pep48 (OD = 0.431), and mix (OD = 0.331). (**c**) The ROC curves were constructed with the results of the control noninfected, and *L. infantum*-infected control serum samples tested by each assay. (**d**) Results of sensitivity, specificity, negative predictive values (NPV), positive predictive values (PPV), and accuracy from the canine visceral leishmaniasis positive and negative animals tested by the peptides No. 45, No. 48, the mix, and SLiA.

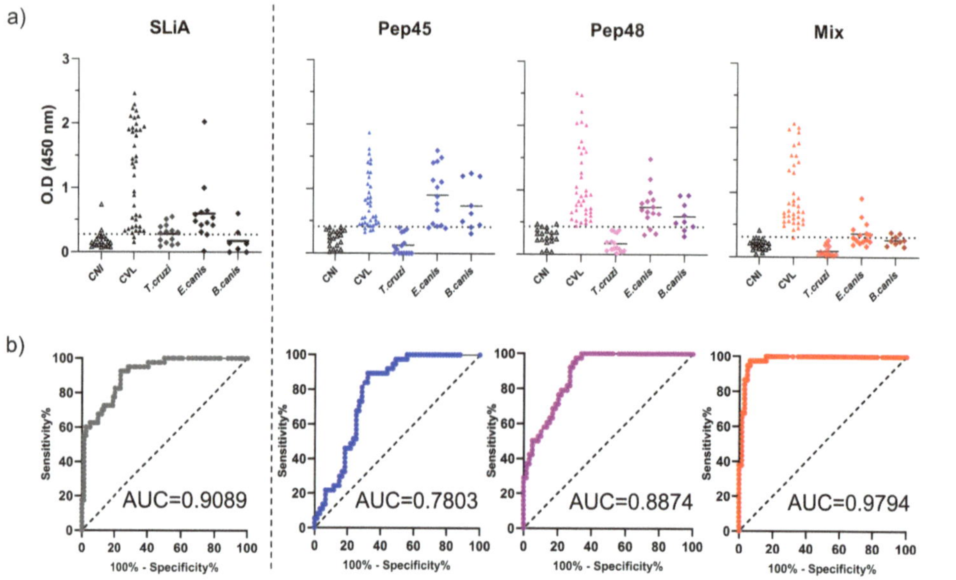

c)

Antigen	Results (% [95% CI])[a]				
	Sensitivity	Specificity	NPV	PPV	Accuracy
SLiA	94.6 (82.3 - 98.5)	50 (37.1 - 62.8)	93.1 (78 - 98)	56.4 (44.1 - 68)	68.1 (58 - 76.8)
Pep 45	86.4 (72 - 94)	66.1 (53.3 - 76.8)	88.6 (76 - 95)	61.5 (47.9 - 73.5)	74 (64.4 - 81.7)
Pep 48	100 (90.6 - 100)	64.4 (51.6 - 75.4)	100 (90.8 - 100)	63.8 (50.9 - 74.9)	78.1 (68.8 - 85.2)
Mix	97.2 (86.1 - 99.5)	84.7 (73.5 - 91.7)	98 (89.7 - 99.6)	80 (66.1 - 89.1)	89.5 (81.8 - 94.2)

[a] CI,confidence interval.

Figure 4. Cross-reactivity of the peptides No. 45, No. 48, and mix with samples from dogs infected with other pathogens of medical and diagnostic importance. (**a**) Distribution of the individual optical density results from serum samples from dogs infected with *Leishmania infantum*, *Trypanosoma cruzi*, *Ehrlichia canis*, and *Babesia canis* using the peptides No. 45, No. 48, mix, and soluble *Leishmania infantum* antigen (SLiA). The dotted lines within the graphs represent the cut-offs calculated by the ROC curve between the negative and positive results of SLiA (OD = 0.273), Pep45 (OD = 0.41), Pep48 (OD = 0.431), and mix (OD = 0.331). (**b**) The ROC curves were constructed with the results of serum samples from control noninfected dogs, *L. infantum*-infected dogs, and dogs infected with other pathogens tested by each assay. (**c**) Results of sensitivity, specificity, negative predictive values (NPV), positive predictive values (PPV), and accuracy from noninfected control dogs, *L.infantum*-infected dogs, and dogs with other canine pathogens tested by the peptides No. 45, No. 48, mix, and SLiA.

3.5. Peptides No. 45, No. 48, and Combination Present No Reactivity with Immunoglobulins from the Serum Samples of Vaccinated Dogs

We evaluated the serologic reactivity of serum samples from dogs vaccinated with two vaccines against CVL, Leish-Tec®, a commercially available vaccine in Brazil, and LBSap, a potential vaccine candidate that will be commercially available (Figure 5a). We observed that the isolated peptides demonstrated low reactivity for these samples, presenting higher sensitivity, specificity, predictive values, and accuracy compared to that shown by SLiA. However, our data indicate that the mixture of peptides demonstrated a better performance (Figure 5b), high sensitivity (97.2%), high specificity (96.8%), and high accuracy (97.1%), as none of the vaccinated dogs presented with any reaction to the peptide mix, whereas 70.5% of vaccinated dogs were detected positive by SLiA. These results indicated that no reactivity or false-positive diagnosis was observed for the vaccinated dogs when the peptide mix was employed.

a)

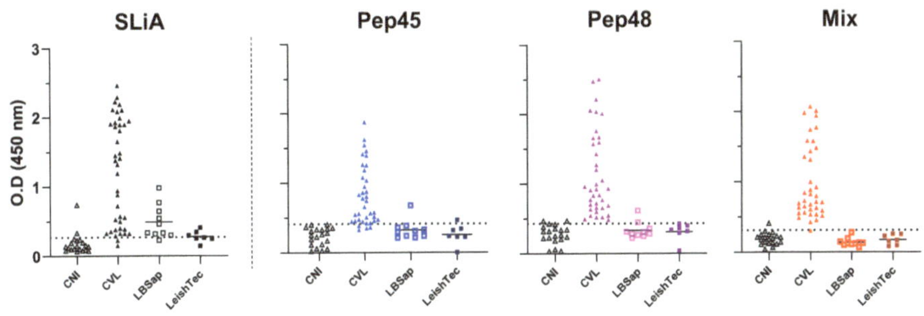

b)

Antigen	Results (% [95% CI])[α]				
	Sensitivity	Specificity	NPV	PPV	Accuracy
SLiA	94.6 (82.3 - 98.5)	43.7 (28.1 - 60.6)	87.5 (63.9 - 96.5)	66 (52.6 - 77.3)	71 (59.4 - 80.3)
Pep 45	86.4 (72 - 94)	93.7 (79.8 - 98.2)	85.7 (70.6 - 93.7)	94.1 (80.9 - 98.3)	89.8 (80.5 - 95)
Pep 48	100 (90.6 - 100)	87.5 (71.9 - 95)	100 (87.9 - 100)	90.2 (77.4 - 96.1)	94.2 (86 - 97.7)
Mix	97.2 (86.1 - 99.5)	96.8 (84.2 - 99.4)	96.8 (84.2 - 99.4)	97.3 (86.1 - 99.5)	97.1 (90 - 99.2)

[α] CI,confidence interval.

Figure 5. Performance of the peptides No. 45, No. 48, and mix in the discrimination of *L. infantum* infected dogs from vaccinated dogs. (**a**) Distribution of the individual optical density results from serum samples from dogs infected with *L. infantum* and vaccinated dogs using the peptides No. 45, No. 48, mix, and soluble *Leishmania infantum* antigen (SLiA). The dotted lines within the graphs represent the cut-offs calculated by the ROC curve between the negative and positive results of SLiA (OD = 0.273), Pep45 (OD = 0.41), Pep48 (OD = 0.431), and mix (OD = 0.331). (**b**) Results of sensitivity, specificity, negative predictive values (NPV), positive predictive values (PPV), and accuracy from noninfected control dogs, *L. infantum* infected dogs, and vaccinated dogs tested by the peptides No. 45, No. 48, mix, and SLiA.

4. Discussion

Bioinformatics has already shown promise in identifying potential antigens for vaccines or diagnosing infectious parasitic diseases, presenting a reduction in the time needed for this exploration and in the costs, in addition to gaining in diagnostic performance, considering the increase in specificity, sensitivity, and reproducibility indexes [12,29]. In this study, bioinformatics tools were used to select linear epitopes for B-cells based on the combination of three prediction algorithms, BepiPred, AAP12, and BCPred12, based on the assumption that the combination of different algorithms has greater accuracy, especially when dealing with protozoa [30]. Our selected peptides show good score values, including algorithms that are considered more restrictive, which is the case for BepiPred [31,32]. Furthermore, using the PPI networks, it was possible to evidence the biological importance of the proteins that contained the selected peptides, as they can participate in the modulation of cellular activities like virulence, metabolism regulation, or latency behavior in some species of the parasite [33–36]. The similarity assessment was performed to obtain a previous result of possible cross-reactions between pathogens that can be found in a co-infection situation [10,37]. In this context, a high similarity can be observed with members of the same genus, such as *L. infantum* and *L. donovani*. However, one must pay attention to the similarity in the treatment of the disease when caused by species of the same genus, in addition to the geographic distribution of certain species, such as *L. donovani*, which is not considered prevalent in the New World [38,39]. These observations, added to the lower similarity noted between different genera of parasites, made all the peptides selected by the B-cell prediction algorithms promising for further steps. Thus, all ten selected peptides were used for the standardization steps of the ELISA technique and the assessment of diagnostic capacity.

All 10 peptides showed potential results in the prediction and similarities scores; however, the screening for in vitro evaluation was based on the standardization tests. In this context, we selected peptide No. 45, with a greater difference in reactivity between the mean of infected and uninfected animals that showed the highest reactivity index (7.94), and peptide No. 48, which had the lowest OD mean of the negative control; this threshold can be used to define the lower limits of detection with a reduced cross-reaction between uninfected and infected animals.

Our data indicated that Pep45, Pep48, and a mix of these peptides showed excellent performance, with high sensitivity (86.4%, 100%, and 97.2%, respectively), high specificity (100%, 90%, and 95%, respectively), and elevated accuracy (91.2%, 96.5%, and 96.5%, respectively) in detecting IgG from *L. infantum*-infected dogs presenting different clinical forms. Moreover, these peptides were subjected to a proof-of-concept experiment to test their ability to distinguish leishmaniasis from other infections (*T. cruzi*, *B. canis*, *E. canis*) or vaccinated dogs. Similar studies that employed immunoinformatic tools demonstrated that the peptides, alone or in combination, were reactive to the serum of infected dogs, with accuracy values ranging from 99.6% to 100%; their assay was highly sensitive and specific when compared to the soluble *Leishmania* antigen, which showed low sensitivity and specificity [40]. These peptides, selected using immunoinformatic tools, display superior performance when compared to other studies that have used crude *Leishmania antigen* preparations such as the commercial kit for canine leishmaniasis diagnosis in Brazil, which uses antigens prepared from *Leishmania major*-like promastigotes [41–43]. The antigens used in these assays display variability in accuracy parameters owing to antigen preparation and antigenic differences among *Leishmania* species. These considerations were consistent with the present study's results, in which the statistical parameters of ELISA using the soluble antigen of *L. infantum* (SLiA) displayed inferior performance.

Our results for Pep48 and the peptide mix indicate an efficiency in the detection of most of the infected dogs, including the asymptomatic ones, increasing the odds of detecting infected dogs that might not be detected using other serological tests. The current protocol recommended by the Brazilian Ministry of Health for CVL diagnosis has several limitations. The specificity of the immunochromatographic assay (DPP®-CVL) presented specificities varying from 70% to 91.7% [44,45], whereas ELISA (EIE®-LVC) presented specificity values varying from 87.5% to 91.76% [44,46].

In this study, we demonstrate that the peptides No. 45, No. 48, and mix presented no reactivity in samples of *T. cruzi*-infected animals, indicating that these peptides may not be shared between *Leishmania* and *T. cruzi* species. In human VL, several studies have shown issues related to cross-reactivity with Chagas disease, owing to the sharing of antigens associated with phylogenetic proximity [47,48]. However, we demonstrated that the mixture of peptides increased the specificity and accuracy compared to that shown by isolated peptides, as confirmed by the higher performance values and lower cross-reactivity in *T. cruzi*, *E. canis*, and *B. canis* pathogens [49], and have previously reported that the association of antigens can lead to better performance because of the use of a single platform for the simultaneous serological measurement of antibodies against different antigens of *L. infantum*. This fact can be associated with the binding of antibodies with higher avidity to the antigens, and the peptide mix could select these antibodies more effectively, favoring a more specific interaction. Despite the lack of concrete evidence of genomic homology between the genus *Ehrlichia* and *Leishmania*, false-positive results are usually found in most conventional serological tests [43,50]. However, some studies do not support serological cross-reactivity among *Leishmania* species, *E. canis*, and *B. canis* (10), based on the phylogenetic differences in such species, since the microorganisms of the genus *Ehrlichia* are intracellular bacteria, whereas *Leishmania* and *Babesia* are protozoa of the orders

Kinetoplastida and Piroplasmida, respectively [10,51,52]. Other studies have associated leishmaniasis infection with an increasing co-infection with *E. canis* and *B. canis*. Attipa et al. (2018) reported that dogs with clinical leishmaniasis are 12 times more susceptible to *E. canis* infection when compared to healthy dogs, indicating a suppression of the immune system caused by *Leishmania* infection, which could allow reactivation of a subclinical *E. canis* infection that was present previously or enable the establishment of a new *E. canis* infection, and recommend the simultaneous treatment of both diseases [11,53]. Although, in the present study, each infection was previously characterized using specific serology (ELISA) and PCR-positive results and confirmed to be PCR-negative for *L. infantum*, new mechanisms must be elucidated to understand the cross-reactivity between such species.

The development of vaccines against CVL has been stimulated as an effective and cost-effective strategy for controlling the spread of leishmaniasis in endemic and expanding areas [54]. Although vaccine-induced anti-*Leishmania* antibodies can be detected for several months after immunization, it is essential to evaluate whether a vaccine-induced seroconversion might cause unnecessary culling of noninfected dogs and, consequently, generate problems in surveillance and control programs [55,56]. Both vaccines (Leish-Tec® and LBSap) described in the present study can induce a humoral response [28,57,58]. We observed that peptides No. 45 and No. 48 presented low reactivity with antibodies obtained from dogs immunized with the commercial vaccine, Leish-Tec®, and dogs immunized with the LBSap vaccine. The peptide mix showed no reactivity, whereas high reactivity was observed (70.5% of the samples) in vaccinated dogs when SLiA was employed. Our findings indicate that the peptides evaluated had better performance compared to other studies that failed to show low reactivity in samples from vaccinated dogs [59].

In summary, our findings showed that the peptide mix presented a higher performance compared to that shown by SLiA and the isolated peptides, with an excellent capacity for the detection of asymptomatic animals and for discriminating infections with other pathogens or vaccinated dogs, which could improve CVL diagnosis, especially in endemic areas. These data also support the establishment of this approach as the gold standard and open new possibilities for employing these peptides in the construction of chimeric proteins, to improve the accuracy of this test or for the development of a rapid test that could be used as a versatile tool in control programs or in human diagnosis disease.

Author Contributions: G.M. and R.M. conducted, designed, and equally contributed to this study. G.M., R.M., N.S. and T.O. performed the experiments. W.C.-V., R.A.-S., J.R., D.R., R.d.B., A.R. and B.R. contributed to the statistical analysis and designed. Writing—review and editing, and funding acquisition B.R., R.A.-S. and A.R. All authors have read and agreed to the published version of the manuscript.

Funding: This research was funded by CNPq (grant number 420912/2023-1), FAPEMIG (grant number, FAPEMIG APQ-02511-22, APQ-02494-22, APQ-02045-23, APQ-02286-21, and RED-00032-22/FAPEMIG—Rede Imunobioleish-Minas), CAPES (this study was financed in part by the Coordenação de Aperfeiçoamento de Pessoal de Nível Superior-Brasil (CAPES)—Finance Code 001), FINEP and Universidade Federal de Ouro Preto—UFOP for funding. A.R., B.R. and R.S. are also grateful to CNPq for fellowships.

Data Availability Statement: The raw data supporting the conclusions of this article will be made available by the authors on request.

Acknowledgments: The authors are also thankful for the use of the facilities at Animal Center Science/UFOP and Centro de Controle de Zoonoses de Governador Valadares, MG (CCZ/GV).

Conflicts of Interest: The authors declare no conflicts of interest.

References

1. Elmahallawy, E.K.; Sampedro Martinez, A.; Rodriguez-Granger, J.; Hoyos-Mallecot, Y.; Agil, A.; Navarro Mari, J.M.; Gutierrez Fernandez, J. Diagnosis of leishmaniasis. *J. Infect. Dev. Ctries.* **2014**, *8*, 961–972. [CrossRef] [PubMed]
2. WHO. Global leishmaniasis surveillance, 2017–2018, and first report on 5 additional indicators. *Wkly. Epidemiol. Rec.* **2020**, *95*, 265–280.

3. PAHO. *Leishmanioses: Informe Epidemiológico nas Américas*; PAHO: Washington, DC, USA, 2020.
4. Brazil. Departamento de Vigilância Epidemiológica. *Manual de Vigilância e Controle da Leishmaniose Visceral*; Departamento de Vigilância Epidemiológica: Rio de Janeiro, Brazil, 2014; ISBN 9788533407428.
5. Brodskyn, C.I.; Kamhawi, S. Biomarkers for Zoonotic Visceral Leishmaniasis in Latin America. *Front. Cell. Infect. Microbiol.* **2018**, *8*, 245. [CrossRef] [PubMed]
6. Coura-Vital, W.; Ker, H.G.; Roatt, B.M.; Aguiar-Soares, R.D.O.; Leal, G.G.D.A.; Moreira, N.D.D.; Oliveira, L.A.M.; de Menezes Machado, E.M.; Morais, M.H.F.; Corrêa-Oliveira, R.; et al. Evaluation of Change in Canine Diagnosis Protocol Adopted by the Visceral Leishmaniasis Control Program in Brazil and a New Proposal for Diagnosis. *PLoS ONE* **2014**, *9*, e91009. [CrossRef] [PubMed]
7. Travi, B.L.; Cordeiro-da-Silva, A.; Dantas-Torres, F.; Miró, G. Canine visceral leishmaniasis: Diagnosis and management of the reservoir living among us. *PLoS Negl. Trop. Dis.* **2018**, *12*, e0006082. [CrossRef] [PubMed]
8. Pessoa-e-Silva, R.; Vaitkevicius-Antão, V.; de Andrade, T.A.S.; de Oliveira Silva, A.C.; de Oliveira, G.A.; Trajano-Silva, L.A.M.; Nakasone, E.K.N.; de Paiva-Cavalcanti, M. The diagnosis of canine visceral leishmaniasis in Brazil: Confronting old problems. *Exp. Parasitol.* **2019**, *199*, 9–16. [CrossRef] [PubMed]
9. Dantas-Torres, F.; de Brito, M.E.F.; Brandão-Filho, S.P. Seroepidemiological survey on canine leishmaniasis among dogs from an urban area of Brazil. *Vet. Parasitol.* **2006**, *140*, 54–60. [CrossRef]
10. Krawczak, F.D.S.; Reis, I.A.; da Silveira, J.A.; Avelar, D.M.; Marcelino, A.P.; Werneck, G.L.; Labruna, M.B.; Paz, G.F. Leishmania, Babesia and Ehrlichia in urban pet dogs: Co-infection or cross-reaction in serological methods? *Rev. Soc. Bras. Med. Trop.* **2015**, *48*, 64–68. [CrossRef]
11. Attipa, C.; Solano-Gallego, L.; Papasouliotis, K.; Soutter, F.; Morris, D.; Helps, C.; Carver, S.; Tasker, S. Association between canine leishmaniosis and Ehrlichia canis co-infection: A prospective case-control study. *Parasites Vectors* **2018**, *11*, 184. [CrossRef]
12. De Brito, R.C.F.; Aguiar-Soares, R.D.D.O.; de O. Cardoso, J.M.; Coura-Vital, W.; Roatt, B.M.; Reis, A.B. Recent advances and new strategies in Leishmaniasis diagnosis. *Appl. Microbiol. Biotechnol.* **2020**, *104*, 8105–8116. [CrossRef] [PubMed]
13. De Brito, R.C.F.; Ruiz, J.C.; Cardoso, J.M.D.O.; Ostolin, T.L.V.D.P.; Reis, L.E.S.; Mathias, F.A.S.; de O. Aguiar-Soares, R.D.; Roatt, B.M.; Corrêa-Oliveira, R.; Resende, D.D.M.; et al. Chimeric vaccines designed by immunoinformatics-activated polyfunctional and memory T cells that trigger protection against experimental visceral leishmaniasis. *Vaccines* **2020**, *8*, 252. [CrossRef]
14. Brito, R.; Guimarães, F.; Velloso, J.; Corrêa-Oliveira, R.; Ruiz, J.; Reis, A.; Resende, D. Immunoinformatics Features Linked to Leishmania Vaccine Development: Data Integration of Experimental and In Silico Studies. *Int. J. Mol. Sci.* **2017**, *18*, 371. [CrossRef] [PubMed]
15. Larsen, J.E.P.; Lund, O.; Nielsen, M. Improved method for predicting linear B-cell epitopes. *Immunome Res.* **2006**, *2*, 2. [CrossRef] [PubMed]
16. Chen, J.; Liu, H.; Yang, J.; Chou, K.C. Prediction of linear B-cell epitopes using amino acid pair antigenicity scale. *Amino Acids* **2007**, *33*, 423–428. [CrossRef]
17. El-Manzalawy, Y.; Dobbs, D.; Honavar, V. Predicting linear B-cell epitopes using string kernels. *J. Mol. Recognit.* **2008**, *21*, 243–255. [CrossRef] [PubMed]
18. El-Manzalawy, Y.; Dobbs, D.; Honavar, V. Predicting flexible length linear B-cell epitopes. In *Computational Systems Bioinformatics: (Volume 7)*; World Scientific: Singapore, 2008; pp. 121–132.
19. Horton, P.; Park, K.-J.; Obayashi, T.; Fujita, N.; Harada, H.; Adams-Collier, C.J.; Nakai, K. WoLF PSORT: Protein localization predictor. *Nucleic Acids Res.* **2007**, *35*, W585–W587. [CrossRef] [PubMed]
20. von Heijne, G. A new method for predicting signal sequence cleavage sites. *Nucleic Acids Res.* **1986**, *14*, 4683–4690. [CrossRef] [PubMed]
21. Almagro Armenteros, J.J.; Salvatore, M.; Emanuelsson, O.; Winther, O.; von Heijne, G.; Elofsson, A.; Nielsen, H. Detecting sequence signals in targeting peptides using deep learning. *Life Sci. Alliance* **2019**, *2*, e201900429. [CrossRef] [PubMed]
22. Krogh, A.; Larsson, B.; Von Heijne, G.; Sonnhammer, E.L.L. Predicting transmembrane protein topology with a hidden Markov model: Application to complete genomes. *J. Mol. Biol.* **2001**, *305*, 567–580. [CrossRef] [PubMed]
23. Szklarczyk, D.; Franceschini, A.; Wyder, S.; Forslund, K.; Heller, D.; Huerta-Cepas, J.; Simonovic, M.; Roth, A.; Santos, A.; Tsafou, K.P. STRING v10: Protein–protein interaction networks, integrated over the tree of life. *Nucleic Acids Res.* **2015**, *43*, D447–D452. [CrossRef]
24. Szklarczyk, D.; Gable, A.L.; Nastou, K.C.; Lyon, D.; Kirsch, R.; Pyysalo, S.; Doncheva, N.T.; Legeay, M.; Fang, T.; Bork, P.; et al. The STRING database in 2021: Customizable protein–protein networks, and functional characterization of user-uploaded gene/measurement sets. *Nucleic Acids Res.* **2021**, *49*, D605–D612. [CrossRef] [PubMed]
25. Kanehisa, M.; Sato, Y. KEGG Mapper for inferring cellular functions from protein sequences. *Protein Sci.* **2020**, *29*, 28–35. [CrossRef] [PubMed]
26. Altschul, S.F.; Gish, W.; Miller, W.; Myers, E.W.; Lipman, D.J. Basic local alignment search tool. *J. Mol. Biol.* **1990**, *215*, 403–410. [CrossRef] [PubMed]
27. Reis, A.B.; Teixeira-Carvalho, A.; Giunchetti, R.C.; Guerra, L.L.; Carvalho, M.G.; Mayrink, W.; Genaro, O.; Corrêa-Oliveira, R.; Martins-Filho, O.A. Phenotypic features of circulating leucocytes as immunological markers for clinical status and bone marrow parasite density in dogs naturally infected by Leishmania chagasi. *Clin. Exp. Immunol.* **2006**, *146*, 303–311. [CrossRef] [PubMed]

28. Aguiar-Soares, R.D.D.O.; Roatt, B.M.; Mathias, F.A.S.; Reis, L.E.S.; de O. Cardoso, J.M.; de Brito, R.C.F.; Ker, H.G.; Corrêa-Oliveira, R.; Giunchetti, R.C.; Reis, A.B. Phase i and ii clinical trial comparing the lbsap, leishmune®, and leish-tec® vaccines against canine visceral leishmaniasis. *Vaccines* **2020**, *8*, 690. [CrossRef]
29. Heyduk, E.; Hickey, R.; Pozzi, N.; Heyduk, T. Peptide ligand-based ELISA reagents for antibody detection. *Anal. Biochem.* **2018**, *559*, 55–61. [CrossRef] [PubMed]
30. Resende, D.M.; Rezende, A.M.; Oliveira, N.J.D.; Batista, I.C.A.; Corrêa-Oliveira, R.; Reis, A.B.; Ruiz, J.C. An assessment on epitope prediction methods for protozoa genomes. *BMC Bioinform.* **2012**, *13*, 309. [CrossRef] [PubMed]
31. Dhusia, K.; Kesarwani, P.; Yadav, P.K. Epitope prediction for MSP119 protein in Plasmodium yeolii using computational approaches. *Netw. Model. Anal. Health Inform. Bioinform.* **2016**, *5*, 19. [CrossRef]
32. Wang, X.; Ren, Z.; Sun, Q.; Wan, X.; Sun, Y.; Hua, Y.; Fu, M.; Shao, N.; Du, Y.; Zhang, Q.; et al. Evaluation and Comparison of Newly Built Linear B-Cell Epitope Prediction Software from a Users' Perspective. *Curr. Bioinform.* **2018**, *13*, 149–156. [CrossRef]
33. Cordeiro-Da-Silva, A.; Borges, M.C.; Guilvard, E.; Ouaissi, A. Dual role of the Leishmania major ribosomal protein S3a homologue in regulation of T- and B-cell activation. *Infect. Immun.* **2001**, *69*, 6588–6596. [CrossRef] [PubMed]
34. Iborra, S.; Parody, N.; Abánades, D.R.; Bonay, P.; Prates, D.; Novais, F.O.; Barral-Netto, M.; Alonso, C.; Soto, M. Vaccination with the Leishmania major ribosomal proteins plus CpG oligodeoxynucleotides induces protection against experimental cutaneous leishmaniasis in mice. *Microbes Infect.* **2008**, *10*, 1133–1141. [CrossRef] [PubMed]
35. Muñoz, C.; San Francisco, J.; Gutiérrez, B.; González, J. Role of the ubiquitin-proteasome systems in the biology and virulence of protozoan parasites. *BioMed Res. Int.* **2015**, *2015*, 141526. [CrossRef] [PubMed]
36. McConville, M.J.; Naderer, T. Metabolic pathways required for the intracellular survival of Leishmania. *Annu. Rev. Microbiol.* **2011**, *65*, 543–561. [CrossRef] [PubMed]
37. Troncarelli, M.Z.; Camargo, J.B.; Machado, J.G.; Lucheis, S.B.; Langoni, H. Leishmania spp. and/or Trypanosoma cruzi diagnosis in dogs from endemic and nonendemic areas for canine visceral leishmaniasis. *Vet. Parasitol.* **2009**, *164*, 118–123. [CrossRef] [PubMed]
38. Jamjoom, M.B.; Ashford, R.W.; Bates, P.A.; Chance, M.L.; Kemp, S.J.; Watts, P.C.; Noyes, H.A. Leishmania donovani is the only cause of visceral leishmaniasis in East Africa; previous descriptions of *L. infantum* and "*L. archibaldi*" from this region are a consequence of convergent evolution in the isoenzyme data. *Parasitology* **2004**, *129*, 399–409. [CrossRef] [PubMed]
39. Carnielli, J.B.T.; Crouch, K.; Forrester, S.; Silva, V.C.; Carvalho, S.F.G.; Damasceno, J.D.; Brown, E.; Dickens, N.J.; Costa, D.L.; Costa, C.H.N.; et al. A Leishmania infantum genetic marker associated with miltefosine treatment failure for visceral leishmaniasis. *EBioMedicine* **2018**, *36*, 83–91. [CrossRef] [PubMed]
40. Toledo-Machado, C.M.; De Avila, R.A.M.; Nguyen, C.; Granier, C.; Bueno, L.L.; Carneiro, C.M.; Menezes-Souza, D.; Carneiro, R.A.; Chávez-Olórtegui, C.; Fujiwara, R.T. Immunodiagnosis of canine visceral leishmaniasis using mimotope peptides selected from phage displayed combinatorial libraries. *BioMed Res. Int.* **2015**, *2015*, 401509. [CrossRef] [PubMed]
41. Paltrinieri, S.; Gradoni, L.; Roura, X.; Zatelli, A.; Zini, E. Laboratory tests for diagnosing and monitoring canine leishmaniasis. *Vet. Clin. Pathol.* **2016**, *45*, 552–578. [CrossRef]
42. Teixeira, A.I.P.; Silva, D.M.; Vital, T.; Nitz, N.; De Carvalho, B.C.; Hecht, M.; Oliveira, D.; Oliveira, E.; Rabello, A.; Romero, G.A.S. Improving the reference standard for the diagnosis of canine visceral leishmaniasis: A challenge for current and future tests. *Mem. Inst. Oswaldo Cruz* **2019**, *114*, 1–9. [CrossRef] [PubMed]
43. Zanette, M.F.; de Lima, V.M.F.; Laurenti, M.D.; Rossi, C.N.; Vides, J.P.; Vieira, R.F.D.C.; Biondo, A.W.; Marcondes, M. Serological cross-reactivity of Trypanosoma cruzi, Ehrlichia canis, Toxoplasma gondii, Neospora caninum and Babesia canis to Leishmania infantum chagasi tests in dogs. *Rev. Soc. Bras. Med. Trop.* **2014**, *47*, 105–107. [CrossRef]
44. Ribeiro, V.M.; Miranda, J.B.; Marcelino, A.P.; de Andrade, H.M.; Reis, I.A.; Cardoso, M.S.; Gontijo, C.M.F.; Paz, G.F. Performance of different serological tests in the diagnosis of natural infection by Leishmania infantum in dogs. *Vet. Parasitol.* **2019**, *274*, 108920. [CrossRef] [PubMed]
45. Figueiredo, F.B.; de Vasconcelos, T.C.B.; de F. Madeira, M.; Menezes, R.C.; Maia-Elkhoury, A.N.S.; Marcelino, A.P.; Werneck, G.L. Validation of the Dual-path Platform chromatographic immunoassay (DPP® CVL rapid test) for the serodiagnosis of canine visceral leishmaniasis. *Mem. Inst. Oswaldo Cruz* **2018**, *113*, e180260. [CrossRef]
46. Lira, R.A.; Cavalcanti, M.P.; Nakazawa, M.; Ferreira, A.G.P.; Silva, E.D.; Abath, F.G.C.; Alves, L.C.; Souza, W.V.; Gomes, Y.M. Canine visceral leishmaniosis: A comparative analysis of the EIE-leishmaniose-visceral-canina-Bio-Manguinhos and the IFI-leishmaniose- visceral-canina-Bio-Manguinhos kits. *Vet. Parasitol.* **2006**, *137*, 11–16. [CrossRef] [PubMed]
47. Santos, M.F.; Alexandre-Pires, G.; Pereira, M.A.; Marques, C.S.; Gomes, J.; Correia, J.; Duarte, A.; Gomes, L.; Rodrigues, A.V.; Basso, A.; et al. Meglumine antimoniate and miltefosine combined with allopurinol sustain pro-inflammatory immune environments during canine leishmaniosis treatment. *Front. Vet. Sci.* **2019**, *6*, 362. [CrossRef]
48. Aguiar, J.C.; Mittmann, J.; Caetano, P.C.; Raniero, L. Using FT-IR spectroscopy for the identification of the T. cruzi, T. rangeli, and the L. chagasi species. *Exp. Parasitol.* **2018**, *192*, 46–51. [CrossRef] [PubMed]
49. Campos, F.M.F.; Repoles, L.C.; de Araújo, F.F.; Peruhype-Magalhães, V.; Xavier, M.A.P.; Sabino, E.C.; de Freitas Carneiro Proietti, A.B.; Andrade, M.C.; Teixeira-Carvalho, A.; Martins-Filho, O.A.; et al. Usefulness of FC-TRIPLEX Chagas/Leish IgG1 as confirmatory assay for non-negative results in blood bank screening of Chagas disease. *J. Immunol. Methods* **2018**, *455*, 34–40. [CrossRef]

50. Ker, H.G.; Coura-Vital, W.; Valadares, D.G.; Aguiar-Soares, R.D.O.; de Brito, R.C.F.; Veras, P.S.T.; Fraga, D.B.M.; Martins-Filho, O.A.; Teixeira-Carvalho, A.; Reis, A.B. Multiplex flow cytometry serology to diagnosis of canine visceral leishmaniasis. *Appl. Microbiol. Biotechnol.* **2019**, *103*, 8179–8190. [CrossRef]
51. Schnittger, L.; Rodriguez, A.E.; Florin-Christensen, M.; Morrison, D.A. Babesia: A world emerging. *Infect. Genet. Evol.* **2012**, *12*, 1788–1809. [CrossRef]
52. Vinícius, M.; Silva, M.; Fernandes, R.A.; Ambrósio, C.E.; Erliquiose, C.E. Erliquiose canina: Revisão de literatura. *Arq. Ciências Veterinárias Zool. UNIPAR* **2011**, *14*, 139–143.
53. Harrus, S.; Waner, T.; Friedmann-Morvinski, D.; Fishman, Z.; Bark, H.; Harmelin, A. Down-regulation of MHC class II receptors of DH82 cells, following infection with Ehrlichia canis. *Vet. Immunol. Immunopathol.* **2003**, *96*, 239–243. [CrossRef] [PubMed]
54. Dantas-Torres, F.; Nogueira, F.D.S.; Menz, I.; Tabanez, P.; da Silva, S.M.; Ribeiro, V.M.; Miró, G.; Cardoso, L.; Petersen, C.; Baneth, G.; et al. Vaccination against canine leishmaniasis in Brazil. *Int. J. Parasitol.* **2020**, *50*, 171–176. [CrossRef] [PubMed]
55. Solano-Gallego, L.; Mirá, G.; Koutinas, A.; Cardoso, L.; Pennisi, M.G.; Ferrer, L.; Bourdeau, P.; Oliva, G.; Baneth, G. LeishVet guidelines for the practical management of canine leishmaniosis. *Parasites Vectors* **2011**, *4*, 86. [CrossRef] [PubMed]
56. Lima, C.; Santarém, N.; Nieto, J.; Moreno, J.; Carrillo, E.; Bartholomeu, D.C.; Bueno, L.L.; Fujiwara, R.; Amorim, C.; Cordeiro-da-Silva, A. The Use of Specific Serological Biomarkers to Detect CaniLeish Vaccination in Dogs. *Front. Vet. Sci.* **2019**, *6*, 373. [CrossRef] [PubMed]
57. Gillespie, P.M.; Beaumier, C.M.; Strych, U.; Hayward, T.; Hotez, P.J.; Bottazzi, M.E. Status of vaccine research and development of vaccines for leishmaniasis. *Vaccine* **2016**, *34*, 2992–2995. [CrossRef]
58. De Mendonça, L.Z.; Resende, L.A.; Lanna, M.F.; Aguiar-Soares, R.D.D.O.; Roatt, B.M.; Castro, R.A.D.O.E.; Batista, M.A.; Silveira-Lemos, D.; Gomes, J.D.A.S.; Fujiwara, R.T.; et al. Multicomponent LBSap vaccine displays immunological and parasitological profiles similar to those of Leish-Tec® and Leishmune® vaccines against visceral leishmaniasis. *Parasites Vectors* **2016**, *9*, 472. [CrossRef] [PubMed]
59. Carvalho, A.M.R.S.; Costa, L.E.; Salles, B.C.S.; Santos, T.T.O.; Ramos, F.F.; Lima, M.P.; Chávez-Fumagalli, M.A.; Silvestre, B.T.; Portela, Á.S.B.; Roatt, B.M.; et al. An ELISA immunoassay employing a conserved Leishmania hypothetical protein for the serodiagnosis of visceral and tegumentary leishmaniasis in dogs and humans. *Cell. Immunol.* **2017**, *318*, 42–48. [CrossRef] [PubMed]

MDPI AG
Grosspeteranlage 5
4052 Basel
Switzerland
Tel.: +41 61 683 77 34

Microorganisms Editorial Office
E-mail: microorganisms@mdpi.com
www.mdpi.com/journal/microorganisms

www.ingramcontent.com/pod-product-compliance
Lightning Source LLC
Chambersburg PA
CBHW041355210926
51850CB00013B/3088

* 9 7 8 3 7 2 5 8 4 4 2 2 7 *